A Wandering Dance through the Philosophy of Graham Parkes

Also available from Bloomsbury:

How to Think about the Climate Crisis, by Graham Parkes
Adorning Bodies, by Marilynn Johnson
Aesthetics and Nature, by Glenn Parsons
Applying Aesthetics to Everyday Life, edited by Lisa Giombini and Adrián Kvokacka
The Social Aesthetics of Human Environments, by Arnold Berleant
The Philosophy of Creative Solitudes, edited by David Jones

A Wandering Dance through the Philosophy of Graham Parkes

Comparative Perspectives on Art and Nature

Edited by David Jones

BLOOMSBURY ACADEMIC
LONDON • NEW YORK • OXFORD • NEW DELHI • SYDNEY

BLOOMSBURY ACADEMIC
Bloomsbury Publishing Plc, 50 Bedford Square, London, WC1B 3DP, UK
Bloomsbury Publishing Inc, 1359 Broadway, 12th Floor, New York, NY 10018, USA
Bloomsbury Publishing Ireland, 29 Earlsfort Terrace, Dublin 2, D02 AY28, Ireland

BLOOMSBURY, BLOOMSBURY ACADEMIC and the Diana logo are
trademarks of Bloomsbury Publishing Plc

First published in Great Britain 2024
This paperback edition published 2026

Copyright © David Jones and Contributors, 2024

David Jones has asserted his right under the Copyright,
Designs and Patents Act, 1988, to be identified as Editor of this work.

The Cover design: Louise Dugdale
Cover image @ *Wanderer Dancing on the Nietzsche Rock (after Friedrich)*
by Setsuko Aihara

All rights reserved. No part of this publication may be: i) reproduced or transmitted in any form, electronic or mechanical, including photocopying, recording or by means of any information storage or retrieval system without prior permission in writing from the publishers; or ii) used or reproduced in any way for the training, development or operation of artificial intelligence (AI) technologies, including generative AI technologies. The rights holders expressly reserve this publication from the text and data mining exception as per Article 4(3) of the Digital Single Market Directive (EU) 2019/790.

Bloomsbury Publishing Inc does not have any control over, or responsibility for,
any third-party websites referred to or in this book. All internet addresses given
in this book were correct at the time of going to press. The author and publisher
regret any inconvenience caused if addresses have changed or sites have
ceased to exist, but can accept no responsibility for any such changes.

A catalogue record for this book is available from the British Library.

Library of Congress Cataloging-in-Publication Data
Names: Jones, David Edward, editor.
Title: A wandering dance through the philosophy of Graham Parkes :
comparative perspectives on art and nature / edited by David Jones.
Description: 1. | London: Bloomsbury Academic, 2024. | Includes
bibliographical references and index. | Summary: "Inspired by the philosopher Graham Parkes,
this collection provides a distinctive study of aesthetics and the climate crisis. Engaging with
Continental European and East Asian traditions, it challenges our definition of self in the
West and asks us to re-evaluate our conventional perspectives. Through a valuable and systematic
treatment of the thought of Parkes, The Wandering Dance in the Philosophy of Graham Parkes makes
the case that a restoration of the intimate relation of self and nature is indispensable in understanding our
place in the order of things and achieving balance in the world"– Provided by publisher.
Identifiers: LCCN 2023055210 (print) | LCCN 2023055211 (ebook) | ISBN 9781350291300 (hardback) |
ISBN 9781350291348 (paperback) | ISBN 9781350291324 (epub) | ISBN 9781350291317 (ebook)
Subjects: LCSH: Environment (Aesthetics) | Parkes, Graham, 1949-
Classification: LCC BH301.E58 W36 2024 (print) | LCC BH301.E58 (ebook) |
DDC 111/.85–dc23/eng/20240321
LC record available at https://lccn.loc.gov/2023055210
LC ebook record available at https://lccn.loc.gov/2023055211

ISBN: HB: 978-1-3502-9130-0
PB: 978-1-3502-9134-8
ePDF: 978-1-3502-9131-7
eBook: 978-1-3502-9132-4

Typeset by Integra Software Services Pvt. Ltd.

For product safety related questions contact productsafety@bloomsbury.com.

To find out more about our authors and books visit www.bloomsbury.com
and sign up for our newsletters.

For Graham Parkes, philosopher, teacher, friend, and wandering dancer.

Contents

Once upon a Signature Event   1

Part One   The First Dance

1   The Wandering Dance: *Zhuangzi* and *Zarathustra*
     Graham Parkes   13

Part Two   The Aesthetic Dance

2   Graham Parkes, Musical Ecologies, and the Daoist Dance
     Roger T. Ames   31
3   On the Chariot of Earth
     Meilin Chinn   49
4   The Role of Aesthetics in Nietzsche's Philosophy and Japanese Culture
     Yuriko Saito   69
5   Heidegger's Evasion of Music
     Kathleen Higgins   87
6   Improvising with Soul: Philosophizing in an Interpretative Key
     Peter D. Hershock   101
7   Image Thinking: Cinematographic Flâneurs and Twenty-First-Century Philosophy
     Andrew K. Whitehead   111

Part Three   The Practice of Dance

8   Rubble and the Philosopher's Stone: The Practice of Philosophy and the Philosophy of Practice
     Jason M. Wirth   125
9   Philosophy as Petromania: Graham Parkes, Jane Bennett, and the (Not So) New Materialisms
     Leah Kalmanson   137
10   A Mind Possessed: The Rhythmic Body and Entraining the Social Mind
     Bradley Douglas Park   149

Part Four Saving the Last Dance

11 Staying True to the Earth in Zarathustra, Zhuangzi, and Zen
    Timothy J. Freeman   179
12 Downbeats and Upbeats in the Meters of Nietzsche, Zhuangzi, and Parkes
    David Jones   215

Part Five Keeping on Dancing

13 Gratitude of a Wondering Wanderer
    Graham Parkes   249

Contributors   253
Subject Index   257
Name Index   264

Once upon a Signature Event

Dance is where we must begin, for dancing is the eco-centric activity of signature events. Signature events are like a dance because they do not happen in a vacuum—there is music, a place to dance, and there are other dancers. At times a dancer will emerge with a new step, an innovative *mudrā*, or a novel movement of the hips that changes the dance. With signature events, something new appears in the world when the participation of a self enters and metamorphoses with and within the world. This kind of self attains its participation from within rather than recurrently being an effect from without. There can be no conquering of nature in this convergence, no mastery over a system, and no ruling over others, whether those others are people, animals, plants, rocks, or landscapes. For this self makes its movements along with the forces of wind, water, geologic arrangements, and the destructive and creative forces of fire that make, remake, and sustain all that is in the dance of life and death. The planet we call earth will certainly fare well-enough on its own without the kinds of selves that act as effects from without—those non-earthbound creatures called *Homo sapiens*. But there are some earthbound *Homo sapiens* who can create something positive that amplifies into our world system, something we might call a signature event. A world without signature events that are created without this kind of participatory self will look differently; this world will lose its special significance and will evaporate into utter universal indifference. Signature events are significant, for they become legendary forces in nature that affect the ebbs and tides of possible futures—signature events mark time. The biologist John L. Culliney and I have referred to this kind of self as a fractal self, a Daoist sage in training.[1] Early in his career, Graham Parkes unwittingly, as it is usually the case, became such a self for comparative philosophy. This book, in so many ways through the dance of its contributing intimate voices, is about this named self and the event attached to him that transformed comparative philosophy, and by extension philosophy in our time.

In 1982, a relatively unnoticed signature event occurred, and like many signature events its unfolding effects took some time to become realized and what emerged spawned a perceptual transformation in the coursing of comparative philosophy. This event opened new vistas for engaging in cross-cultural philosophy. I happened to be one who witnessed this event and unlike many others (perhaps even the signatory of the event himself), I sensed that we were on another verge in comparative philosophy, and as Heraclitus said in his fragment on the soul, this verge was a sign of "its deep being."[2] My sense now is that I was positioned in such a way through my admiration for this signatory and that this positioning gave me a privileged place of viewing its verge—perhaps even a better view of the event itself, even beyond that of the signatory. It was, as these types of things go, just a matter of perspective.

In honesty, I merely happened to be in the most advantageous position to see this; although I was open to the reevaluation of comparative philosophy, I just happened to have a perspective that others simply could not occupy because as time's turnings would have it, I was simply the first, the random first student who felt the need to have this non-tenured professor direct his dissertation. My being first was purely fortuitous—there were no divine interventions, no teleological plans, for nothing but fate brought me into his fold. It was all much like Heraclitus's filings that reap themselves into this or that arrangement. I had done nothing to deserve being first, and for the honor of being the one putting this book together. I just emerged from the mix, but it is true though that I was less rule bound than some others; I was by nature rebellious and loved Nietzsche, who was my entrée into philosophy as a seventeen-year-old freshman—this was a time before I came to better understand Confucius and became tempered by his aspiration for harmony. This event for me was just as random as being the first person Timothy Freeman encountered on his arrival to Sakamaki Hall when I talked him into taking a seminar on *Thus Spoke Zarathustra*. In this seminar we read the text line by line as Parkes made his nuanced corrections to the most currently used translation of the day. The time of the seminar ultimately swelled into another signature event of what many think as the definitive translation of *Thus Spoke Zarathustra* that Parkes published with Oxford University Press in 2005. Freeman is in this book and occupies an additional prominence as historically being the second PhD student of Graham Parkes—we were, and continue to be, transformed by his work, his being, and his life's dance.

What appeared on the horizon in 1982 was a "prolegomena to a wider and deeper study of Nietzsche and Zhuangzi as psychologically acute philosophers intent on effecting a transformation of our ideas of self and world—and thereby of ourselves."[3] This was the first work in English, and likely in any language of which I am aware, that gestured comparative philosophy in this direction. Never had Nietzsche and Zhuangzi been brought together in conversation, especially in a dance with each other. It all seems so natural and logical to us now, but this was the first occurrence that demonstrated a resonance between Zhuangzi and Nietzsche, and this moment transformed comparative philosophy. The first part of the title of this book bears the original title of that landmark article, *A Wandering Dance*. And the second part discloses more specifically about the who and what of this book: *A Wandering Dance through the Philosophy of Graham Parkes: Comparative Perspectives on Art and Nature*. The subtitle indicates the aesthetic and ecological dimensions of the subsequent impact of his work and this natural coupling brings forward the necessary way in which we must approach the ecological challenges facing our planet and her diversity of species today. It is of little surprise that Parkes's latest book is titled *How to Think about the Climate Crisis: A Philosophical Guide to Saner Ways of Living*, which is also published by Bloomsbury.

We begin with the original publication of "The Wandering Dance: Zhuangzi and Zarathustra" that was published in *Philosophy East-West* in 1982. Keeping the article in its original form was a conscious decision, but the author has converted the Wade Giles form of romanization into the pinyin for readers' convenience. I am grateful to Franklin Perkins, the editor of *Philosophy East-West*, for allowing its republication in this book.

Dancing requires music; music is the background of dancing; music gives dancing its soul. Good dancing demands that we be in step with music's rhythms, its underlying pulses, syncopations, and the patterns and arrangements of beats, which are music's constant heartbeats. We should think of music in its broadest sense—inclusive of the songs birds and crickets sing, the movement of the wind through the trees, the lapping of an ocean's waves on a rocky or sandy shoreline, or the torrent of a rushing river or the more silent sounds of pooling water—for the wandering dance is all about this, moving in step with the multifarious perspectives throughout the earth's disclosing dance of its unfolding. But the wandering dance is more than just this, for it is also about the shadow side of its being and its becoming. This music of life includes the sounds of the lion's jaws on the crunching bones of a wildebeest or the dance of wolves bringing down a caribou and the sounds of their devouring the newly transformed carcass into a dinner and then the sounds of others coming to the feast—this is the Feast of Logos. And the sound of a scorpion's sting and the rattlesnake's rattle is also included in this same dance—they too have a part, and there are many other parts in the dance of the Great Earth. Hobbes was not entirely mistaken about the so-called nasty side of nature—but from nature's side there is no nasty perspective, for it is beyond good and evil—and the failure to hear this part of the composition and to learn to dance its dance and be able to dance with it is a profound negligence, for one will be out of step.

Nietzsche and Zhuangzi were not saccharin sages; they understood the beauty of all the movements of the Great Dance; they understood that without death, there is no life, and that without life, there is no death—we must die at the right time to participate in the Great Dance. This comes with an earthbound acceptance—philosophy is the preparation for death, albeit a bit differently imagined from how Socrates might have intended, for it is an acceptance of life. To be earthbound, to welcome and accept being earthbound, and to live as being earthbound is the only hope we can manifest for our species. And by preserving the signature events that transform our beloved planet that we have come to inhabit, our decisions about how we live, how we accept and promote diversity, and how we ultimately define ourselves are the only hope, for this is the Great Hope.

Graham Parkes has always been a purveyor of this hope as he has whispered gleefully in our ears throughout his professional and personal life. The music of his whisper has taken many forms that find the necessary silence between the movements of his concerto—his solo musical compositions with his accompanying instruments are accompanied by the greatest of orchestras of the largest of scales—the Great Earth. His solo instruments of composition sing to others for emulation, inspiration, and call us to go beyond, farther beyond, but not beyond the Great Earth; for this is what signature events do when a participatory self melds into and with the Great Dance. It is at this juncture where emergence gives rise to itself.

A Wandering Dance begins with an aesthetic dance of nature, for the project of becoming earthbound requires us, like no other species, to create and develop a profound sense of the aesthetic. Aesthetic awareness is necessary and vital for the sustainability of the earth. To realize and enhance the beauty of the flowing tide of the earth's becoming through the creation and execution of signature events by an intimate and participatory self gives new meanings and accelerates the process of change across scale. Aesthetic expression, in all its forms, cuts through the illusion that reasoned arguments, driven by either science or philosophy, can change human behavior on our planet.[4]

Graham Parkes's longtime colleague and friend, Roger T. Ames, appropriately begins our collection. Ames is one of the world's most eminent Chinese philosophy scholars and has transformed the landscape of Chinese philosophy by making the philosophy of China relevant to Western audiences while still retaining its originary relevance for the Chinese themselves. Ames's chapter is titled "Graham Parkes, Musical Ecologies, and the Daoist Dance," and in it he argues for a different conception of responsibility from the Chinese worldview. His attention focuses on the *Daodejing* and a contrapuntal responsiveness that entails accountability and authenticity of character and being in the world. In Daoist philosophy, creativity is understood as co-creativity that privileges situation over agency, significance over originality, and, as Ames states, "is expressed in terms of a collaborative, meaning-producing relationality." In contrast to a more rational ordering that is characteristically marked by its reductionism and emphasis on causality, Daoism gives primacy to a dynamic relationality of a holistic and aesthetic sense of ordering. Drawing upon and endorsing Parkes's *How to Think about the Climate Crisis: A Philosophical Guide to Saner Ways of Living* where Parkes offers collaborative relationality as the foundation for a global environmental ethic, Ames suggests this ethic redefines the human role within the Anthropocene and that this advocacy, at least to some extent, is inspired by Parkes's intimate familiarity with the early Daoist canons. As co-creator of Confucian Role Ethics along with Henry Rosemont, Ames suggests that "Parkes has laid the groundwork for what might come to be called a Daoist 'role ethics,' that is, the human capacity and responsibility in our role as cosmic co-creators for 'extending the way' for our own time and place."

Drawing on several sources, Meilin Chinn's "On the Chariot of Earth" follows the many routes Parkes's work takes back to the earth, especially his investigations of inner and outer landscapes of soul, including rocks. Chinn maintains that a common thread throughout Parkes's diverse body of work is how we must correct our understanding of nature so we can readjust our relationship with the natural world and its layered sense of soul, that is, nature is not dead matter. Viewing nature as soulless matter leads to environmental degradation. There is, however, a more appropriate way of understanding nature, and this understanding consists of learning how to read nature's text more properly. The routes of *feng shui*, *anima mundi*, and *homo natura* all converge, according to Chinn, "in a place where all things are vital and perspectives are agile: the world of *qi* 氣 energies in the Chinese tradition." By following the deeper meanings of these routes back to the earth, we gain an enhanced perspective that can assist us to remedy the extinctions and ecological disasters we face today. The primary sources engaged with and emphasized by Chinn are Zhuangzi, Nietzsche, and the depth psychologist James Hillman. These sources, often used by Parkes himself, point to the way back into a nature that is alive, vital, and ever transforming.

In the next chapter, we move more directly into the importance of comparative aesthetics, especially the magnitude of Japanese aesthetics in understanding the human place in the natural sphere vis-à-vis the aesthetic sense emerging from Nietzsche's philosophy. In "The Role of Aesthetics in Nietzsche's Philosophy and Japanese Culture," Yuriko Saito first turns to eighteenth-century modern Western aesthetics, which was primarily concerned with the aesthetic experience of the viewer, audience, and reader that had an emphasis on a disinterestedness aesthetic attitude and experience removed from other areas of the concerns of morality, politics, religion, science, and other practical concerns. This contemporary model of aesthetic discourse still dominates today with its emphasis on the spectator who derives their aesthetic experience from an object for its own sake. Saito finds solace in the work of Nietzsche who observed that Western aesthetics focused on the point of view of the receiver who considered art as from only the point of view of the spectator. Nietzsche's aesthetics focused more on issues generated from the creator's viewpoint, but for Nietzsche his notion of creator encompassed human beings as artists in a grander sense, that is, of creators of the self and their lives. Saito concludes that aesthetics is inseparable from all life concerns.

Paralleling Nietzsche's emphasis is the array of what constitutes traditional Japanese aesthetics, which is primarily shaped by the writings of its practitioners. The various arts comprising Japanese art—painting, tea ceremony, flower arrangement, haiku, linked verse, Noh theatre, and martial arts—are often linked to the life of the everyday. By its nature, discussion of the aesthetic experience relies less on the work as object, the audience, or about literary or artistic criticism and more on the convergence of the moral, spiritual, and practical aspects of what is constitutive of the good life and how to live it. Saito explores the resonances and divergences between the role of aesthetics in Nietzsche's philosophy of life and the main themes of traditional and pre-modern Japanese aesthetics.

In her chapter, Kathleen M. Higgins pursues her continued perplexity of "Heidegger's Evasion of Music," which is the title of her chapter. Focusing on *Being and Time* and "The Origin of the Work of Art," her primary question is why Heidegger ignores the topic of music while being more interested in other artistic topics, such as the visual arts and poetry. One possibility is this is something Heidegger knew little about or lacked the depth to discuss music in detail. Nevertheless, Heidegger knew little about Daoism but was willing to borrow heavily from it, as Parkes has pointed out in his translation of Reinhard May's *Heidegger's Hidden Sources*.[5] Avoiding speculation on Heidegger's motives, Higgins indicates various features of Heidegger's project that might have given him philosophical pause for the inclusion of music in his phenomenological philosophy. Although music might have been beneficial and even consistent with some of Heidegger's aims, these goals may have nevertheless promoted his omission of music. Higgins offers readers three possible reasons for this omission. The first reason is that music, unlike the visual representational arts and language, does not bring forth a world. One can pick out, for example, an aspect of a painting that has the capacity to explain the experiential field of the world's flux—language has this same capability. A musical note, or individual chord, may lack this capacity and need more notes, chords, and so forth to accomplish Heidegger's objectives. Higgins's second point is concerned with Heidegger's awareness—driven by Nietzsche's awareness of music's capacity to "forge solidarity"—that, as she writes, "music can draw attention to what people qua human beings have in common, or more aptly, it can make us *feel* this connection. No particular identification with a given community is required." For Nietzsche, the Dionysian meant that all political situatedness could be transcended. A third possibility noted by Higgins—that Heidegger subsumes musical qualities under language with the power of rhythm—is encompassed by poetic language. She points out that these possibilities are not necessarily exclusive, but that with respect to music, Heidegger is closer to the Confucian position on music, and not the Daoist philosophers who emphasized the subjective and Dionysian side of music.

In the penultimate chapter of this section, Peter D. Hershock discusses philosophizing as an embodied practice in his "Improvising with Soul: Philosophizing in an Interpretative Key" by developing an alternative to the compositional metaphor used by Parkes in his *Composing the Soul*. The approach taken by Hershock is anticipated by Parkes in his exchanges with others that are "dynamic kin to the practice of 'trading fours' and interweaving soloing explorations in improvisational jazz and blues." According to Hershock, improvising with soul takes precedence over composing the soul and in its antecedence gains more purchase in resolving the value conflicts arising from, as he says, "climate change, persistent global hunger, and the new global attention economy and surveillance capitalism." Following Buddhist sources, Hershock argues that Buddha-nature consisted of creating an intentional disposition on realizing a relational dynamics that expressed the meaning and the means to liberating all things from conflict, trouble, and suffering. In other words, for Hershock, "the means to and meaning of uncompelled presence" is what is at issue. He continues by building a bridge between musical and ethical improvisation. In freely improvised music a "sonic/aesthetic ecology" emerges from the emptiness of a "*jointly evoked, wholly shared* world" where there is "an immediate coordination of values or modalities of appreciation in the absence of representation and reflection." An implication of this when applied to ethics is that ethics begins to expose its aesthetic dimensions and becomes profoundly historical. Ethical performance is not about achieving some goal of the good or the avoidance of doing bad things, but to realize an exemplary relational dynamics with other performers. Hershock concludes that ethics is "improvisational art or way (*dao*) of virtuosic human course correction."

Andrew K. Whitehead's chapter on "Image Thinking: Cinematographic Flâneurs and Twenty-First-Century Philosophy" shifts our aesthetic attention to film and is the concluding chapter in the book's first section. Whitehead's primary interest is to explore the possibility of employing alternatives, especially film, for philosophical expression and even for thinking. In other words, to what extent can philosophy be expressed and ultimately practiced in mediums other than academic writing? What is at issue for Whitehead is the role of language in philosophy and philosophizing and that philosophy is necessarily a linguistic practice. Following recent challenges from phenomenological research, he uncovers through Dieter Lohmar's notions of non-linguistic thinking in Husserl's writing disclosures of "scenic phantasmata" or "images or sketches of objects, persons, events, or situations in both dreams and daydreams." This idea of "phantasmatic" thinking is fruitful for Whitehead in his developing film as a philosophical medium and how film significantly contributes to the practice of philosophy. At bottom is the idea that phantasmatic thinking accelerates thinking and enables syntheses of different aspects more effectively than linguistic and linear thinking. As such, film can be seen as a more suitable medium for the expression of certain thoughts or modes of thought related to philosophical undertakings. Whitehead critically reviews the distinction between philosophy *of* film and film *as* philosophy by placing the writings and cinematographic works of Graham Parkes in dialogue with Paisley Livingston, who Whitehead deems as "a skeptical voice and less-than-sympathetic interlocutor." By doing so, he lines up with Parkes and provides an argument that film can be used in making genuine contributions to philosophy. However, underlying Whitehead's argument is a more substantial question—"What is Philosophy?"

The next section of *A Wandering Dance* focuses more specifically on "The Practice of Nature." This section includes chapters by Jason M. Wirth, Leah Kalmanson, and Bradley Douglas Park. We begin with "Rubble and the Philosopher's Stone: The Practice of Philosophy and the Philosophy of Practice." Jason M. Wirth embraces the work and legacy of Graham Parkes's attentiveness to the strange and nearly incomprehensible voices of rocks and stones as being intrinsically valuable. In doing so, Wirth suggests that by listening to these "inanimate objects" of our experience we are accepting an invitation to "be more philosophical about what matters as philosophy and how it does so." In his translation of François Berthier's text on Ryōanji, Parkes notes that "we don't normally think of rocks as having language, nor of gardens devoid of vegetation as posing questions, nor of stone as something that can be read, and possessing surfaces reflecting something in the depths of human being." Through a meditation on the Kubota Garden in Seattle—a place dear and near to Wirth—he reminds readers that the peaceful state of rocks/stones was born from the earth's violent drama of their coming to be and how they represent good examples of what Japanese Buddhists call "form" in their slow evolution of becoming what they are. In other words, rocks are masters of holding to their form even though they also move. Rocks too change, albeit slowly and deliberately from their birth to the constant weathering they experience before their disintegration—they are connected to the interdependent vitality of all beings; they are "empty" (*kū* 空), for according to the *Heart Sutra*, "form is emptiness and emptiness is form." Rocks teach us to be empty and to embrace this emptiness is to embrace impermanence, without which there is no change, no evolving, no becoming who we are. Wirth concludes that "by transforming the icons of density and heaviness—the lithic realm—we can detect an avenue to a new sense of philosophy as a practice of overcoming. Philosophy is simultaneously an explication of that practice and its enactment."

Our discussion of lithophilia continues with Leah Kalmanson's "Philosophy as Petromania: Graham Parkes, Jane Bennett, and the (Not So) New Materialisms." In this chapter, she poses the question: Is "new materialism" a type of *qi*-studies (*qixue*)? By bringing together Bennett's work of how to enact ecological reform around her new materialism with the *qixue* and *lixue* (*li*-studies—*li*, principle or emergent pattern, which is woven out of *qi* studies), Kalmanson suggests that Graham Parkes and the Song-Ming philosophers point out in their distinct ways that *habits* cannot be countered and reforming our habits requires the adoption of better practices. She reminds us how the Chinese tradition is a rich repository full of practices for cultivating persons who have the potential to cultivate better worlds. Parkes has often referred to the general idea that all forms of physical matter are condensed *qi* and that rocks especially represent, as Kalmanson says, "vibrant and fantastical concentrations" of *qi*. As we see in Wirth's chapter, but now in Kalmanson's words that "the heft and rigidity of the stone does not indicate inertness but rather dynamism and power—the stone is not immobile but *unyielding*; its form is *decisive*; its living force is demonstrated precisely by its resolute materiality." The vital stuff of things, or *qi*, infuses bodies, mental awareness, and the physical environment that are mutually influencing and responsive to each other. Hence, there is no difference between, as Kalmanson writes, "sentient and insentient, organic and inorganic, physical and biological, and so forth; there are only different configurations of *qi* at microcosmic and macrocosmic levels, which can, under optimal circumstances, be productively attuned to each other." *Qi*-realism is dissimilar to and can be understood as differing from animism because there is no attribution of animate qualities to inanimate objects. From the outset, *qi*-realism helps avoid the dualistic distinction between animate/inanimate—everything is *qi*-real, that is, stones and rocks, along with everything else, are all different configurations of the "vital stuff" of the world.

Bradley Douglas Park's "A Mind Possessed: The Rhythmic Body and Entraining the Social Mind" ends this section and serves as a prelude to the last. In the wake of Parkes's work on the body, Park pursues the body as a field of resonance "in an open and pulsing dance with its environment," one that "solicits entrainment and then calibrates itself through this history." Thus, our being is rhythmic and our body-mind dances as it generates its own time through its own intrinsic timing as it affects what Park identifies as "new horizons of sensitivity—new edges of influence and perturbation." These horizons of sensitivity are our openness to the world and those other temporalities with whom we share the world. We thus entangle ourselves with other beings in the world. Park sets up all of this through Nietzsche's insistence on the "great Reason" of the body that represents the Dionysian and Heraclitus's commitment to the world's flux and the challenge to Plato's devotion to the "right mind" and self-consciousness. Park believes that interbodily entrainment offers the most rewarding explanatory pathway because of its grounding in direct causal coupling. When one system's movement or frequency of signal, such as in pendulums or the colony phasing of rhythms in bees, entrains the frequency of another system, direct causal coupling occurs. As a universal phenomenon, through entrainment we train new bodies, or as Park imparts "to have a world, and thereby a mind." Park insists that we are primed to dance in the world and that by learning "to dance" with other bodies in the world, we become minded. Thus, the rhythmic body is fundamental to the possibility of being-in-the-world.

We wander to the last (and longest dances) of *A Wandering Dance* with chapters by Timothy J. Freeman and David Jones. With our succession as students reversed, this section opens with Freeman's "Staying True to the Earth in Zarathustra, Zhuangzi, and Zen." In this chapter he engages Parkes's work by drawing further attention to the resonances between Nietzsche's thought, Daoism, and Zen philosophy and how to rethink these distinct philosophies by playing them off each other. In doing so, Freeman pays special attention to the relevance of the project by emphasizing how this rethinking is significant in a time of ecological crises, a time when the very future of life on earth is at risk. Freeman first focuses on Parkes's suggestion of the experience of "seeing things as they are" in *Zarathustra* and the *Zhuangzi* and then advances his thinking by applying the problem posed by the notion of will to power, as well as the notion of "seeing things as they are" in Zen. In taking up the comparison between Nietzsche's thought and Buddhism, he sharpens the discussion with a critical reflection on Parkes's emphasis on the importance of a "psychical transformation" in Nietzsche's thought, Daoism, and Zen. In his honing of the discussion with Parkes, Freeman reflects directly on the idea of eternal recurrence, the key idea in Zarathustra's call to stay true to the earth, and its possible resonances with Daoism and Zen, especially with "the bodhisattva vow to return to life over and over … to save each and every one of the numberless beings in the universe" and the significance of laughter and joking in staying true to the earth.

I have positioned my contribution to this collection last. I do this not to give myself any special place, nor to have the last word. Perhaps it is appropriate, however, since I was the first that I should now be last, for as Heraclitus reminds us: "In a circling, beginning and end are one and the same." As stated in the beginning of this introduction, being Graham Parkes's first student was a fortuitous occasion, even though I remain honored to have that divergent distinction. There is, however, a poetic appropriateness in the placement of "Downbeats and Upbeats in the Meters of Nietzsche, Zhuangzi, and Parkes" in this book by way of its nascent conception appearing first in my PhD thesis and then in print as "Crossing Currents: The Over-flowing/Flowing-over Soul in *Zarathustra and Zhuangzi*."[6] Spawned by Parkes's article "The Wandering Dance: Chuang Tzu and Zarathustra," I pondered not only the resonances between Zhuangzi and Nietzsche's thinking, as Parkes had so aptly accomplished, but the felt difference between the two and what could emerge, what kind of dance could arise, between these two disparate philosophers coming from inherently different worldviews and cultural contexts. Using the metaphor of water, their streams seemed to wander in divergent ways. "Downbeats and Upbeats in the Meters of Nietzsche, Zhuangzi, and Parkes" looks more closely and plays out this divergence while retaining the folding into itself of their resonances and by releasing their effusing dissonances.

* * *

A Wandering Dance through the Philosophy of Graham Parkes: Comparative Perspectives on Art and Nature is a rich collection of essays that not only celebrate the oeuvre of Graham Parkes's contributions to comparative and world philosophy, but critically engage and use his work's corpus as a point of departure for their own work and the evolution of philosophy—for this is what signature events do; they amplify into the future, mutate in their own ways, and transform the very system in which they originated. This book represents only a fraction of what Graham Parkes has spawned in his career and life. Our contributors are just some of the dancing stars to whom he has given birth.

In gratitude, I dedicate this book to Graham Parkes and give deep thanks to its contributors. I am most grateful to Setsuko Aihara for use of the image of her painting

Wanderer Dancing on the Nietzsche Rock (after Friedrich) that graces the cover. In addition, I express my gratitude to Colleen Coalter for her support of this project and her steadfast patience that spanned the years of the pandemic that postponed its completion. My gratitude also extends to the Bloomsbury staff, especially to Suzie Nash, and our anonymous reviewers. Without Graham Parkes, this superb collection and the splendid magnificence of its authors could not exist in their phenomenally interesting ways. And there are so many other wandering dancers out there who have listened and continue to listen to his music, well beyond the covers of this *Wandering Dance*.

—David Jones, Kīlauea Summit

Notes

1. *The Fractal Self: Science, Philosophy, and the Evolution of Human Cooperation*, John L. Culliney and David Jones. University of Hawai'i Press, 2017.
2. Heraclitus translations are my own.
3. "The Wandering Dance: *Chuang Tzu* and *Zarathustra*," Graham Parkes. *Philosophy East-West* 1982.
4. Statistics link climate change to the carbon emissions and we have been slow to act. For example, see https://education.nationalgeographic.org/resource/global-co2-emissions/.
5. As Parkes states in his complementary essay "Rising Sun over Black Forest," "in view of the conclusions drawn by Reinhard May, one is forced to entertain the possibility that this harmony may have been occasioned by some quiet appropriation on Heidegger's part" (84).
6. I am grateful to Huang Yong for permission to recast the article in its current and most divergent form that first appeared in *Dao: Journal of Comparative Philosophy* Volume IV, No. 2 (June 2005).

Part One

The First Dance

1

The Wandering Dance: *Zhuangzi* and *Zarathustra*

Graham Parkes

Nietzsche's *Thus Spoke Zarathustra* and the collection of chapters known as the *Zhuangzi* are philosophical texts from quite different times and places, written in absolutely unrelated languages and stemming from totally disparate historical and cultural circumstances. The *Zhuangzi* dates from the end of the Warring States period in ancient China and is, along with the *Laozi*, the central text of Daoist philosophy, a source ever returned to and a lasting formative influence on Chinese thought. While *Zarathustra* prophetically ushers in the twentieth century, it more ominously finishes off two and a half millennia of Western metaphysics. But in spite of these differences, the style of these two texts is so similar and their philosophical content so uncannily congruent that a careful comparison is called for, on several grounds. Both texts have been rather poorly understood, owing in part to the radical nature of the philosophy they embody. Moreover, to place the obscure alongside the obscure sometimes can effect clarification on both sides. As awareness of congruence develops, an idea on one side which at first appears to lack a counterpart may bring to our attention a hitherto unnoticed element on the other. And in the case of these texts the parallels will turn out to be so striking that we shall be impelled to search all the more keenly for the essential differences—an undertaking that will enhance our understanding of both philosophies.

What follows is intended as prolegomena to a wider and deeper study of Nietzsche and Zhuangzi as psychologically acute philosophers intent on effecting a transformation of our ideas of self and world—and thereby of ourselves. Since no work has yet (to my knowledge) appeared in English in this area, the following paper will stake out the ground, ranging widely to uncover the appropriate topics for comparison, rather than delving deeply into any one of them. Such more detailed work will remain to be done—by these or others' hands.

If the first phase of such an extensive comparison is to stand on its own, its scope must be restricted—and we find a ground for such restriction in the disparate bulks of the corpora of our two thinkers. Little is known about the historical person Zhuangzi—no more than about his near-contemporary and kindred spirit Heraclitus the Obscure (a figure who will, by the way, be shadowing the course of the comparison all along). The text of the *Zhuangzi* is not uniform but composite, put together in such a way as to suggest a number of authors from different periods. However, it is generally agreed that the so-called "inner chapters" (which comprise seven of the thirty-three extant sections) are composed by a single author, whom we call "Zhuangzi," and that these embody the central ideas of the work as a whole. In what follows, I shall draw primarily from the inner chapters, using passages from the "outer" and miscellaneous chapters (which represent the earliest commentary upon the text) where they articulate a theme or elaborate an idea merely implicit in the core of the work. Whereas the thirteen thousand or so characters of the inner chapters are all that is known to us of Zhuangzi's writings, Nietzsche's literary output was enormous, amounting to some ten thousand pages in the latest critical edition. Since we have more text of Nietzsche than of almost any other great philosopher, I shall focus on one text in particular, *Thus Spoke Zarathustra*, since it is closest in style to the *Zhuangzi*, and because Nietzsche considered it his best work and the fullest embodiment of his mature thought.

A Correspondence of Style

The Zhuangzi is a patchwork of anecdotes and dialogues, often with no apparent order, and only loosely unified thematically in each of the seven chapters under consideration. Zarathustra's speeches are similarly episodic and yet are held together by a relatively coherent narrative that chronicles his alternating engagements with society and withdrawals into isolation, culminating in an eventual transformation of the speaker's psyche. Just as Zarathustra and the book's other characters are more or less closely fitting masks of their author, so the philosophy of Zhuangzi is imparted by a large cast of characters, with Confucius as the most prominent—and including on occasion Zhuangzi himself. As one might expect from an ironist in the same class as Nietzsche, the roles played by Confucius (who for Zhuangzi is the equivalent of Socrates/Plato for Nietzsche) range from unenlightened straight-man to straightforward proponent of Zhuangzi's own views. Both works are supremely poetic and stem from a deep understanding—philological as well as psychological/philosophical—of their respective traditions. The language of the *Zhuangzi* is of a quality unparalleled amongst ancient Chinese texts; and even if one finds the philosophy of Zarathustra distasteful, it is impossible to deny that the text is stylistically one of the most rich and powerful in the Western philosophical tradition. While the elements of parody in the *Zhuangzi* do not have the force or the weight of Zarathustra's frequently oracular/Biblical tone, both works are deeply humorous—each constituting perhaps the most *amusing* philosophy of its tradition—emphasizing laughter as an often necessary concomitant of insight into the ways things are.[1]

Most important of all, *Zarathustra* and the *Zhuangzi* are first and foremost works of *imagery*. This has given rise to a reluctance to consider them as genuine philosophy at all, but such a judgment stems from too parochial an understanding of the nature of philosophy. It is easy (and convenient) for the short sighted to forget that philosophy, now in the Anglo-American tradition mostly purged of images and under the exclusive tyranny of concepts, began as poetry, with the poetic utterances of Xenophanes, Heraclitus, and Parmenides. Forgetting that the philosophic treatise was not invented until Aristotle, one fails to notice that Plato (although by no means lenient toward poets) himself complements concept with image, argument with anecdote, and frequently, as he runs up against the limits of rational dialectic, has recourse to the vivid imagery of ancient myth. And just as in Plato the poetic style and dramatic form are integral to the substance of his philosophy, so, too, in *Zarathustra* and the *Zhuangzi* the imagistic *way* the ideas are presented comprises the philosophical import.

Beyond being works of the philosophical imagination, both texts share the same *kinds* of images. The primary source of imagery is the natural world: the elements—sky, earth, fire, and water; the sun, moon, and stars; the climate, weather, and seasons; and the realms of plant and animal. I think the reason for the predominance of images from the natural world is significant; how both philosophers say what they have to say reflects what they have to say. They are concerned to promote a particular way of being in the world—a mode of involved yet reflective *participation* in the world rather than of detached *observation* of the cosmos.[2]

The common enemy is anthropocentrism, the unquestioning prejudice in favor of the human perspective on the world at the expense of all others, and both thinkers seek to foster an appreciation of alternative perspectives by weighting the imagery on the side of the nonhuman cosmos. Their motivations are, however, somewhat different. Zhuangzi is reacting against the tendency of many of the Hundred Schools which preceded him, and of Confucianism and Mohism in particular, to place man at the center of the cosmos, though still acknowledging his participation in it. Nietzsche has a more serious imbalance to redress, the result of two millennia of Platonic/Christian denial of the body and man's animal nature: the tendency, encouraged by Cartesianism, for man to identify himself as being essentially mind, or spirit, and to ignore or devalue his participation in the physical world. (The aim would then be, in Wallace Stevens's words, "the body quickened and the mind in root."[3]) But in spite of such differences in philosophical milieu, Nietzsche's later understanding of will to power as an interpretive energy inherent in all things comes close to the pan psychism that informs the *Zhuangzi*.

Since both Zhuangzi and Nietzsche are philosophers of *flux* (with the *Yijing* and the fragments of Heraclitus as their precursors), their predilection amongst the elements lies with water, as the most obviously mutable. The *Laozi* established water as a primary image for the psychical fluidity at which Daoist philosophy aims, and the imagery of the *Zhuangzi* draws heavily from lakes and rivers and seas (see, especially, chapters 1 and 17). The *Zhuangzi* often evokes water to illustrate relativity of perspective, and we might hear its "A fish by staying in the water lives, a man by staying in the water dies" (Zz 18, 189) as an echo, from half-way around the globe, of Heraclitus's fragment 61 of two centuries earlier: "Sea water is at once very pure and very foul: it is drinkable and healthful for fishes, but undrinkable and deadly for men." *Zarathustra* plays frequently with water in its possibilities of being contained and containing, and presents the flow from lake through river to sea as a metaphor for the transformation of the encapsulated self, through the overflowing of libido or psychic energy, into a state of open participation in the cosmic play of will to power.[4]

The predominant image from the realm of vegetation in both texts is the *tree*. The human being is like the tree insofar as its natural unfolding comes from, and psychically mirrors, the interplay between the four elements (in the Chinese tradition wood is itself a fifth element, or rather *process*). And as the tree in the form and direction of its growth spans the dimension between the primal powers of sky and earth, so man lives and dies (as Blake so vividly reminds us) at the interface between the upper and lower realms, and between the powers of light and darkness. The Daoist idea of the interdependence of good and evil is consummately exemplified in the section of *Zarathustra* entitled "By the Tree on the Mountainside": "It is the same with man as the tree. The more he strives upwards into height and light, the more strongly do his roots strive earthwards, downwards, into the dark, the depths-into evil" (*Za* 1.8). It is a law of psychological development for Nietzsche (as for depth psychologists like Freud and Jung after him) that over concentration on the "higher" pole of a continuum of opposites leads to a toppling over of the psychical tree, and such one-sidedness would, for Zhuangzi, be like "taking heaven as your authority and doing without earth" (*Zz* 17, 147).

However, the abundant flora in both texts are by far outweighed by the mass of *fauna*. The most common device in the *Zhuangzi* used to point up the arbitrary limitations of the anthropocentric perspective is the poetic evocation of the perspective of a variety of species of animal. Some of the most telling and famous passages invoke fishes and birds of mythical proportions, frogs in wells, and swarms of different insects. Just as the *Zhuangzi* contains a richness of fauna unrivalled by any other Oriental philosophical text, so the bestiary of *Zarathustra* is unique in Western philosophical literature. Not since Aristotle's magnificent treatises on animals has such a vast and varied menagerie crawled, soared, trotted, and slithered across the pages of a philosophical text: over seventy different species are mentioned by name. Zarathustra's own animals, the eagle and the serpent (representing the extremes of the aerial and the chthonic), are privy to his most secret thoughts and are far more vocal than he in the second major attempt (in "The Convalescent") to give voice to the central idea of the book—the eternal recurrence. Although Zhuangzi's stories are designed to draw the reader into a variety of animal perspectives, whereas Nietzsche's employment of beasts is more in metaphor and simile than in invitational anecdote, there is a significant parallel in the kinds of animals that predominate.[5] In both works mammals, while abundant, are remarkably few in comparison with reptiles, insects, birds, and fish. The significance of this emphasis is, I think, this: mammals are the animals closest to human beings and those with whom it is easiest to fall into the pathetic fallacy; it is far less easy to project human feelings and emotions onto creatures such as insects and fish. The relative " otherness" of non-mammals serves Nietzsche's and Zhuangzi's purposes of knocking us off our anthropocentrism, and yet they both manage at the same time to employ this otherness to enhance the sense of our participation in the realms of the nonhuman.

Radical Perspectivism

A major difference between conceptual and imaginative philosophies is that while the former are generally articulated in a framework independent of the reader that can be assented to or rejected at a safe distance, a philosophy presented in images works on the reader's psyche by inviting the kind of participation in their play that effects a psychical transformation more radical than just a change of mind. Whereas conceptual thinking is predicated upon the idea of opposites (concepts tend to grasp by excluding their opposites), images play along a continuum of gradations and embrace opposites in their mutual interdependence.

A philosophy that acknowledges the relativity of opposites tends to be a perspectivism (how things appear depends on your point of view, your place on the continuum) as well as a philosophy of flux. While such philosophies generally have been less popular than those that assert the existence of unchanging and absolute values, the dynamic perspectivisms of both Zhuangzi and Nietzsche have important and illustrious precursors. The *Yijing* is the archetypal work of flux: there, to borrow a couple of fragments (12 and 57) from Heraclitus, "upon those who step in the same rivers, different and again different waters flow. [...] In changing, it [the ever-living fire] is at rest." Such a philosophy of flux leads naturally to a perspectivism: the opposites of yin and yang are intimately linked, each depending on the other in order to be what it is and having the germ of the other immanent in it; what is going on depends on what has been going on and where the process is heading; the value of a line in a hexagram depends on what is above it and what is below. The perspectivism of the *Zhuangzi* is all the more radical since it is reacting against what it understands as the moral absolutism of Confucianism. Similarly, Nietzsche's perspectivism is a reaction against the essentialism of Platonism and Christian philosophy and also harks back to the beginnings of Western philosophy with Heraclitus. The central, anti-essentialist message of both our texts is well summed up in fragment 111 of that darkly seminal thinker: "It is by disease that health is pleasant, by evil that good is pleasant, by hunger satiety, by weariness rest."

The most general way in which the Zhuangzi expresses the interdependence of opposites is through the primal forces of heaven and earth and yin and yang.

> If then we say "Why not take the right as our authority and do without the wrong, take the ordered as our authority and do away with the unruly", this is failing to understand the pattern of heaven and earth, and the myriad things as they essentially are. It is as though you were to take heaven as your authority and do without earth, take the Yin as your authority and do without the Yang; that this is impracticable is plain enough.
>
> (*Zz* 17, 147)

Zhuangzi makes the same claim in terms of the relatively neutral pronouns "it" and "other":

> No thing is not "other," no thing is not "it." If you treat yourself too as "other" they do not appear, if you know of yourself you know of them. Hence it is said: "'Other' comes out from 'it,' 'it' likewise goes by 'other,'" the opinion that "it" and "other" are born simultaneously. However, "Simultaneously with being alive one dies," and simultaneously with dying one is alive [...].
>
> (*Zz* 2, 52)

Here Zhuangzi is doing two things: he is suggesting that since the opposites "it" and "other" have generated one another, there is the possibility—through seeing how this happens—of their being annulled, and he is also introducing the opposition whose interdependence is most emphasized in the Zhuangzi—that between life and death.[6] (It is appropriate here to be reminded of Socrates's arguments concerning the interdependence of life and death in the *Phaedo*, and of the fragments of Heraclitus—numbers 62, 77, and 88—which lie behind them.)

Now while the mutual interdependence of life and death does not receive explicit elaboration in Zarathustra, the theme runs throughout the text in the constant interplay between *übergehen* (to go over, across) and *untergehen* (to go under, to die, to perish) and *überwinden* (to overcome). In order to go across, to undergo the transition to the overman (*Übermensch*), Zarathustra must overcome himself—and this he does by going under, by dying away from the self as an encapsulated ego, separate from the cosmic play of will to power. One of his frequent voicings of the dictum "The human is something that is to be overcome" is echoed a few lines later by the eyes of the pale criminal: "My I is something that is to be overcome" (*Za* 1.5, 1.6). And for Zhuangzi, "The utmost man is selfless" (*Zz* 1, 45).

Valuing

Zhuangzi and Nietzsche both agree on the genealogy of opposites: these arise from evaluation undertaken from a particular perspective, from perspectival value judgments. For both thinkers, the most common perspective from which value judgments are made is the perspective of utility. Nietzsche constantly argues (more in works other than *Zarathustra*) that the motive for valuing is control, that we simplify the manifold of experience by discriminating into opposites for the sake of power over the environment and other people. "No people could live without first valuing; [...] A tablet of the good hangs over every people. See, it is the tablet of their overcomings; see, it is the voice of their will to power" (*Za* 1.15).

Just as for Nietzsche every being manifests will to power—primarily through interpreting and construing the world in terms of values—so, too, for Zhuangzi every being has its own perspective, determined by the conditions particular to it. The emphasis may be on spatial conditions (the Peng bird is so enormous that it needs an exceptionally strong wind in order to take off; the well-frog's perspective is limited by the space he lives in) or on temporal limiting factors (the morning mushroom knows nothing of the moon, the summer cicada knows nothing of spring and autumn) (*Zz* 1, 44). The value judgments may be utilitarian, gastronomic, or sexual/aesthetic (*Zz* 2, 58), but, whatever they are, they work only in a particular context and are, beyond a specific perspective, invalid.

Utility is a value that is especially context-dependent, since things of use are used for a particular function. In the "Autumn Floods" chapter of the *Zhuangzi*, the spirit of which is especially close to that of the first two inner chapters, we read: "That a large beam can be used to batter down a city wall but can't be used to plug up a small hole speaks of a difference in function" (*Zz* 17, 147). The anti-essentialist message here is that there is nothing inherently useful about a battering ram: it depends on what it is to be used for. The context-dependence of the useful is brought out at its most fundamental level in the delightful encounter between Zhuangzi and Huizi in chapter 26 concerning a topic dear to Nietzsche's heart—the relation between ground and lack of ground (*Zz* 26, 100). The point here is that any footprint-sized piece of the broad earth's surface is not grounding per se: it is so only in relation to the ground that surrounds it.

It is not that the perspective of utility is in itself wrong or a "bad thing": what both thinkers are concerned to emphasize (and Nietzsche does this with respect to the practical scientific attitude in general) is that it is only one perspective amongst many. The problem arises when we become fixated in a particular perspective, as demonstrated in the interchange between Zhuangzi and Huizi concerning the enormous gourds, at the end of the first chapter. The gist of this passage could be paraphrased in Heideggerian language by saying that fixity of perspective leads to an obsessive concentration on actuality and a corresponding blindness to the myriad possibilities in every situation. Both Nietzsche and Zhuangzi would, I think, agree that the motivation for our discriminating continuums of gradations into pairs of polar opposites and our taking a stand with a fixed perspective at one or other end is a strong feeling of malaise in the face of the perpetual flux of existence.

However, the phenomenon that most radically relativizes all perspectives of utility is the dream.

Dreaming

While philosophers have generally had remarkably little to say about dreaming, Nietzsche and Zhuangzi are exceptions, and their ideas on the topic are again quite congruent. Dreaming comes up in both texts in two ways: firstly, we are presented with reports of particular dreams, and secondly, it is suggested that dreaming is a universal condition of our being here—that we are always dreaming.

We receive reports of four dreams dreamed by Zarathustra, and they play a crucial role in the development of his relations with his disciples and the world in general, and also in his increasingly successful attempts to voice the thought of the eternal recurrence. Part Two opens with a dream which stimulates him to go down once again from his mountain-top retreat and to re-engage the world of men (*Za* 2.1); a second dream, in which Zarathustra finds himself "a guardian of tombs in the lonely mountain-castle of death," enables him to undertake the redemption of his personal past (2.19); and a third dream, in which his "stillest hour" assures him that he knows the thought of recurrence, brings Part Two to a close by prompting him to leave his disciples and withdraw once again into solitude (2.22). In a fourth dream, Zarathustra stands "beyond the world" with a pair of scales and weighs it. The world presents itself to him "as if a tree waved to me, broad-branched, strong-willed, bent as a support, even as a footstool for one weary of his way [...]" (3.10).

In an important dream in the inner chapters, an enormously broad oak appears to carpenter Shi, just after he has in waking life condemned it as worthless, and berates him for being stuck in the perspective of utility and failing to appreciate "the usefulness of being useless." As a parting shot, the tree reminds the carpenter that he too is about to die (*Zz* 4, 73). Whereas Zarathustra's death dream opens up a new perspective on his past, Zhuangzi's death dream, in which he is addressed by a skull he is using as a pillow, overturns his perspective on life by portraying existence in the underworld as much preferable (*Zz* 18, 124–5).

The most celebrated dream in the text, in which Zhuangzi dreams he is a butterfly, has further implications for his perspectivism: since when we are in the dream-world it can present itself as fully real and our sense of the reality of the day-world is lost, that reality is put into question and the day-world perspective radically relativized.[7] The story of the dream makes the further point, relevant also to Nietzsche's perspectivism, that when one is in a certain perspective it is impossible to see it *as a perspective*. Only when we are placed in a different perspective can we appreciate the limitations of our former standpoint.

It is a short step from saying that we are always in some perspective or other to saying that we are always in some dream or fantasy or other. In chapter 6, Confucius asks Yan Hui: "Is it just that you and I are the ones who have not yet begun to waken from our dream?" (*Zz* 6, 90). This should shake our confidence, he continues, in thinking that we know the true nature of the "I" who supposedly "does this and does that." The point is made more emphatically in chapter 2, where Chang Wuxi says: "While we dream we do not know that we are dreaming […]; not until we wake do we know that we were dreaming. Only at the ultimate awakening shall we know that this is the ultimate dream. […] You and Confucius are both dreams, and I who call you a dream am also a dream" (*Zz* 2, 59–60). In addition to establishing the universality of dreaming for Zhuangzi, this passage points up an important feature of his perspectivism and its further congruence with Nietzsche's. Like Nietzsche, who emphasizes that experience is always necessarily perspectival, Zhuangzi does not believe that we could ever attain a kind of "perspective-less seeing." What we wake up to is the realization that we are always bound by some perspective: this awakening is itself a perspective—but one that acknowledges and embraces the multiplicity of all possible perspectives, and is thus "open in every direction" (*Zz* 17, 148).

From his first published book, *The Birth of Tragedy*, in which he spoke of the necessity for the Greeks of interposing the Apollonian dream-world of beautiful illusion between themselves and the Dionysian abyss, Nietzsche has emphasized the way in which deep-level fantasy activity conditions all our experience.[8] While Zarathustra elaborates (and itself exemplifies) the idea of existence's being a product of creative/interpretive will to power, we find a more concise presentation of this idea in *The Gay Science*, the book Nietzsche published just before *Zarathustra*. Speaking of his discovery that "the whole past of humanity and animality" conditions our present experience at an unconscious level, he writes:

> I suddenly woke up in the midst of this dream, but only to the awareness that I am dreaming and that I must go on dreaming in order not to perish […] that amongst all these dreamers I too, the "one who recognizes," am dancing my dance, that the one who recognizes this is a means for prolonging the earthly dance and so belongs to the masters of ceremonies of existence […] in order to preserve the

universality of dreaming and the mutual comprehension of all these dreamers and thereby to preserve the continuation of the dream (aphorism 54).

As in the *Zhuangzi*, it is not that there is anything wrong with dreaming while awake; being in a perspective conditioned by unconscious fantasy is an essential way human beings are in the world. What is blameworthy is the refusal to admit that we are dreamers, to become aware of the extent to which the "real world" is projected by human needs and desires, and to celebrate this creative activity by both seeing through and playing with it at the same time.

Wandering

It is significant that both our texts, in order to impart a sense of a philosophically more healthy response to the perspectival nature of human existence, employ the same image—that of wandering. The first chapter of the Zhuangzi is entitled *Xiao yao you*: "free and easy wandering," "going rambling without a destination," or "wandering in unconditioned freedom." The anecdotes of this section, which conduct the reader through a variety of perspectives ranging from the vegetative through the animal to the human, all point up the limitations of adopting a fixed standpoint. They, and many of the anecdotes which follow, offer the alternative of a perspectival fluidity and flexibility, a free and easy wandering through a multiplicity of possible points of view. Such wandering is often contrasted with a fixed moral (generally Confucian) position of benevolence or righteousness.

Zarathustra is established as a wanderer at the very beginning of the Prologue where, on coming down from the mountain, he meets the old saint, whose first words to him are: "No stranger to me is this wanderer [...] Zarathustra. [...] Does he not walk like a dancer?" (*Za*, Prologue 2). The subsequent narrative follows Zarathustra's career as he wanders from place to place, trying out the perspectives of mountain top and valley, underworld and ocean. Part Three opens with a section entitled "The Wanderer," in which, anticipating his first major attempt at imparting the thought of the recurrence, and reaffirming his awareness of the interdependence of opposites on the psychological path, Zarathustra says: "Before my highest mountain I stand and before my longest wandering; to that end I must first go down deeper than I ever descended" (*Za* 3.1).

The paths of self-transformation in both texts are not straightforward but rather crooked, and often lead backwards or round in circles. They thus reflect a predilection of both philosophers for the curved and bent over the straight and upright (favored by most paths of transcendence). "My walk goes backward and goes crooked," says the madman of Chu, who is generally an advocate of good sense in the *Zhuangzi* (*Zz* 4, 75). "All good things approach their goal crookedly," says Zarathustra (*Za* 4.13.17); and at the end of the book, one of the "higher men" asks Zarathustra: "Isn't the perfect sage fond of walking on the most crooked ways? Your appearance is the evidence!" (4.18.1).

The word for "wander" in the *Zhuangzi*, *you*, also has the connotation of "dance," being derived from a word for the way the pendants of a banner dance in the wind and cognate with a term meaning "to dance, float, swim about in water." We could therefore translate Chang Wuxi's final advice to Ju Quezi as: "Dance in—let yourself be moved by—the limitless, and so find things their lodging-places there" (*Zz* 2, 60). This corresponds to the dance as a central image in *Zarathustra* and an indispensable capability of the overman. The overman must be a dancer because, through realizing the relativity of all perspectives, he knows that there is no longer any firm ground on which to take a stand. Every apparently firm ground (*Grund*) is, for Nietzsche, an abyss (*Abgrund*): "Where does man not stand at the edge of abysses? Is to see not itself—to see abysses?" (*Za* 3.2.1). For Zhuangzi, too, the appropriate response to the realization of the relativity of all standpoints is to develop lightness of foot and learn to dance over the abyss: "[the enlightened ruler] keeps his foothold in the immeasurable and wanders where there is nothing at all" (*Zz* 7, 96).

Crippling

Both texts are distinguished by a number of images of the pathological, the deformed, and the grotesque. In the inner chapters alone we meet a hag/leper, a cripple, a madman (twice), three amputees, three hunchbacks, and a man "ugly enough to astound the whole world" (*Zz* 2, 53; 3, 64; 4, 74; 5, 77–81; 6, 88). *Zarathustra* does not harbor as large a cast of bizarre characters, although the foaming fool ("Zarathustra's ape") and the ugliest man are significant in being aspects of Zarathustra's own psyche (*Za* 3.7, 4.7). One of his most important speeches, "On Redemption" (the major themes of which—the redemption of the past and the idea of "willing backwards"—form a bridge between the ideas of will to power and eternal recurrence), is delivered to a hunchback and a crowd of cripples and beggars—by a bridge. Thus speaks Zarathustra: "A seer, a willer, a creator, a future himself and a bridge to the future—and, alas, also, as it were, a cripple at this bridge: all this is Zarathustra" (*Za* 2.20). This touches on a central theme in the book: the necessity, and even desirability, of suffering on the path toward self-transformation. "I love him whose soul is deep, even in being wounded," says Zarathustra in the Prologue. While the foaming fool in *Zarathustra* does not exhibit the crazy wisdom of the madman of Chu, we must remember that by conventional standards the overman is bound to appear demented. "Where is the madness with which you should be inoculated?" asks Zarathustra at the end of his first speech to the people. "See, I teach you the overman: [...] he is this madness" (Prologue 3).

The plethora of deformed characters in the *Zhuangzi* serves, I think, two purposes, both of which are in harmony with the ideas behind *Zarathustra*. The presence of ugliness, deformity, or disease is what alone gives to beauty, integrity, and health their meaning, and these, as with all opposites, are harmoniously embraced by the *dao*. And just as the beautiful is seen to be beautiful only from a more or less arbitrarily fixed perspective, so being deformed is not necessarily the drawback that it appears to be—since crippled Shu is able to live out his years in comfort precisely because of his deformity (*Zz* 4, 74). The characters who have had a foot amputated are philosophical about it—"When Heaven gave me life, it saw to it that I would be one-footed"[9]— ascribing their misfortune to fate in a manner consonant with, though less impassioned than, Nietzsche's *amor fati*. Amputation was a common punishment in China, and we should read the remarks in this context about "crippled virtue" in connection with the following passage from chapter 6: "When Yao has already branded your hide with Goodwill and Duty, and snipped off your nose with his 'That's it, that's not,' how are you going to wander on that free and easy take-any-turn-you-please path?" (*Zz* 6, 91). Here it is clear that rigid moral prescriptions "cripple the soul," as Nietzsche might say, and impair the natural unfolding of one's talents.

Stages on the Way

I wish now to set up a schema against which a variety of spiritual and psychical transformations of the self can be understood. It takes the form of a quasi-Hegelian triad of phases, which I shall call immersion, detachment, and reintegration. (Nietzsche imagines these as the stages of the camel, the lion, and the child.) In the first phase the self is "not yet a self," being interfused with the world, participating unconsciously in the social group and the phenomena of nature. It is a stage of relative innocence; the Hegelian sock has not yet been torn. Then the self withdraws, detaches itself as a self-conscious ego, over against a world of objects and other people, saying no to the way society has set things up, to tradition as it is handed down. This detachment can take the extreme form of a spiritual withdrawal or a transcendence of spatiotemporal experience toward the Absolute (compare the soul's ascent to the intelligible realm for Plato, union with Brahman in Hinduism, or the attainment of *nirvāṇa* for early Buddhism). In the final phase a reintegration with the world is effected, a return to participation, but now reflective and self-conscious. The self re-engages the world without being totally taken in by it. The innocence and spontaneity of the child are joined with the archaic wisdom of the animals.[10]

Now much of the literature on Zhuangzi has taken him as an advocate of the second phase, of detached, quietistic mysticism, of a wandering beyond the temporal affairs of men, in constant communion with the eternal *dao*.[11] While there are passages in the inner chapters that appear to advocate a dispassionate detachment from the world, the close parallels so far with Nietzsche should make us wary of such an interpretation. Looked at carefully, these passages are seen to reflect only an intermediate stage on the path of free and easy wandering. Detachment is a necessary stage on the way toward self-transformation "whoever cannot loose himself other things bind still tighter" (*Zz* 6, 88)—but more important is reengagement, since to keep on forcefully rejecting the world would betray a residuum of attachment to it (compare *Zz* 5 and 6, 79–85). This last point is brought out clearly in the following passage: "By abandoning the world one can be without entanglements. Being without entanglements one can be upright and calm. Being upright and calm one can live again with other things [the world]. Living again one can come close [to *dao*]. You return to become the 'helper of Heaven'" (*Zz* 19, 182).

The idea that one returns to become "a helper of Heaven" is central to the *Zhuangzi*. One returns to a participation in the ongoing processes of change in the natural *dao*. This participation is most evident in the passages in which death is discussed as just another phase in the succession of transformations of yin and yang (*Zz* 6 and 7, 88–97): here the dissolution of the self is expressed by images of parts of the body turning into roosters, crossbow pellets, cartwheels, and willow trees. It is important to realize that this "going along with the *dao*," "letting things be," the nonassertive activity of *wuwei*, is—as with the stage of the child in Zarathustra—by no means a mere regression to the first stage of unconscious participation. For at this stage there is no freedom: the self is unreflectively involved in the processes of change. The term "helper of Heaven" suggests that the engagement is now active and conditioned above all by freedom—a central idea in the thought of the *Zhuangzi* and an indispensable feature of the wandering dance.

The major question underlying all that has gone on so far is this: what relationship is there between the central idea of the *Zhuangzi*, the *dao*, or Way, and the two major ideas of *Zarathustra*, the will to power and eternal recurrence? This question is too deep for me to do more here than to venture some quick reflections into its obscurity. Somehow, the darkness of these ideas tends toward an equivalence. On one level *dao* and will to power are the same in being the totality of existence—present, past, and future. Just as for Nietzsche everything in existence is will to power, so in the *Zhuangzi* there is a correspondence between the universal *dao* and the idea of *de*, or power, which is the manifestation of the *dao* in particular existents.[12] And if will to power is what everything is, eternal recurrence is how the way all things are. For a philosophy of flux, such as Nietzsche's or Zhuangzi's, "that everything recurs is the closest approximation of a world of becoming to a world of being."[13]

But let us now try to see how the two major ideas of Zarathustra are connected with the principal ideal that the work projects—the overman. The way to the overman, "the bridge to the highest hope and a rainbow after long storms" (*Za* 2.7), involves abandoning the egoistic will that is impotent against the past and so wreaks revenge by branding its passing as deserved and all temporal existence as nugatory. To redeem the past by overcoming the "spirit of revenge" is to learn to "will backwards," to be able to say to one's entire past, and especially to things apparently "fated" and beyond the range of will power, "Yes—thus I will it" (*Za* 2.20).

Now here we come upon a point where we see the value of the comparative approach in illuminating hitherto unnoticed aspects of one or the other side of the comparison. It is central to Zhuangzi's philosophy of organism and to his understanding of the *dao* that everything in existence is related to everything else. This idea tends to be associated almost exclusively with Oriental thought, the philosophy of Huayan Buddhism being perhaps its consummate elaboration, and one would not normally expect to find such an idea in Nietzsche—unless prompted to look for it by a comparison with Zhuangzi. Now it turns out that this very idea is the link between the transformation of the will that makes possible the overman and the affirmation of the idea of recurrence. The first hint of this connection comes at the end of "The Convalescent," where Zarathustra's eagle and serpent put these words into his mouth: "Now I die and dwindle away and suddenly I am nothing. Souls are as mortal as bodies. But the knot of causes in which I am entangled recurs—and will recreate me."[14]

If we look closely at the difficulties Zarathustra has in expressing the idea of recurrence, we see that the major stumbling block on his ways toward the overman is nausea at the realization that the "smallest man" must also recur eternally. (Let us remember that this must now refer to the smallest man, the stinking rabble, and the ugliest man within Zarathustra's own psyche as much as to those in the "external" world.) It is easy enough to affirm the recurrence by willing the eternal return of the "good" parts of the past (whether of one's personal history or of the history of the race); heavier and more difficult is the realization that the good and the bad are indissolubly linked, and that to will the recurrence of a single good thing is to will the recurrence of everything bad.

Ultimately, then, the idea of the interdependence of all things is applied to the cosmos as a whole—as expressed in the magnificently Dionysian culmination of the penultimate section of the book, "The Sleepwalker-song," which Zarathustra addresses to the "higher men."

> Just now my world became complete, midnight is also midday,—
> Pain is also a joy, cursing is also a blessing, night is also a sun,—go away or you will learn: a wise man is also a fool.
> Did you ever say Yes to a single joy? Oh, my friends, you thereby said Yes to all woe. All things are linked together, intertwined, enamoured,—
> —if ever you wanted one thing twice, if ever you said, "You please me, happiness! Fleeting moment!" then you wanted everything back!
> —All anew, all eternally, all linked together, intertwined, enamoured, oh then you loved the world. (*Za* 4.19.10)

This passage expresses an understanding in harmony with *dao*, which unites all opposites and by virtue of which "the ten thousand things are one." There is a distinctly Daoist tone to the implication of the idea of recurrence that one is to learn to accept even the most vile aspects of existence and to say to them, "Yes—this too belongs." But it is on this issue that we begin to touch upon what seems to me to be the central difference between the two philosophies. Whereas the Daoist sage cultivates an acceptance of the darker aspects of existence, the goal of the overman is an exuberant affirmation—"a Dionysian Yes-saying to the world, as it is, without any subtraction, exception or selection."[15] But let us approach this difference in tone from the "negative" side first.

While the perspective of death is paramount in both texts, and going under a major move on the way to the transformed self, we do not find in the *Zhuangzi* the terror in the face of the abyss, the undergoing of the great suffering required for the crossing to the overman, the nausea at the prospect of the eternal recurrence of the smallest man. While in the *Zhuangzi*, in the light of the interconnectedness of all things, to attempt always to avoid pain and strive after pleasure would be like opting for yin and dismissing yang—the way of the perfecting man winds through vales of sorrow as well as over plateaus of joy—there is nevertheless more emphasis in *Zarathustra* upon the necessity and even desirability of suffering for man to become who he is. This difference stems, I think, from the disparate historic-philosophical backgrounds of the two thinkers. With Nietzsche we have two and a half thousand years of Platonism and Christian thought to contend with, which have engendered an enormous tension in the bow of the European spirit, culminating in the radical Cartesian dichotomy that has totally cut off the self from the world and in the Copernican revolution and the ensuing "death of God." The existential situation in nineteenth-century Europe (as now) gives grounds enough for Angst in the face of the abyss. With the *Zhuangzi*, stemming as it does from a tradition of Chinese thought in which self and world remain organismically bound together, there is no such abysmal split and so less suffering to be undergone in the attempt at healing.

Correspondingly, there seems to be in the *Zhuangzi* less intensity of passion at the other pole, less exuberance. The images of the overman's fiery solar will and of the leaping dance are more vital than the Daoist's mirror-like lunar reflections and the harmonious participation of *wuwei*. While the *Zhuangzi* encourages man to emulate the natural way (and promotes an unromanticized view of the natural congruent with Nietzsche's, of a nature embracing extremes of cruelty and the grotesque), there is a strong sense in Zarathustra that the creativity of the overman is an *opus contra naturam*, which goes beyond the attainment of harmony with the natural and in some respects works against nature.

This difference in mental temperature—the *Zhuangzi*'s cool harmony as against Zarathustra's friction-generated heat—is less a result of residual egoism on the part of the overman than of a disparity in the degrees of encapsulation of the self that is to be overcome. The self with which Zhuangzi was confronted had begun, under the same kinds of pressures of self-interest that Confucius strove to reduce, to shrink from being a relational matrix into a coagulation around a nodal point; and yet the process was not so far along that a great amount of energy was needed to dissolve the solidifying self back into the network of the world. But in Nietzsche's tradition the self had congealed into a rigidly encapsulated ego, and therefore much more intense heat was necessary for undertaking the alchemical work of burning it out. But the difference in intensity should not obscure the isomorphism between the underlying transformations.

I hope, in conclusion, that the losses from considering these two thinkers apart from their historical contexts have been offset by the gains in clarity about their ideas yielded by the comparative approach. This approach has been somewhat tentative, because it treads new ground, but at least a few steps have been made along the way of the wandering dance.

Notes

We agreed that I should leave the essay in its original form rather than revising it, except for changing Wade-Giles romanization to Pinyin and updating the references. I have retained the translations of quotations from Nietzsche that I did in 1982, but passages in Nietzsche's *Thus Spoke Zarathustra* are now referred to by "Za" followed by the part and chapter number (and section number, where applicable). I added numbers to the chapters in my later translation of the book (Oxford World's Classics, 2005), but the chapters can be found easily enough in any edition by simply counting. I refer to Angus Graham's translation of the *Zhuangzi* (*Chuang-tzŭ: The Inner Chapters*, London: George Allen & Unwin, 1981) by Zz followed by the chapter number and the page number of *The Inner Chapters*. I occasionally modified the translation with the help of my colleague Roger Ames, for whose willingness to share his knowledge of the Chinese language and of Daoist philosophy I remain grateful.

I now also recommend the translation by Brook Ziporyn, *Zhuangzi: The Essential Writings* (Indianapolis/Cambridge: Hackett, 2009). The numbering of the fragments of Heraclitus follows the Diels/Kranz edition, and the translations are my own from the German.

Editor's Note : We are grateful to *Philosophy East and West* and its editor Franklin Perkins for permission to reprint the original "The Wandering Dance: Chuang Tzu and Zarathustra," by Graham Parkes, which appeared in *PEW* 33.3 (1983) 235-50. This article was apparently the first study of Nietzsche and Daoist philosophy in English and was a significant step forward for comparative philosophy and remains an instructive contribution and inspiration to this day. To retain purity of the original article, we have retained the author's older translations of *Thus Spoke Zarathustra*. Nonetheless, I recommend consulting his more recent 2005 translation published by Oxford University Press. This translation is superior and more inclusive than what is presented here. Since this is the original version of the article, readers will encounter some language designations that would not be used today.

1. In addition to being full of jokes, the *Zhuangzi* emphasizes laughter as an important stage on the way: "Rather than go towards what suits you, laugh: rather than acknowledge it with your laughter, shove it from you. Shove it from you and leave the transformations behind; then you will enter the oneness of the featureless sky" (Zz 6, 91). The most important of the numerous outbursts of laughter in *Zarathustra* arises from the first full vision of the overman, when the shepherd who has bitten off the head of the black snake of nihilism leaps up—"No longer shepherd, no longer human—one transformed, enlightened, *laughing!* [...] Oh, my brothers, I heard a laughter that was no human laughter" (Z 2.2.2).
2. See Owen Barfield's insightful discussion of participation in his *Saving the Appearances: A Study in Idolatry* (Barfield 1965).
3. Stevens 1978, 527.
4. As I suggest in the last section of this paper, Nietzsche's conception of will to power as an all pervading cosmic force approximates Zhuangzi's notion of *dao*. For a discussion of water as a psychical metaphor, see my essay, "The Overflowing Soul: Images of Transformation in Nietzsche's *Zarathustra*," *Man and World* 16 (3) (1983): 335-48.
5. There are also a few differences in emphasis: for instance, the horse plays a very minor role in *Zarathustra* but is the predominant mammal in the *Zhuangzi*, and there is a greater preponderance of fish in the latter than in *Zarathustra*.
6. See, especially, Zz 2, 59-60; 6, 85-91; and 18, 123-4.
7. See Zz 2, 61. It is interesting that butterflies appear in connection with Zarathustra's second and fourth dreams. The butterfly is an ancient image for the soul—one meaning of the Greek *psuchē* is "butterfly"—connected, perhaps, with the soul's tendency to "flit around" during sleep.
8. Concerning the role of creative fantasy in constituting the world, see especially the discussions in *Dawn of Morning* 119 and *The Gay Science* 54 and 57.
9. Zz 3, 64. Many of Zhuangzi's utterances concerning "Heaven" sound very un-Nietzschean unless one realizes that he uses the term *tian* in two ways: in the narrower sense "heaven" is a counterpart of "earth"; but in the broad sense, for which I shall use "Heaven," *tian* denotes the unity of the powers of heaven and earth, and so its meaning comes close to "nature" in the sense of "the Way of nature."
10. The sage is often likened to the child in Daoist philosophy: for example, Yan Hui speaks of the sage in his relation to Heaven as "childlike" (Zz 4, 68).

11   Compare Watson (1968, 3) where he considers the question of how to live in a world "dominated by chaos, suffering, and absurdity," and writes: "Zhuangzi's is the answer of a mystic … : free yourself from the world."
12   While *de* is commonly translated as "virtue," by analogy with the Latin *virtus*, Graham translates it, more aptly, as "power." For an insightful treatment of the relationship between the ideas of *de* and will to power, see Roger Ames, "Coextending Arising (*De*) and Will to Power: Two Doctrines of Self-Transformation," *Journal of Chinese Philosophy* 11 (2) (1984): 113–38.
13   Nietzsche, *Sämtliche Werke: Kritische Studienausgabe* 12: 7[54] (= *The Will to Power*, sec. 617).
14   *Za* 3.13.2. Compare the stories in the *Zhuangzi* concerning death as a dissolution of self and body and a transformation into cosmic participation (*Zz* 6, 87–91; 18 and 32, 123–5).
15   Nietzsche, *Sämtliche Werke* 13: 16[32] (= *The Will to Power*, sec. 1041).

References

Ames, Roger T. 1984. "Coextending Arising (*De*) and Will to Power: Two Doctrines of Self-Transformation." *Journal of Chinese Philosophy* 11 (2): 113–38.
Ames, Roger T. and Rosemont, Henry, Jr. 1998. *The Analects of Confucius: A Philosophical Translation*. New York: Ballantine Books.
Barfield, Owen. 1965. *Saving the Appearances: A Study in Idolatry*. New York: Harcourt, Brace & World.
Chai, David. 2020. *Daoist Encounters with Phenomenology*. London: Bloomsbury Academic.
Diels, Hermann. 1912. *Die Fragmente der Vorsokratiker. Griechisch und Deutsch*, Vol. 1. Berlin: Weidmannsche Buchhandlung.
Nietzsche, Friedrich. 1980. *Sämtliche Werke: Kritische Studienausgabe*, 15 vols. Munich: dtv.
Parkes, Graham. 1983. "The Overflowing Soul: Images of Transformation in Nietzsche's Zarathustra." *Man and World* 16 (3): 335–48.
Stevens, Wallace. 1978. "The Rock." In *Collected Poems*. New York: Knopf.
Watson, Burton. 1968. *The Complete Works of Chuang Tzu*. New York: Columbia University Press.
Zhuangzi. 1981. *Chuang Tzu: The Inner Chapters*. Translated by A. C. Graham. London: George Allen & Unwin.

Part Two

The Aesthetic Dance

2

Graham Parkes, Musical Ecologies, and the Daoist Dance

Roger T. Ames

> *Music ... frees me from myself, it sobers me up from myself, as though I survey the scene from a great distance ... It is very strange. It is as though I had bathed in some natural element. Life without music is simply an error, exhausting, an exile.*[1]
> —Friedrich Nietzsche, *Correspondence* 1888 (Young, 458–9)

Graham Parkes, late and soon, has lived the life of an impassioned public intellectual. This being said, I want to suggest that Parkes is not just anyone's public intellectual, but one of a particularly Daoist stripe. That is, having been liberated by the rhythms of the Daoist dance that draws upon the natural order itself, he is able to address our environmental issues by widening his perspective to "survey the scene from a great distance" and to "see the world as the world" (*yitianxia guantianxia* 以天下觀天下) (*Daodejing* 54).

While such exemplary public intellectuals are properly defined by their shared sense of responsibility for the human condition in its broadest sense, we must allow that "responsibility" assumes different forms in different philosophical contexts. Indeed, it is Parkes's own particular sense of responsibility—perhaps eccentric in his own world—that has drawn him to the texts of Chinese philosophy broadly and, more particularly, to those of a decidedly Daoist lineage. What then is this Daoist sense of responsibility that has won Parkes over to its side?

Within the context of familiar liberal values as they come to be invested in the notion of free, self-determining, unencumbered, self-sufficient, and autonomous individuals, "responsibility" is often expressed in the language of liability and obligation. Such responsibility is a kind of accountability—while being autonomous, we still have the burden of having a duty to other persons or some existing law, and thus being answerable to this obligation in our actions.[2] As freely choosing individuals, however, the alternative to such accountability is the duty we have to ourselves to live our lives authentically. In both cases we have responsibility for those behaviors over which we exercise some degree of control, whether for acts of compliance with some assumed moral order that is not of our own making, or for those deliberate, self-determining actions that we deem truly conducive to the kind of persons we want to become. In the case of compliance, the dividend we reap for this sense of responsibility is preserving our integrity as persons of principle who do the right thing; in the second case as agents of our own freely chosen actions, we have the existential integrity of striving to become true and authentic human beings.

Where responsibility is conceived as a kind of accountability to some external order, it would seem to set boundaries on the degree of freedom we might have to exercise our imagination and create our own way forward in the world. On the other hand, responsibility understood in the more existential sense of being responsible for who we will ultimately become would seem to be the opposite—a demand to exercise our own personal freedom to create ourselves as fully authentic beings. While accountability is expressed as deference in our relation to some independent standard, authenticity by contrast would seem to require that we sometimes daringly and audaciously subordinate any such relations that might constrain our freedom in order to become our own best thoughts. We thus have a clear contrast between the responsibility as defined by our transcendental idealists such as Kant who would offer us imperatives that are categorical (duty as the obligation of the will to comply with the moral law), and the radically different responsibility understood by our extreme particularists such as Nietzsche who would exhort us to say yes, to say no, and then to live dangerously along the straight line that would take us to our authentic selves.

Both of these senses of responsibility—to be accountable and to be authentic—are also at play in the Chinese tradition, but because the conception of radically contextualized and irreducibly social persons is so fundamental and pervasive, we find that the notions of both agency and personal authorship are construed in a profoundly different way. The familiar tension between compliance and existential creativity remarked upon above has a significantly different configuration. In fact, I will argue that responsibility as it is expressed in the classical Chinese worldview, taking the classical Daoist cosmology as it is proffered in the *Daodejing* as my specific example, is for the most part a kind of creative, contrapuntal responsiveness that requires from us both accountability and authenticity. Daoism offers us an understanding of creativity—or better, co-creativity—that, in privileging situation over agency and aggregating significance over originality, is expressed in terms of a collaborative, meaning-producing relationality.[3] Further, a corollary to the primacy given to vital relationality is the privileging of a holistic, aesthetic sense of order in which all competing orders are included in the unsummed totality of the effect—that is, in the *dao* that makes this tradition "Daoism." This Daoist aesthetic sensibility contrasts markedly with the reductionistic rationalizing of order that would select out one particular iteration of order deemed to be causal and foundational and raise it over all of the subordinated others to give it privilege as the single, defining order.[4]

In formulating this Daoist sense of responsive responsibility, I am drawing upon and endorsing Parkes's *Befriending the Dragon: How Chinese Philosophy Can Help Save Humanity and Enhance the Lives of Millions on the Way*.[5] In this monograph, he appeals to this kind of collaborative relationality to advocate for a global environmental ethic that seeks to redefine our role within the Anthropocene while it is still possible, an advocacy that I would suggest is at least in some degree inspired by his intimate familiarity with the early Daoist canons. In this new book, and in his earlier work on

Daoism too, Parkes has laid the groundwork for what might come to be called a Daoist "role ethics"—that is, the human capacity and responsibility in our role as cosmic co-creators for "extending the way" for our own time and place.

In Confucian role ethics, each generation must with resolution aspire to a cultivated virtuosity that would enable us to move straight ahead together as the road forward is laid, thus being "true" (*zhi* 直 and *de* 惪) to each other in staying the course and extending the cultural tradition. This image of human beings with eyes and face forging resolutely straight ahead is depicted in the component elements of the pre-stylized Chinese graph, *dao*, itself: 𢕟.

By contrast the Daoist "focus-field" (*dedao* 德道) approach to laying a connector for our own time and place is captured in the *wu* 無-forms—"non-coercive acting" (*wuwei* 無為), "objectless desiring" (*wuyu* 無欲), and "unprincipled knowing" (*wuzhi* 無知)—that are formulated in the *Daodejing* as its technical vocabulary for prescribing the optimal relational disposition that can be achieved between one's person and its various environs.[6] Within the process cosmology that serves as the interpretive context for the *Daodejing*, the cultivation and expression of such a *wu*-form disposition become a rhythmic musicality. It is certainly no more than a coincidence that historically the character for "dancing" *wu* 無 (now written as 舞) became a phonetic loan character for the *wu* 無-"negative, none, without, lacking" that we find in these binomial *wu*-forms such as 無為 *wuwei* "non-coercive acting." But even so, the cultivated *wu*-form disposition is perhaps best captured by the image of the expressive rhythms and gestures of a deferential cosmic dance that draws its energy of generation and transformation from the natural world.[7] Indeed, the title of the *Daodejing* itself can be read as recommending the cultivation of a disposition wherein *dao* as the environing field is brought into its clearest resolution through the virtuosity of the insistent particular—the *de*—that is able through its *wu*-form disposition to express optimal relationality in this cosmic dance. Such an expressive virtuosity is certainly revealing of our authentic, insistent selves (*de*), but this can only be achieved as our persons, through a cultivated responsiveness, are able to find full coalescence, not with but within, some persistent and yet always evolving, natural and moral order (*dao*).

Indeed, this appeal to a holistic, holographic understanding of gerundive persons-in-context as well as to the inseparability of agents and their actions brings immediately to mind an excerpt from the Yeats poem, "Among School Children":

> O chestnut-tree, great-rooted blossomer,
> Are you the leaf, the blossom or the bole?
> O body swayed to music, O brightening glance,
> How can we know the dancer from the dance?

And again, we might recall the first line of the *Daodejing* that announces the inability of language to do justice to the unceasing process of human world-making: "Way-making that can be put into words is not really way-making."[8] The expectation that it is the expressiveness of dance that can elevate the discourse above and beyond the spoken word recalls the confessional ascent from speech to dance we find in the *Great Preface* to the *Book of Songs*:

> Our songs are whither our purposes carry us. While still resident within our heartmind they remain our purposes; expressed in words they become our songs. When feelings stir within us they take shape in our speech, but when speech cannot do these feelings justice, we have recourse to sighs and laments. When our sighs and laments are inadequate, we intone and sing them. And when our chant and song are not up to the task, we unconsciously dance out these feelings with our hands and feet.[9]

In this antique Daoist tradition, it is the contrapuntal responsiveness of dance that provides the medium through which our personal thoughts and feelings that suffuse our experience in the world can find their highest and most revealing expression.

When we reflect on Parkes's own long and eventful career in philosophy, we cannot but admire its inimitable choreography. The complexity and compass of his dance has been shaped in part by his unrelenting resistance to being constrained by philosophical limits of any kind. In his work, Parkes has been an "*intra*-cultural" philosopher for whom the subject of philosophy itself, far from by being fragmented by focusing on the comparison among, or the conjoining of erstwhile discrete elements, is one complex thing. Having no outside, philosophy can be reconnoitered only from within. Philosophizing so conceived is a kind of Wittgensteinian "criss-crossing": the selecting and correlating of some episodes of insight from among the boundless many within the wholeness and continuity of our ever-evolving personal narrative. And Parkes has his philosophical heroes—Emerson and Nietzsche among them—whom he has learned from and emulated closely in this respect. As Parkes observes in *Composing the Soul*:

> For both thinkers [Emerson and Nietzsche] all conditions of the human being, not just certain privileged states, are potential candidates for philosophical transformation—and especially those that have traditionally been neglected.

And then, bringing these two friends in history together, Parkes cites Nietzsche who uses a passage from Emerson (in German translation) as the epigraph to the first edition of *The Joyful Science*:

> To the poet and sage all things are friendly and hallowed, all experiences useful, all days holy, all human beings divine.[10]

And Parkes as a matter of style has certainly, like most of us, availed himself of the printed word, but, with Emerson and Nietzsche as his models, has used language better than most of us with much elegance and aplomb. Beyond his many books and articles, however, he has also traversed the boundaries of different media in his philosophical work, making use of film and imagery and music and theater and dialogue, not to mention tide pools and rock gardens, and yes dance too, as well as so many other discursive strategies, including daughter Helen, in compiling his own colorful philosophical album.

Parkes together with Hans-Georg Moeller were partners in crime in cultivating a formidable University College Cork program in comparative philosophy that in its time produced a crop of some of our most promising comparative avatars. Moeller in an essay published in a volume entitled *Zhuangzi and the Happy Fish* uses "rambling" (*you* 遊) as a technical term found throughout the *Zhuangzi* to describe a peculiarly Daoist mode of productive philosophizing—what he calls a kind of "philosophical musing."[11] And Parkes on many occasions has happily experimented with such musing. But his philosophical style, although certainly Daoistic, is perhaps better represented by the *Daodejing* than the *Zhuangzi*, a text that is profoundly normative in its attempt to formulate its *wu*-form strategy for optimizing the human experience. In the remainder of this essay, then, I want to join in common cause with Parkes not only in exploring the nature of this *Daodejing* normativity, but in endorsing its *wu*-forms as an inspiration for formulating an environmental ethics in our role as responsible collaborators within the natural world.

We might begin from the observation that approximately one-third of the *Daodejing* is rhymed, underscoring the text's respect for counterpoint as a high value in producing musicality within the important literary dimension of the human experience. From an analysis of this rhyme structure as well as from the documents recovered in the recent Guodian archeological find, we can speculate with some confidence that the compilation of the *Daodejing* as an evolving process began some time early in the fourth century BCE.[12] Another counterpoint to be remarked upon found not only in the received *Daodejing* text as it emerges from these beginnings, but also in the Guodian manuscripts as our earliest redaction of this text, is the clear anti-Confucian polemic that levels a sustained attack on both the artificiality and the limited, specifically human focus that is defining of this earliest statement of the Confucian project. An example of this polemic is the *Daodejing*'s critique of the Confucian moralist way of construing appropriate family relations, with the text first setting the problem and then offering its own solution:

> It is only when grand way-making is abandoned
> That we find appeal to the Confucian language of "consummate conduct" (*ren*) and "optimizing appropriateness in our roles and relations" (*yi*).
> It is only when "wisdom" (*zhi*) and "erudition" (*hui*) have arisen
> That we are saddled with great duplicity.
> It is only when the six family relationships are in disarray
> That we find appeal to "family reverence" (*xiao*) and "parental affection" (*ci*).
> It is only when the state has fallen into troubled times
> That we have recourse to upright ministers.[13]
> ********
> Cut off "sagacity" (*sheng*) and get rid of "wisdom" (*zhi*)
> And the benefit to the common people will be a hundredfold.
> Cut off "consummate conduct" (*ren*) and get rid of "optimizing appropriateness in our roles and relations" (*yi*)
> And the common people will return to real family reverence and true parental affection.
> Cut off cleverness and get rid of the search for personal profit
> And there will be no more brigands and thieves ….
> But these three suggestions as they stand are still lacking

And need to be supplemented by the following counsel:
Display a genuineness like raw silk and embrace a simplicity like unworked wood,
Lessen your concern for yourself and reduce your desires.[14]

By way of commentary on this portion of the *Daodejing*, we might observe that when the authentic way of being human prevails in the world, the family-based natural morality of the community takes care of itself and, in so doing, enables its members to prosper and flourish. It is only in a period of moral decline that "cough in ink" philosophers appear to pronounce upon the obvious and, in so doing, ironically exacerbate the problem by formulating and institutionalizing a prescriptive regimen that only serves to suffocate what were our natural, unmediated moral sentiments. Indeed, the argument in these two chapters is first that the imposition of an ossified catechism of Confucian moral terms offends against the natural human feelings that fund our basic human institutions. And secondly, the celebration of a narrowly construed human erudition obstructs the wisdom available to us were we to instead coordinate our own lives with the natural rhythms of the cosmos. The *Daodejing* offers us an argument for authenticity over formal prescription in both our ethics and our practical wisdom.[15]

Although the *Daodejing* is explicitly critical of the terms of art that frame the Confucian ethical discourse, the Daoist text itself is still clearly prescriptive in appealing as it does to authenticity in family and communal relations over a morality of contrivance. For the *Daodejing*, the classical Confucian ethic that we find in the *Analects* and other such texts is half right in asserting that the cultivation of our vital relationality is the proper starting point within the project of living morally. But the *Daodejing* would argue that the focus on human relations within these early Confucian canons at the cost of relationality more broadly construed—the exclusive way of the human being expressed as either *rendao* 人道 or *rendao* 仁道—is not sufficiently capacious to do justice to the broad environmental compass of what it means to become fully human. Instead, the *Daodejing* appeals to an unbounded cosmic way of living (*tiandao* 天道) that is inclusive of the world of human beings and of the heavens and the earth as the cosmic context within which our lives are lived.

This same anti-Confucian polemic continues with a vengeance in another chapter of the *Daodejing* in which the text provides its own lapsarian genealogy of morals. It avers that we humans have fallen from a virtuosity defined by an unmediated, natural, and unencumbered expression of our feelings and, now in embracing a decadent, prescriptive Confucian morality, have entered a spiraling decline.

Persons of the highest virtuosity do things non-coercively and unintentionally,
While those of lesser virtuosity always have their motives in working at what they do.
Persons who are exemplary in their consummatory conduct (*ren*) are always unintentionally working at what they do,
While those who are optimally appropriate in their roles and relations (*yi*) always have their motives;
Persons who are exemplars of propriety in their roles and relations (*li*), in

discovering that no one is paying them any heed, yank up their sleeves and drag others along with them ....
As for propriety, it is the thinnest veneer for doing your best and making good on your word, and is the first sign of real trouble.
Foreknowledge is tinsel to decorate the proper way, and is the first inkling of real ignorance.
It is for this reason that persons of consequence are discriminating:
They set store by the substance rather than the veneer,
And are invested in the fruit rather than the flower.[16]

There is a clear contrast set here between effortless actions that are inspired by and conduce to deeply rooted habits of moral virtuosity, and those that are purposely contrived as an external show of the tinseled conduct that others deem to be proper conduct. The process of a technical morality being imposed from without is nothing more than a record of our steady moral deterioration. The more elaborate the terms of such norms, the clearer is the indication that morality has slipped from what is done spontaneously and unconsciously as the spirit moves us, to what we do in a calculated way for some self-conscious and self-serving reason. Morality in becoming increasingly instrumentalized is reduced to an expedient means to some ulterior end. For the Daoist, the more that the choreography of our conduct has become the burlesque of Confucian ritualized living, the thinner and more diluted moral sentiments become.

The fact that "family" (*jia* 家) as the governing metaphor for Confucian philosophy broadly is a considered and deliberate strategy for optimizing the human experience is not lost on the Daoists. Indeed, that the sustained tensile strength of the social fabric of China from earliest times has been derived from the institution of family or clan lineage (*jiazu* 家族 or *shizu* 氏族) is nothing less than the history of the Chinese social order. The *Daodejing* too clearly understands that family is that singularly powerful human institution wherein we as its members are ready to give ourselves to our familial nexus utterly and without reserve—we gladly give of our time, of our money, and our body parts, and even our lives, when the need arises.

Indeed, in this *Daodejing* corrective on a contrived Confucian normativity, it advocates for its own alternative model of family and community. As we have seen above, the *Daodejing* appeals to the family metaphor—albeit to celebrate authenticity over artifice. The text itself is replete with references to familiar activities we associate with family and family lineage: autumnal sacrifices to ancestors, progeny protecting their mother, and children and grandchildren conducting the customary rites to those who have come before. In its reproductive function, the sire of the many is no match for the female who, being properly underneath, is able to use her stillness to best the male. There are images of newborn babes cherishing their mother's milk and the infant male at the height of his potency with penis erect.

But the *Daodejing* extends this family metaphor from the more narrowly construed human institution to a broad and inclusive cosmic compass and focuses on the unceasing and wholly natural process of procreativity itself. There are seminal concentrations of *qi*, and a dark female, with the wispy and delicate gates on her fecund emptiness, swinging open and close. The text offers explicit images of a kind of cosmic sexuality in which waters gush through cavernously deep and receptive valleys as the fetal beginnings of a cosmic birthing, and *dao* as "mother" gives birth to all things. The point being made with the *Daodejing*'s extension of family to include nature's procreativity is that real normativity like cosmic order is capacious and accommodating of all things, with none of them being privileged over the others, and with each of them having their appropriate time and role in the cosmic cycle:

> The cosmos is not partial to human morality;
> It treats everything as so many straw dogs.
> Sages too are not partial to human morality;
> They treat the common people as so many straw dogs.[17]

The Song dynasty scholar Su Zhe 蘇轍 offers us a seminal commentary on this passage:

> The heavens and the earth show no partiality, and allow things to follow their own natural course. All things are born and die away, where the heavens and earth are neither motivated by kindness in giving birth to things nor by cruelty in taking their lives. They are just like us when we tie up stalks of straw to make dogs to use in our sacrifices. We fashion them and place them on our altars not because of our love for them, but because it is the right time. And when the sacrifice is over, we discard them without ceremony for passers-by to trample under foot not because of some animus towards them, but also because it is the right time.[18]

Parkes too offers his own commentary on this passage, emphasizing the need not simply to emulate nature, but to coordinate our human values with those of the natural order in seeing "worlding" as an inclusive "worlding," thereby taking cosmic order on its own terms:

> The worldview is trans-humanist and un-anthropocentric, insofar as humans are understood as being irrevocably subject to the powers of Heaven and Earth. And so the Daoist sage-ruler empties himself of personal likes and dislikes to become a medium for the impartiality of those greater powers, letting them inform the social realm in its continuity with the natural.[19]

This image of "straw dogs" can be read in three importantly different and yet complementary ways. First, nature does not participate in the pathetic fallacy of human exceptionalism wherein the human being is singled out for special treatment. Nature treats all things, humans and otherwise, with the same degree of care and concern. Secondly, both nature and those human sages, who properly emulate nature, treat all things and all people with parity, and, when it is their time, have the ingenuity to get the most out of even the least among them. And thirdly, "straw dogs" as sacrificial artifacts like all things are celebrated according to their proper season. Even a clutch of straw is entitled to reverence at the proper time and place. In the grand transformations of the natural cycle, all things have their moment, and when that moment passes, things must give way to other things. There is nothing in nature, high or low, that is singled out to be revered in perpetuity. And importantly, the natural cycle for the Daoist, far from entailing the counterintuitive notions of *ex nihilo* birth and utter annihilation, is described as a ceaseless cycle wherein the ebb and flow of the presencing of things, and their ineluctable transformation into other things, are a constant stream:

> I do not know what to call it.
> Were I to give it a style-name, I would call it "way-making" (*dao*),
> And forced to give it a name, I would call it "expansive."
> Being expansive means that it passes on,
> And passing on means that it travels far
> Only to return again.[20]

Of course, this passage has to be read together with a second description of how this cycle of transformation proceeds as one thing lives on in becoming something else:

> Returning is how way-making unfolds,
> And ever weakening is how it functions.
> The things of the world are born of the determinate,
> And the determinate are in turn born of the indeterminate.[21]

The normativity of the *Daodejing* simply put—that is, its highest value—is to optimize the creative possibilities of any particular situation within this generative process of transformation, wherein each situation is boundless and holographic, having implicated within it the entire cosmos. It is for this reason that only those persons who cherish themselves and all that is implicated within them—that is, the world itself—can be trusted with the governance of the world:

> Thus only those who value the care of their own persons more than running the world can be entrusted with the world. And only those who begrudge their own persons as though they were the world can be put in charge of the world.[22]

Of course, a corollary to this holistic, aesthetic value is the perception that the kind of coercion that would necessarily follow from rationalizing a situation by the privileging of one among the many orders would be a gross diminution of these same creative possibilities. Such forced imposition and redefinition would be nothing short of an impoverishing violence. We might appeal to another representative chapter of the text that summarizes this inclusive, aesthetic aspiration in a concrete way:

> Sages are never of a constant mind,
> Taking the many minds of the common people as their own.
> To not only treat the able as able,
> But to treat the inept as able too

> Is to have a quantum gain in ability.
> To not only treat the credible as credible
> But to treat those you do not trust as credible too
> Is to have a quantum gain in credibility.
> As for the presence of sages in the world, in their efforts to draw things together:
>> They make of the world one muddled mind.
>> The common people all fix their eyes and ears on the sages,
>> And the sages treat them as so many children.[23]

In this passage, implicated within and constitutive of the narratives of the sages are the diverse lives of the common people. The ordinary people—some more able and credible than others—certainly look to these sages in finding their bearings and setting their different directions. But the life trajectories of these same common people, far from being governed by the sages, are given the space to retain and relish their own spontaneity—their own "self-so-ing" (*ziran* 自然)—that is needed to live their own unique lives on their own terms. The text expresses the sage's inclusive pluralism in which all have their place through the image of the people being treated by the sage as so many children. For the sages to treat the people like children, far from suggesting condescension, is respecting their potency and their as-yet undetermined possibilities. For children, blithely unaware of the seemingly scripted actions of adults, never walk in a straight line, but like rhizomes, favor sideways and verticality in their movements and growth. Their disjunctive laughter as a profoundly social activity animates everyone and nourishes the happy growth of their own uniqueness while at the same time enlivening their families and communities. It is all too common that in the schoolroom education of our happy children, their first order, natural wonder is smothered as we foist upon them a structured, methodical "curriculum" (literally, Latin for a "racecourse"). Thus, being marshaled to run the course, our children all too soon come to wear the stern and serious faces of little adults. By contrast, sages resist imposing any kind of specific regimen on their people and, like the happy chaos and cacophony of recess on the playground, allow the prevailing order—the *dao*—to evolve as the unsummed totality of the many different orders of these children. By allowing all of their people to enjoy the diversity of participating whole-heartedly in a happily muddled mind, the copious differences among them are activated in making a difference for each other, and for the sages as well.

This *Daodejing* passage with its holographic understanding of persons and their fields of experience—the shared mind of sages and their people—is explicit in calling into question our familiar distinction between the erstwhile separate domains of an inner self and an outer world. This gestalt shift is of course immediately relevant to Parkes's ecological sensibilities, where persons and their natural environments are "aspectual" as coterminous and mutually entailing perspectives on the same phenomenon. Just as, with their swaying bodies and brightening glances, we cannot know the dancers from their dance, so neither too can we know the avid gardeners of a refulgent natural environment from their garden.

The Chinese term "mind" or better "minding" (*xin* 心) serves in this passage and elsewhere as a metonym for person(s) person-ing. It has conventionally been translated as "heartmind" by our clever sinologists who, in so doing, challenge the familiar separation of the cognitive and the affective, connoting as *xin* does both thinking and feeling. In order to make sense of this *Daodejing* passage on the *xin* of the sages—a passage that is reminiscent of the Mencian claim that "the myriad happenings of the world are all implicated here in me"[24]—we need to invoke an alternative to our common sense dichotomous understanding of a "subjective self" and an "objective world." Most obviously, as noted above, it is a commonplace that *xin* does the work of both cognizing and feeling in a life experience that includes both felt thoughts and cognitively informed feelings. But further and importantly, there is no strict dichotomy between intellection and sensation, between body and mind, between structure and function, between thinking and doing, between center and context, between nature and culture. Indeed, *xin* or "heartmind" might be better read gerundively (if ungrammatically) as something like "lived bodyheartminding within our field of experience"—that is, the embodied, embedded, affective and cognitively informed, narratives-within-narratives that constitute our lives as lived. These various aspectual distinctions are nonanalytic and mutually entailing; they do not serve to separate and isolate different components within this "lived bodyheartminding" nor fragment the activities that are defining of our lived experience, but rather underscore the complexity of the continuing, holistic human narrative.

In reflecting on an environmental ethic, the ecological focus-field notion of person assumed in this *daode* and *qi* cosmology stands in stark contrast to a metaphysical realist conception of an inner, private domain and a shared outer world. It begins from a doctrine of internal, constitutive relations and requires a fundamentally different understanding of persons in which their particular identities and the unsummed totality—their foregrounded focal identities and the field of their life narratives—are two holographic and thus mutually entailing ways of perceiving the same phenomenon. That is, any particular phenomenon in our field of experience can be focused in many different ways: on the one hand, it is a unique and persistent particular (the sage "sage-ing"), and, on the other, it has the entire cosmos and all that is happening implicated within its own particular pattern of relationships (the people living their lives). Just as each note in a symphony has implicated within it the entire performance, so persons as live focal events have implicated within them their entire field of experience. And just as the symphony is the complex totality of the effect as it is construed seriatim from the perspective of each unique note without the privileging of any particular one among them, so persons are anarchic in construing the entire field of experience from their own unique perspective without the regulation of some invisible hand.

William James with his own alternative to the old "inner-outer" psychology that was the inspiration for the social psychology of John Dewey and George Herbert Mead provides us with a helpful image immediately resonant with this Chinese notion of *xin* that might help us de-exoticize this focus-field notion of person to serve as the basis for this Daoist environmental role ethics. In his *Pluralistic Universe*, James uses a phenomenology of consciousness to reflect upon and to sketch a rather vivid picture of what he calls "the pulse of inner life," a pulsation that, in being both holistic and vitally specific at the same time, like *xin*, requires that we abandon any notion of "inner" and "outer" as exclusive domains. And as with the notion of *xin*, we must reconceive the relationship between inner and outer in focus-field, holographic terms where different foregrounded perspectives are simply opportunities to shift among different aspects of the same phenomenon. In James's own words:

> In the pulse of inner life immediately present now in each of us is a little past, a little future, a little awareness of our own body, of each other's persons, of these sublimities we are trying to talk about, of the earth's geography and the direction of history, of truth and error, of good and bad, and of who knows how much more? Feeling, however dimly and subconsciously, all these things, your pulse of inner life is continuous with them, belongs to them and they to it …. The real units of our immediately felt life are unlike the units that intellectualist logic holds to and makes its calculations with. They are not separate from their own others, and you have to take them at widely separated dates to find any two of them that seem unblent … my present field of consciousness is a centre surrounded by a fringe that shades insensibly into a subconscious more.

James uses the language of "center" and "field" to give us his alternative to what we have been describing above as the holography of focus and field. He brings into contrast the foregrounding of a single impulse that expands outward to encompass the boundless field, and the foregrounding of the continuous ecological field that is construed from our own perspective with our own unique focal meaning:

> Which part of it properly is in my consciousness, which out? If I name what is out, it already has come in. The centre works in one way while the margins work in another, and presently overpower the centre and are central themselves. What we conceptually identify ourselves with and say we are thinking of at any time is the centre; but our full self is the whole field, with all those indefinitely radiating subconscious possibilities of increase.[25]

Again, James in his *Principles of Psychology* gives us insight into the existential aspect of the lived musicality that animates such pulsations from the inside of this "full self" that "is the whole field" by quoting a passage at length from Adolf Horwicz's seminal work *Psychologische Analysen auf physiologischer Grundlage* (§11):

> A piece of music which one plays one's self is heard and understood better than when it is played by another. We get more exactly all the details, penetrate more deeply into the musical thought …. And this is also surely the reason why one's own portrait or reflection in the mirror is so peculiarly interesting a thing to contemplate… not on account of any absolute "c'est moi," but just as with the music played by ourselves. What greets our eyes is what we know best, most deeply understand; because we ourselves have felt it and lived through it. We know what has ploughed these furrows, deepened these shadows, blanched this hair; and other faces may be handsomer, but none can speak to us or interest us like this.[26]

Perhaps a first key to understanding this holographic relationship between persons and their world is the doctrine of internal relations that follows from and is corollary to ecological insight into the primacy of vital relationality. While the substance ontology of early Greece establishes a doctrine of external relations among discrete "things" that each have their essential integrity, the processual cosmology, as it is expressed in the "Great Commentary" on the *Book of Changes* (易經大傳) and as it is implicated in the early Confucian and Daoist texts as their interpretive context, treats phenomenon as conterminous ecological events that are constituted by their internal relations. In envisioning this relational "event" alternative to the "being" of substance ontology, Peter Hershock looks to a doctrine of intrinsic, constitutive relations that makes "objects" simply the product of a mental abstraction from lived relations.

Hershock diagnoses our problem—a problem that can be traced back at least to as early as Aristotle's ontology—as our culturally specific, recalcitrant habit of seeing the world as being comprised of pre-existing, discrete, and isolatable "things" that then enter into external relations among themselves. Challenging our penchant to give primacy to these "things," Hershock observes:

> Autonomous subjects and objects are, finally, only artifacts of abstraction …. What we refer to as "things"—whether mountains, human beings, or complex phenomena like histories—are simply the experienced results of having established relatively constant horizons of value or relevance ("things"). They are not, as commonsense insists, natural occurring realities or [things]. Indeed, what we take to be *objects* existing independently of ourselves are, in actuality, simply a function of habitual patterns of relationships.[27]

Hershock goes on to offer us what might serve as an intellectual cure for our culturally bound, default assumption that such discrete "things" are primary, allowing us to see "through the conceit that relations are second-order realities contingent upon pre-existing actors." A doctrine of internal relations requires of us a different common sense:

> This amounts to an ontological gestalt shift from taking independent and dependent actors to be first order realities and relations among them as second order, to seeing relationality as first order (or ultimate) reality and all individual actors as (conventionally) abstracted or derived from them.[28]

What this doctrine of internal relations means for environmental ethics is that while personal cultivation is itself the cultivation of our natural world, activities that result in environmental degradation carry with them our own self-abnegation and impoverishment. We within our natural environments either win together with these environs, or we all lose big time.

A second key to understanding a Daoist environmental role ethics that follows from this doctrine of internal, constitutive relations and the Daoist focus-field holography is summarized in the contemporary philosopher, Zhao Tingyang's notion of "All-under-

Heaven" (*tianxia* 天下). Zhao summarizes his geopolitical "worlding" thinking by appealing to a passage from the *Daodejing*:

> It is by cultivating virtuosity in your person that your virtuosity will then be made genuine,
> And by cultivating it in the family that the family's virtuosity will then be made ample,
> By further cultivating it in the village that its virtuosity will then be made enduring,
> By cultivating it in the state that its virtuosity will then be made copious,
> And by cultivating it in the world that its virtuosity will then be made all-pervasive.
> View the person as a person,
> The family as a family,
> The village as a village
> The state as a state
> And the world as the world.
> How will I know how the world is faring? By taking it as it is.[29]

A starting point for understanding Zhao's "worlding" thinking is the observation made by Hershock that the possibilities of the one and the many are much diminished if treated separately. That is, maintaining the disintegrating values of equality and individualism in our families and communities guarantees that difference will only be variation (variety) rather than a compounding ecological diversification (diversity) in which we differ *for* each other rather than just *from* each other. As Hershock remarks, "something that is good for each of us, considered individually, may not be good for all of us."[30] Zhao in this same spirit is advocating for a new concept of international relations that goes beyond competing nation-states inclined to see the world only in terms of their own interests. Zhao is trying to formulate a new inclusive geopolitics in which we see the world as the world—that is, as one single ecology. To manage such an ecology, we need governance that applies a "relational rationality" (*guanxilixing* 關係理性) that, having the reach and inclusiveness to correlate authority, power, and diverse interests in determining future policies, can serve as an alternative to the irrationality of contestation motivated by self-interest.[31]

Parkes has made good use of Zhao's work in his *Befriending the Dragon*. In appropriating this *tianxia* position for his environmental ethics, Parkes describes it as being

> open to diversity and oriented toward a global power structure that's multipolar and polycentric .... It's a matter of adapting the appropriate perspective and "scale." ... [T]he good ruler is able to manage All-under-Heaven on the basis of a genuinely global perspective, looking not from the perspective of his person or state, but through the world itself.[32]

Zhao Tingyang, true to his own tradition, draws his "worlding" thinking from the process cosmology made explicit in the "Great Commentary" appendix of the *Book of Changes*. *Tianxia* is not "*the* world" or "*a* world" that can be objectified by a definite or indefinite article, but is rather a boundless "worlding." In such an ecology, there is no outside—as Parkes observes, only a "multipolar" and "polycentric" inside.[33] And the many interdependent states that make up *tianxia* constitute each other through their patterns of internal interactive activities that function as organic "*intra*-national" relations like organs within a body, rather than as fragmented "*inter*-national" relations among discrete, sovereign polities. We might think of these geopolitical relations in terms of what contemporary philosopher Tang Junyi has offered us as a defining postulate of this early process cosmology: "the inseparability of one and many, of particularity and continuity, of uniqueness and multivalence" (*yiduobufenguan* 一多不分觀). That is to say, the identity and value of each of the states, like members of a family, is achieved through and dependent upon what they do for each other. Said another way, states are always unique events in a continuing stream rather than marbles in a jar, and their varied interests are to be interpreted and responded to in terms of their shared narratives rather than their discrete identities.

Of course, Zhao Tingyang's holistic conception of "*intra*-national" relations is a geopolitical model that expands immediately into Parkes's geo-environmental model. Particular persons, families, communities, and states are our focal, eventful identities that can, through our commitment and resolution, enable the unbounded organic totality within which we are all embedded to thrive. And the operative question within the flow of such mutually implicated and interpenetrating events is not "who are we?" as distinct aspects within the ecology, but rather, "whence" do we come and "whither" do we go on the road (*dao*) we are traveling together? And the wisdom is that if our collaborators do better within our family, community, political and natural ecologies, so do we.

Notes

1 From *Friedrich Nietzsche: A Philosophical Biography* (New York: Cambridge University Press). Authored by Julien Young © 2010. Reprinted by arrangement with Cambridge University Press.

2 Michael Sandel's *Democracy's Discontents: America in Search of a Public Philosophy* (Cambridge, MA: Harvard University Press, 1992) rehearses a narrative over the past fifty years that provides an example in which shared obligations have been mitigated by Supreme Court decisions. According to Sandel, the enforcement of the liberal definition of autonomous person has had the direct consequence of liberating disaffected husband-fathers from their marital responsibilities, and of driving wife-mothers who are given sole responsibility for children into abject poverty.

3 See Roger T. Ames, "'The Way Is Made in the Walking': Responsibility as Relational Virtuosity," in *Responsibility*, edited by Barbara Darling-Smith (Lanham, MD: Lexington Books, 2007).

4 See A. N. Whitehead, *Modes of Thought* (New York: Free Press, 1938), 53–60, for his distinction between a holistic aesthetic order and a reductionistic rational or logical sense of order. This sense of rational order is the basis of systematic philosophy,

and is what John Dewey has called "*the* philosophical fallacy." Such a rationalizing assumption has had many descriptions in the twentieth-century internal critique of foundationalism and objectivism within the Western philosophical narrative. It is what Heidegger, for example, has referred to as "theo-ontological thinking," what Wilfred Sellars has called "the myth of the given," what Whitehead has called "the perils of abstraction" and "misplaced concreteness," what Richard Rorty has called "privileged representation," and so on.

5   Editor's Note: This was the original title of Graham Parkes's *How to Think about the Climate Crisis: A Philosophical Guide to Saner Ways of Living*.

6   For an explanation of this "focus-field" cosmology and the cultivation of the *wu*-forms as a precondition for achieving optimal relationality, see the introduction to Roger T. Ames and David L. Hall, *Daodejing: Making This Life Significant: A Philosophical Translation*. New York: Ballantine, 2003, especially 11–14 and 36–53 respectively.

7   Phonetic loan characters (*jiajiezi* 假借字) are characters that are "borrowed" to write another homophonous or near-homophonous morpheme. The character *wu* 無 as it is found on the oracle bones 𣥂 has a person holding two ox-tails, and means "dancing" or "rain dance." But later inscribed on the bronzes *wu* 無 became a loan character for the abstract idea of "none, without." Subsequently in order to disambiguate the two meanings of "dancing" and "without," an alternative character *wu* 舞 adding two feet 舛 to the character *wu* 無 appears to reference "dancing."

8   *Daodejing* 1: 道可道非常道。.

9   《詩經, 大序》詩者, 志之所之也, 在心爲志, 發言爲詩, 情動於中而形於言, 言之不足, 故嗟歎之, 嗟歎之不足, 故詠歌之, 詠歌之不足, 不知手之舞之足之蹈之也 。.

10 Graham Parkes, *Composing the Soul: Reaches of Nietzsche's Psychology* (Chicago: University of Chicago Press, 1994), 12.

11 Moeller 248–60.

12 Kohn and LaFargue, 231–53.

13 *Daodejing* 18: 大道廢, 有仁義; 智慧出, 有大僞; 六親不和, 有孝慈; 國家昏亂, 有忠臣。.

14 *Daodejing* 19: 絕聖棄智, 民利百倍; 絕仁棄義, 民復孝慈; 絕巧棄利, 盜賊無有。此三者以為文不足。故令有所屬: 見素抱樸, 少私寡欲。.

15 Hans-Georg Moeller makes just such an argument in a provocative and compelling way in his recent advocacy of this kind of Daoist morality as a sanguine alternative to the stiff, unforgiving moralities that usually make up the curriculum in our ethics classes. See his *The Moral Fool: A Case for Amorality* (New York: Columbia University Press, 2009).

16 *Daodejing* 38: 上德無為而無以為; 下德為之而有以為。上仁為之而無以為; 上義為之而有以為。上禮為之而莫之應, 則攘臂而扔之。。。夫禮者, 忠信之薄, 而亂之首。前識者, 道之華, 而愚之始。是以大丈夫處其厚, 不居其薄; 處其實, 不居其華。故去彼取此。.

17 *Daodejing* 5: 天地不仁, 以萬物為芻狗; 聖人不仁, 以百姓為芻狗。.

18 蘇轍《老子解》: 天地無私而聽萬物之自然。故萬物自生自死。死非吾虐之, 生非吾仁之也。譬如結芻以為狗, 設之于祭祀, 盡飾以奉之, 夫豈愛之?時適然也。既事而棄之, 行者踐之, 夫豈惡之?亦適然也。.

19 Graham Parkes, "The Art of Rulership in the Context of Heaven and Earth" (Albany: State University of New York Press). 79.

20 *Daodejing* 25: 吾不知其名, 字之曰道, 強為之名曰大。大曰逝, 逝曰遠, 遠曰反。.
21 *Daodejing* 40: 反者道之動; 弱者道之用。天下萬物生於有, 有生於無。.
22 *Daodejing* 13: 故貴以身為天下, 若可寄天下; 愛以身為天下, 若可託天下。See also *Daodejing* 26: 奈何萬乘之主, 而以身輕天下?輕則失本, 躁則失君。"How could someone be the king of a huge state and treat his own person as less important than the world? If he treats his person lightly, he loses the root; if he becomes agitated, he loses his throne."
23 *Daodejing* 49: 聖人恆無心, 以百姓心為心。善者, 吾善之; 不善者, 吾亦善之; 德善。信者, 吾信之; 不信者, 吾亦信之; 德信。聖人在天下, 歙歙為天下渾其心, 百姓皆注其耳目, 聖人皆孩之。The received text of *Daodejing* 49 has 聖人無常心: "Sages are without a constant mind." On the basis of a Mawangdui text A variant that has 聖人恆無心, Liu Xiaogan 劉笑敢 uses received commentaries to argue for the cogency of this alternative: "Sages are ever without a constant mind." See his *Laozi Past and Present* (老子古今) (Beijing: Zhongguo shehuikexue chubanshe, 2006, Vol. 1, p. 487). I would read the *wuxin* 無心 here as an additional *wu* 無-form that expresses a sedimented habit of engagement: an unmediated "thinking and feeling" or "thinking and feeling immediately." Like *wuwei* 無為, *wuxin* describes an optimal pattern of deferential relationality rather than the absence of activity.
24 *Mencius* 7A4: 孟子曰: 萬物皆備於我矣。反身而誠, 樂莫大焉。強恕而行, 求仁莫近焉。Mengzi said, "Is there any enjoyment greater than, with the myriad happenings of the world all implicated here in me, to turn personally inward and to thus find resolution with these happenings? Is there any way of seeking to become consummate in my person more immediate than making every effort to act empathetically by extending myself into the places of others?"
25 William James, *A Pluralistic Universe* (New York: Longmans, Green and Co, 1909), 286–8.
26 William James, *Principles of Psychology* (New York: Henry Holt & Co, 1890), 326–7.
27 Peter D. Hershock, *Buddhism in the Public Sphere: Reorienting Global Interdependence* (New York: Routledge, 2006), 140.
28 Hershock, *Buddhism in the Public Sphere*, 147.
29 *Daodejing* 54: 修之於身, 其德乃真; 修之於家, 其德乃餘; 修之於鄉, 其德乃長; 修之於國, 其德乃豐; 修之於天下, 其德乃普。故以身觀身, 以家觀家, 以鄉觀鄉, 以國觀國, 以天下觀天下。吾何以知天下然哉?以此。.
30 Peter D. Hershock, *Valuing Diversity: Buddhist Reflection on Realizing a More Equitable Global Future* (Albany: State University of New York Press, 2012), 133.
31 Zhao 31–44.
32 Parkes, "Befriending the Dragon," spiral printout ms., 368–9
33 See Zhao Tingyang, *Tianxia de dangdaixing*, chapter 5 entitled "Without an Outside" 《無外》.

References

Baxter, William H. 1998. "Situating the Language of the *Lao-tzu*: The Probable Date of the *Tao-te-ching*." In *Lao-tzu and the Tao-te-ching*, edited by Livia Kohn and Michael LaFargue, 231–53. Albany: State University of New York Press.

Hershock, Peter D. 2006. *Buddhism in the Public Sphere: Reorienting Global Interdependence*. New York: Routledge.

Hershock, Peter D. 2012. *Valuing Diversity: Buddhist Reflection on Realizing a More Equitable Global Future*. Albany: State University of New York Press.
James, William. 1890. *Principles of Psychology*. New York: Henry Holt & Co.
James, William. 1909. *A Pluralistic Universe*. New York: Longmans, Green and Co.
Moeller, Hans-Georg. 2015. "Rambling without Destination: On Daoist 'You-ing' in the World." In *Zhuangzi and the Happy Fish*, edited by Roger T. Ames and Takahiro Nakajima, 248–60. Honolulu: University of Hawai'i Press.
Parkes, Graham. 1994. *Composing the Soul: Reaches of Nietzsche's Psychology*. Chicago: University of Chicago Press.
Parkes, Graham. 2018. "The Art of Rulership in the Context of Heaven and Earth" in *Appreciating the Chinese Difference: Engaging Roger T. Ames on Methods, Issues, and Roles*, edited by Jim Behuniak, Albany: State University of New York Press.
Parkes, Graham. 2020. "Befriending the Dragon," spiral printout ms., 368–9.
Parkes, Graham. 2021. *How to Think about the Climate Crisis: A Philosophical Guide to Saner Ways of Living*. London: Bloomsbury Publishing.
Young, Julien. 2010. *Friedrich Nietzsche: A Philosophical Biography*. New York: Cambridge University Press.
Zhao, Tingyang. 2016. 趙汀陽, *Tianxia de dangdaixing: Shirjie zhixude shijian yu xiangxiang* 天下的當代性: 世界秩序的實踐與想像. Beijing: China CITIC Press.

3

On the Chariot of Earth

Meilin Chinn

Graham Parkes has shown us many routes back to the earth across time, tradition, and place, and through inner and outer landscapes of soul and rock alike. An important refrain across Parkes's diverse body of work is that we must correct our understanding of nature in order to right our relationship with it and ourselves. As he explains, many proposed solutions to the ecological crisis "overlook the underlying fantasies and prejudices that condition (for the most part unconsciously) our perception of the natural world—and thereby also our interactions with the environment."[1] Misguided ideas such as the view that nature is "dead matter" encourage environmental degradation but can be transformed by learning the "text of nature properly." Here I would like to follow a few of the remarkable routes Parkes has traveled—*feng shui*, *anima mundi*, and *homo natura*—in order to appreciate the importance of his work on place and perspective, both of which can help shift the contemporary drives toward extinction and ecological disaster. These routes notably converge in a place where all things are vital and perspectives are agile: the world of *qi* 氣 energies in the Chinese tradition.

Commonsense *Fengshui* 風水

Parkes has repeatedly reminded his readers that most humans throughout history did not think of the natural world as "dead matter in motion," despite the outsized influence of this view on the development of modern technology.[2] Yet, technological progress on its own does not prove the natural world is inanimate; it's certainly possible we have advanced within a limited domain while still harboring mistaken views about nature. Any account of nature must also reckon with the hermeneutic and phenomenological position of human beings: creatures capable of being "unnatural" or at least at odds with nature and whose representations of nature often reinforce its hidden and enigmatic tendencies.[3] Further, whatever conclusion we draw about nature's vitality still leaves the question of whether this should drive our relationship with it. This is an especially salient point when it comes to the ethical and existential dimensions of the current environmental crises. For example, Parkes has suggested that despite the efficacy with which the Cartesian-Newtonian worldview has allowed us to manipulate the natural world in service of our technological goals, we ought to consider how viewing nature as dead matter encourages environmental degradation.[4] We have good reason to be clear-eyed about the limits of technological solutions to environmental problems and, as he puts it, we ought to consider alternative views with better consequences.

One of the worldviews Parkes suggests is the Chinese practical environmental science and art known as *fengshui* ("wind and water"). In "Winds, Waters, and Earth Energies: *Fengshui* and a Sense of Place," Parkes encourages us to draw from the "commonsense" aspects of *fengshui* as a means for deepening our relationship to nature and the environment. For Parkes, this requires setting aside the aspects of *fengshui* that may intentionally or accidentally encourage mystification, exoticism, and charlantary—including mantic practices and arcane symbolism as well as contemporary, popular versions of *fengshui* that stray from its central insights.[5] Parkes favors treating *fengshui* as "a set of sensible recommendations grounded in sensitivity to the natural environment" that can enhance our relations with a "down-to-earth" sense of place.[6] While fully separating the esoteric elements may turn out to be difficult, there are certainly aspects of *fengshui* practices that translate outside the culturally specific imagery and ideas of its history. As he describes it, the "sensible core" of *fengshui* is quite compatible with the physics and biology of the twenty-first century and "accessible to anyone nowadays with an open mind and opened senses."[7]

Fengshui involves paying attention to the telluric and atmospheric forces of places that shape the landscape as well as the mutual flourishing of humans with their environments and one other. Historically speaking, *fengshui* was predominantly used to choose optimal dwelling and burial sites (*zhai* 宅) and offered means to sustain an ongoing place-based relationship between the living and dead. Its practical benefits also include knowledge applicable to health, ecology, agriculture, and architecture, as well as to existential, aesthetic, and ethical concerns about the human dynamic with nature. Regarding this dynamic, *fengshui* can be seen as an applied practice of the traditional Chinese idea of *tianren heyi* 天人合一 ("harmonious unity of humans and nature") achieved through humanity's understanding of *dili* 地理 ("earth principles" or "earth patterns").[8]

More specifically, *fengshui* is the study of the flowing patterns of *qi* in nature and is akin to Chinese medicine as the study of the flowing patterns of *qi* in the body. This correlation provides the empirical and experiential basis for accessing the commonsense aspects of *fengshui*, the most fundamental of which is the reciprocity between human beings and their environments. *Fengshui* was shaped by the views of a *qi*-based, process cosmology in which the microcosm of the human body parallels the macrocosm of landscape—an isomorphism which Parkes notes at one level "comes simply from an appreciation of environmental influences."[9] For example, the "earth's veins" or *di mo* 地脈 correspond to the *qi* arteries of the human body mapped in acupuncture. As a result of the correspondence *and* reciprocity between the living patterns of embodied human existence, landscape, and atmosphere, sites (*zhai*) are "not merely some location in abstract space, but rather a place defined both by a particular topography and by the kinds of human activities that take place in it."[10] Understanding and working with these flowing patterns of vital energy-matter require learning to perceive them in the first place, which, as Parkes notes, is difficult for those with different background assumptions about the material world.

Seeing the world as comprised of *qi* may be the most challenging aspect of *fengshui* to integrate into contemporary techno-scientific views, yet it is also the aspect most directly accessible to personal, embodied experience. As Parkes remarks, "anyone who is skeptical about the existence of the qi that flows through and energizes the body need only try some qigong exercises (for example, the set known as the (*ba duan jin* 八 段錦, or 'eight seconds of brocade') often used as a warm-up for *taiji* practice) in order to experience the flow of qi in his or her own person."[11] While *qi* alone is difficult to perceive, its flowing patterns are available to witness in our bodies and by extension in our environment, for example, in wind and water, and even in stone. As the classic *fengshui* text *Twenty-Four Difficult Problems* states, "*Qi* is the microcosm of form. Form is the manifestation of *qi*. *Qi* is hidden and difficult to know. Form is manifest and easy to see."[12] To illustrate this in Parkes's beloved mineral realm, the external aesthetics of stone—leanness, surface texture, and foraminate structure—reveal the "vast interior forces that formed the rock."[13] Rocks become "kernels of energy" instead of inanimate lumps of non-biotic matter once our perception is transformed appropriately.[14]

In recognizing the mineral realms and even cultural artifacts as vital forces, *fengshui* is an ecological practice that surpasses the limits of biocentrism.[15] It is also a phenomenological craft reliant on aesthetics as a perceptual practice bridging sensing and embodied meaning. In these methods, our senses are trained on the activities of *qi* within the progenitive reciprocity between human beings and their environments. Here we find another way that *fengshui* is common sense: it generates a shared sense of place or *sensus communis* through which we can collectively communicate and understand our mutual lifeworld at specific locations on earth.[16] This shared sense of place—including the living and their ancestors, plants, animals, minerals, and artifacts—provides a simultaneously personal and communal basis for living in accord with that place and its relations.[17]

Crucially, this *sensus communis* does not operate according to a putative binary between humans and nature even if substantial tensions exist. In fact, *fengshui* is one of a number of practices in the Chinese tradition that aim to ameliorate these tensions by generating what Tu Wei-Ming called an "aesthetic experience of mutuality and immediacy" with nature.[18] The aesthetic emphasis should be noted here. The environmentally damaging worldview in which humans are fundamentally separate from the natural world of "dead matter" ignores perceptual evidence to the contrary, and, as such, invites us to consider the positive consequences of allowing aesthetic experiences to shape our view. The benefits are amplified when these experiences result from the "strenuous and continual effort at self-cultivation" that Tu Wei-Ming argues is necessary for aesthetic mutuality with nature. As he puts it, "it is true that we are consanguineous with nature. But as humans, we must make ourselves worthy of such a relationship."[19] For example, Parkes describes how practicing *taiji quan* cultivates sensitivity to the dynamic forms of our energies in relation to the forces of heaven and earth, in an effort that eventually "brings about precisely what fengshui encourages—a greater awareness of the relations between one's activities and the configurations of the surroundings, whether natural or built."[20] This awareness is yet another example of the *sensus communis* required to build harmonious human sites (*zhai*) on the "chariot of earth" under "heaven's canopy" (*kanyu* 堪輿).[21] What's at stake is not merely overcoming a false binary with nature, but reorienting humans (including their artifacts and technologies) toward the vitality of all things in dynamic arrangements of wellbeing.

Anima Mundi and the Musical Face of the World

The commonsense aspects of *fengshui* that site humans appropriately upon the earth also help develop our sense for *anima mundi* or the "soul of the world." Parkes has linked the world of *qi* energies in the Chinese tradition to *anima mundi* in an innovative move that sidesteps objections to animism as the anthropocentric projection of human qualities onto inanimate objects and nonhuman beings. This view of animism presumes the mind-matter and animate-inanimate dualisms absent in *qi* philosophies, as Parkes points out, and even in the human-centered Confucian way of life "careful attention to interactions with *things*" was central to ritually appropriate and socially harmonious life.[22] Daoists decentered the Confucian focus on human social relations by extending attentive reciprocity into ever-broadening fields of *qi* energies, leading beyond the human world to the great expanses of the earth and heavens. Once our perspective is opened this way, we are better able to experience what Parkes has described as the continuum of *qi* energies transforming themselves "from rarefied and invisible, as in the breath, to condensed and palpable, as in rock" and may discover that *qi* is not just "life energy," it constitutes supposedly "inanimate" matter too.[23]

Mutual affect and response structure this world of *qi* energies much like musical instruments attuned to one another in sympathetic or mutual resonance (*ganying* 感應). Not surprisingly, the idea of *qi* resonance or *ganying* in China drew on musical resonance as the model for how things help animate one another and for how we can know the world whether near or far. This is exemplified in an iconic story from the *Huainanzi* 淮南子 of a *guqin* 古琴 tuner causing two instruments to resonate sympathetically while only striking one: "When the guqin tuner strikes the *gong* note [on one instrument], the *gong* note [on the other instrument] responds: when he plucks the *jiao* note [on one instrument], the *jiao* note [on the other instrument] vibrates. This is because correlated musical notes are in mutual harmony."[24] Parkes rightly emphasizes that while resonance "works most powerfully between things of the same kind, it can also work *across* kinds, between us and 'inanimate' things."[25] This is possible because animate and inanimate things do not differ as metaphysical kinds; they are distinguished by relative degrees and sorts of *qi* processes—or what could be called their *anima*. These include activities such as dissolving or condensing, withdrawing or expanding, and stilling or agitating, which are meaningful according to the continuum of relations in which they occur. Parkes insightfully explains that attending to the *anima* of things requires the "affective aspects of understanding: it is not a matter of knowing things in the framework of an abstract epistemology, but of getting a feel for them, sympathizing with them."[26] As a result, he says, even our interactions with supposedly lifeless things can help generate mutual aliveness in which objects have an almost "spirit-like" quality.

In addition to *qi* philosophies, Parkes's work on *anima mundi* has been influenced by psychologist James Hillman's attempt to remedy the shortcomings of viewing *anima/psyche/*soul as existing inside the human alone and/or as a unifying panpsychic principle separate from physical matter. Hillman proposed that ignoring the "soul of the world" hampered psychotherapy's ability to help individuals and contributed to an increasingly disintegrated and unwell world.[27] In place of traditional notions of *anima mundi*, such as found in Platonic cosmology and Jungian psychology, Hillman suggested the following:

> [L]et us imagine the *anima mundi* as that particular soul spark, that seminal image, which offers itself through each thing … Then *anima mundi* indicates the animated possibilities presented by each event as it is, its sensuous presentation as a face … All things show faces, the world not only a coded signature to be read for meaning but a physiognomy to be faced. As expressive forms, things speak; they show the shape they are in … This imaginative claim on attention bespeaks a world ensouled."[28]

In centering our perspective on how things show themselves—their "faces"—our anthropocentric projection of life onto supposedly lifeless things is diminished in favor of allowing particular things to become presently alive in their uniqueness. This transformation of perspective also reveals the one living world as more fundamental than any two worlds, whether subject and object, phenomenal and natural/scientific, or animate and inanimate. As Hillman writes, the world ensouled replaces "the familiar notion of psychic reality based on a system of private experiencing subjects and dead public objects."[29]

Parkes aptly describes Hillman's shift in perspective as an "aesthetic turn" that emphasizes what things reveal to sense-perception (*aesthesis*) and the "phenomenological inclination to *let them show themselves* rather than imposing our views on what and how they can or should be."[30] Hillman's move here directly ties *aesthesis* to *psychē* in a manner that deepens environmental aesthetics and ecopsychology, and clarifies the commonsense aspects of *fengshui* as both down-to-earth and the *sensus communis* for shared place, in this case the shared place of *anima mundi*. "*Psyche is the life of our aesthetic responses*," he writes, and "those primordial aesthetic reactions of the heart are soul itself speaking."[31] Having defined soul or *psychē* as aesthetic at heart, it is not surprising, Parkes notes, that Hillman saw the healing potential of the arts, creativity at the psychological level, and the power of everyday aesthetics to develop a "fitting ecological response" in which even mundane daily tasks become ways of "participating in the central life of Earth and serving a function in the play of the world."[32] The *anima mundi* of our local world creates a place in which we can practice awareness of the unique presence of each thing and act with fitting integrity toward them, which Parkes compares to the "awakened activity" practiced by Buddhists.

The striking connection Parkes draws between *anima/psychē/*soul and the world of *qi* energies (often portrayed as "vital breath") is reinforced by Hillman's argument for loving the world by "breathing" it in and attending to the "breath" of things. Hillman begins by returning to the ancient ideas that the heart is an organ of perception and the location of the *sensus communis*, whose role is to apprehend images.[33] In other words, the heart's function is aesthetic. *Aisthesis*, he says, is "taking in" or "breathing in," a "gasp" or "ah" in which wonder transfigures matter and lets "each thing reveal its particular aspiration within a cosmic arrangement."[34] Intimacy with a thing grows as we hear it speaking, and in our listening, its soul is enlivened and our imagination activated. In this process of personification, phenomena become "lovable," but Hillman emphasizes, not merely because of human projection. Instead, their own sense and imagination (what Hillman calls their expressed interior image) unfold. Consequently, he argues that phenomena will not be "saved by grace or faith of all-embracing theory, or by scientific objectiveness or transcendental subjectivity"; instead, they are saved "by the *anima mundi*, by their own souls and our simple gasping at this imaginal loveliness. The *ahh* of wonder, of recognition, or the Japanese *shee-e* through the teeth. The aesthetic response saves the phenomenon, the phenomenon that is the face of the world."[35] But the key here is that *aisthesis* is not sense perception alone, for this would not save the phenomenon; the emotional response that arises within *aisthesis* is necessary to the transformation of perspective and behavior, or to borrow Parkes's language, the "awe" and "humility" we feel in the face of things.

In "Awe and Humility in the Face of Things: Somatic Practice in East-Asian Philosophies," Parkes identifies a pivotal aspect to the current environmental catastrophe: "Among the numerous factors driving this current insanity is a lack of awe and humility in the face of the wonders of the world—other people, natural phenomena, and the things we use in living our lives."[36] One way to cultivate the right response to things, Parkes argues, is skilled somatic practices—"disciplined and repetitive activities of the body that have a cumulative effect on its physiology and transform its experiencing"—such as the arts and meditation. Of particular relevance to the discussions here of *anima mundi* and *feng shui* is Zhuangzi's "fasting of the heart" (*xinzhai* 心齋). Parkes describes this practice as: "Such fasting dissolves sedimented judgments and prejudices in the mind, and loosens habitual reactions in the body, so that the *qi* energies of heaven and earth (*tian* or *tiandi*) can flow through unimpeded and keep one on course."[37] Perspectival agility opens up through emptying the self of its preferences, fixations, judgments, and projections; this can be accomplished through sitting meditation or "the consummate practice of crafts, to experience not through the senses and the heart-mind but rather through the *qi* energies."[38]

In the *Renjianshi* 人間世 (In the Human World) chapter of the *Zhuangzi*, Confucius teaches the practice of fasting the heart to Yan Hui as guidance on how to transform in accord with all things. Yan Hui must practice progressively subtler listening, moving from listening with the ears to listening with the heart, and then listening with the "vital breath" (*qi*). Confucius tells Yan Hui: "The ears stop at what they hear. The heart stops at what its thoughts tally. As for the vital breath, it is empty (*xu* 虛) and awaits the arising of things."[39] Whereas in Hillman's depiction, the heart is where phenomena are breathed in, Zhuangzi's fasting of the heart in order to listen with *qi* brings one deeper into open, spacious receptivity and even further away from self-projection and anthrocentrism by bringing one into the emptiness of listening with the vital breath or *qi*. The result is profound adaptability to the vagaries of circumstance and particularity of life, as well as skilled attenuation of ego, which,

Confucius instructs Yan Hui, is the only way to survive the "cage" of politics and avoid accidentally furthering the destructive forces he seeks to transform. Through fasting the heart, a place is made for the images of things themselves, though, as Parkes clarifies, this does not mean completely abandoning the human perspective. Rather, we flexibly interchange our perspective with other beings and keep it in "dynamic interplay with the perspectives of natural phenomena," becoming a "genuine human being from whom 'neither the Heavenly [the natural] nor the human wins out over the other.'"[40]

The somatic practice of listening in the fasting of the heart passage in the *Zhuangzi* is paralleled in the opening of the *Qiwulun* 齊物論 ("Discussion on Smoothing Things Out") chapter where progressively subtler listening allows one to hear the music of the earth and heavens. Ziqi has been discovered in a trance of sorts—"having gone beyond himself"—while listening to the piping of *tian*. He provocatively asks his friend Yancheng Ziyou whether, despite hearing the music of people, he is able to hear the "piping" of the earth: "When the Great Clump blows forth its vital breath (*qi*), it is called the wind. As soon as it arises, the hollows of the ten thousand things sound furiously. Can't you hear them, long and drawn out?"[41] The passage continues by describing an increasingly silent receptivity through which we might hear the piping of *tian*—a process that brings one nearer to things in their naturalness as *ziran* 自然 or "spontaneously so" of themselves. We "wander" (*you* 遊) with and wonder at things in their vital becoming, in what Parkes calls their "awe-inspiring" coming and going, rather than through limited perspectives which hold things in fixed states of being.[42] Just as the transformations of *qi* are more obvious in the winds and waters of *fengshui* or in the rhythmic body movements of *taiji quan*, they are more easily sensed in music and by listening musically.

Given that the fasting of the heart and listening to the music of *tiandi* are enacted through increasingly open and receptive listening, we should expect to find other somatic practices in which the body resonates with the *qi* of things described in musical terms. For example, in the story of Cook Ding discussed by Parkes, when the cook drops his habitual perceptions and allows himself to be guided by the specific, natural patterns of the animal, "the thwacking tones of flesh falling from bone would echo, the knife would whiz through with its resonant thwing, each stroke ringing out the perfect note, attuned to the 'Dance of the Mulberry Grove' or the 'Jingshou Chorus' of the ancient sage-kings."[43] Music's autonomy from representational meaning offers a somatic opportunity to practice resonant listening to each thing's *anima* with less distortion than when we impose the abstractions of language. This transforms the space in which we encounter the music of things into a place. The *Zhuangzi* is generally read as offering radical shifts of perspective, but what may be less obvious is that becoming the free and easy wanderer and returning to our "naturalness" (*ziran*)—our native spontaneity and responsiveness with things—involve spatial shifts as well. The central images in the *Zhuangzi* of wandering and emptying involve liberation from static, homogenous, and abstract conceptions of space in favor of places where one can encounter unique *anima* as "faces," including *anima mundi* as the musical face of the earth.[44] In these places, rather than just learning the "text" of nature, which risks obscuring nature through representation and the innate anthropocentrism of human language, we might learn nature's music.

Overcoming Extinction, Becoming Our Future Ancestors: *Homo Natura*

At the end of "In the Light of Heaven before Sunrise: Zhuangzi and Nietzsche on Transperspectival Experience," in a section titled "Creative Experience," Parkes addresses the tension in the *Zhuangzi* between the ideal of the sage's mind as an empty mirror and the suggestion that humans can assist in the operations of the heavens.[45] At issue is the extent to which egoism and anthropocentrism can be overcome without detached quietism. If the result of fasting and forgetting practices is "just being empty, nothing more" so that we may impartially let things show themselves, how do we assist the creative forces and processes of the cosmos? Silence, one might presume, does not speak. Put in terms of the urgent environmental calamities of our time and the need to halt and reverse the drive toward extinction, how do we address these human-made disasters and build a new world while "rejecting nothing, welcoming nothing: responding but not storing"?[46] For Parkes, a comparison with Nietzsche "reveals a more engaged aspect" of the *Zhuangzi* in which the wanderer is more like Zarathustra's "creators, procreators, or enjoyers of becoming," who are capable of "loving the Earth" and, crucially, *acting* upon this love with the "lightness of foot" that comes from being attuned to the music of the world.[47] Following Parkes along this route, this dancer's path reveals it is a mistake to equate emptied, resonant listening with passivity and inaction, an error that can be corrected by understanding why becoming *homo natura* requires not only "unhumanizing" nature, but equally "naturalizing" humanity.

On the way to the music of the world, it helps to first reflect on Nietzsche's idea of being "without melody." The person Nietzsche describes in §626 "*Without Melody*" of *Human, All Too Human: A Book for Free Spirits* sounds quite like the image of the sage described variously in the *Zhuangzi*: still like a mirror, harmonizing, and at most, responding like an echo without holding on to anything. Nietzsche writes:

> There are people for whom a steady inner repose and a harmonious ordering of all their capacities is so characteristic that all goal-oriented activity is repugnant to them. They resemble music that consists of nothing but long, drawn-out harmonic chords, without ever showing even the start of an articulated, active melody.[48]

Such people are strange and frustrating to those caught in frenetic modern pursuits—think of the endless consumption, productivity, and self-obsessions of our times, including pursuing nature as a goal, whether as capitalist profit, a code to be cracked, or a bucket list of recreation locales. Modern people are impatient toward those without melody, Nietzsche says, because they "*become* nothing without our being able to say that they *are* nothing." And yet, he notices, in certain moods we may ask an unusual question, "Why melody at all?" and "Why isn't it sufficient for us when life mirrors itself peacefully upon a deep lake?" Why does Nietzsche use the absence of melody, rather than another musical element, to make his point? Melody is certainly not opposed to harmony. Echoing Confucius's advice to Yan Hui in the *Zhuangzi*, Nietzsche compares being without melody to a protective "profound stillness" toward life that shelters one even in the midst of great turmoil. Yet isn't rhythm the least "still" element of music? Perhaps melody is singled out by Nietzsche because it is the aspect of music closest to language and subjectivity, and thus to the common world of concepts and the self-projections of anthropomorphic meaning. In music, melody often sounds like the "voice" of the song (even when there is no singer). If we pair this suggestion with Nietzsche's description of how music "sobers" him up by creating distance from himself, as though he had "bathed in some natural element," then music without melody is all the more unfettered by the chronic strivings of the self and common meanings of the herd.[49] In any case, although the person without melody has an air of quietude, they still resemble music. They are free from purpose-driven action and maintain equilibrium or "harmonic euphony" in the face of external disturbance, as Nietzsche puts it, but they are not silent.

We should also read the above passages alongside §109 of *The Joyous Science*, which challenges the interpretation that being "without melody" or fasting the heart leads to detached quietism.[50] Here Nietzsche warns us against a number of views, some obvious, others more subtle and sneaky, all of which anthropomorphically project our aesthetic and moral judgments onto the world, for example, by conceiving of it as an organism, a machine, or by assuming there must be an order and purpose to the universe that can satisfy our desire for these. In each case, we err by extrapolating essentialist and universalist principles from singular phenomena or by turning the human hunger for meaning into an attribute of the cosmos.[51] These "shadows" of a dead God obscure the universe as a "music box that repeats eternally its tune which may never be called a melody."[52] There are no "enduring substances and matter is as much of an error as the God of the Eleatics"—there is only eternally returning music without melody. For Nietzsche, value is created through joyfully affirming the eternal recurrence, this ever-repeating tune, not in unlocking a transcendent purpose to the universe. To be like music without melody, then, is to be like the universe, much like one must empty the fixed, purposive, and judging self before open listening to the piping of nature is possible.

Nietzsche continues in §109 along lines reminiscent of the *Zhuangzi*, saying there is "nobody who commands, nobody who obeys," and what's more, we should not infer from this lack of "design" that the universe is mere "chance"; words like these only have meaning in reference to each other, and so it is only in a designed world that chance would have meaning. In the passages immediately following Zhuangzi's description of listening to the piping of the earth and heavens, he too cautions against searching for the conclusive source or ultimate order of things beyond the rhythms and patterns of their spontaneous emergence. He describes the arising of things as being like "music coming out of emptiness" and warns that as tempting as it is to propose a hidden cause, if there is a grander force at play making everything "sound," it leaves no enduring material evidence, only the sounding which is indistinguishable from what it animates. In fact, there really is no substantial "what" here, just ever transforming, moving *qi*. Habitually categorizing things into "this" and "that" is sometimes provisionally useful, but always at odds with the perpetual flux of the world. Practices such as fasting the heart, listening to the music of the earth and heavens, and being without melody transform our persistent tendency to view the world according to our desire for an ultimate order and purpose—metaphysical, moral, or otherwise.

These perspective shifts reveal the profound interdependent becoming of all things and the related insight that detached quietism cannot fully unfasten one from the world, even if turning away, fasting, forgetting, and other emptying practices are beneficial and, moreover, necessary to becoming Zarathustra's joyful creator or a Daoist assistant to *tian*. As Zarathustra sings, "all anew, all eternally, all chained together, entwined in love, oh then you *loved* the world—."⁵³ Affirming the eternal recurrence teaches that all things must be inextricably intertwined and ever-changing, not separate and enduring, in order to recur "all anew." Thus, the eternally returning tune magnifies our relations with all our kin, including humans, but also plants, mountains, waters, other than human animals, and even weather. Parkes draws our attention to the Overhuman's (*Übermensch*) kinship with nature in Zarathustra's prologue, emphasizing that the Overhuman is "the sense of the earth," "this sea," and "this lightning," and overcoming the human "is to acknowledge and emulate the nonhuman nature—mineral, animal, vegetal—of which we consist and on which we depend."⁵⁴ With Parkes's guidance, the Overhuman as *homo natura* becomes clearer, as does Nietzsche's twofold task of dehumanizing nature *and* naturalizing the human. The second half of the task assures that we do not stop at emptying our habitual projections upon nature; in becoming *homo natura*, we also joyfully affirm our entwined kinships, our love of the world, and the creativity of will to power without ego—without which we could not become like nature.

To put this another way, if translating the human back into nature as *homo natura* means becoming like nature, then quietude is already on its way to engaged activity. As *yin*, any receptive and relatively tranquil process is already beginning to proffer and move as *yang*. Creative vitality can be cultivated from the dynamics of *qi* energies, which Parkes has linked to Nietzsche's will to power as early as his ground-breaking (and waymaking) work in "The Wandering Dance." There he writes, "Nietzsche's later understanding of will to power as an interpretive energy inherent in all things comes close to the pan-psychism that informs the *Chuang Tzu*."⁵⁵ More recently, he revisited this comparison, replacing pan-psychism with *qi* energy and highlighting the more engaged aspects of the *Zhuangzi* where wanderers of many stripes act in the world and "the sage practices non-attachment rather than detachment."⁵⁶ The sage nourishes their vital energy (*qi*) by disentangling themselves from the false opposites of the human world, eventually arriving at the pivotal place from which they can respond easily, spontaneously, and naturally, that is, they can assist in the operations of *tian*: "Such a one will position himself in the measure of his place without overflow and yet hide himself away in the proportions of the endless, roaming through that in which all beings end and begin."⁵⁷ At this place, called the Great Beginning (*tai chu* 泰初) elsewhere in the text, "a way in which all the world can become peaceful and balanced" is revealed.⁵⁸ Why? Because becoming an assistant to *tian* affirms the whole and thus does not exile one's humanity, as Parkes emphasizes with this passage: "Insatiably partake of the Heavenly, but do not neglect the human either—then even the ordinary people around you will soon come to live by what is genuine in them."⁵⁹ Later alchemists in the Chinese tradition heeded the warnings in the *Zhuangzi* that not emulating nature, as *ziran*, eventually brings illness and an unnatural end to life, including one's own, and saw the potential for reversing this drive through the power of the Great Beginning available in each moment.

Zhuangzi's assistant to *tian* and Nietzsche's *homo natura* do not succumb to false oppositions between nature and human, nor between life and death. Following his description of the cosmic music box, in a sentiment echoed throughout the *Zhuangzi*, Nietzsche says, "Let us beware of saying that death is opposed to life. The living is merely a type of what is dead, and a very rare type."[60] The living and the dead share the same mutual dependence revealed in the dynamics of the eternal recurrence and the transformations of *qi*. This means our kin include dead ancestors: animal, vegetal, mineral, and atmospheric. What's more, we are future ancestors to these realms, and in the eternally returning tune, we are our own future ancestors. This astonishing perspective can help develop a sense of place, a *sensus communis*, with our ancestors through time. Parkes points out a passage in the *Zhuangzi* that "seems strangely to anticipate Nietzsche's thought of the eternal recurrence," in which the sage is described as one who "gets through to the intertwining of things so that everything forms a single body around him," and which speaks to the joy of remembering "the old homeland, the old neighborhood" despite "nine out of ten of your old friends" lying beneath the ground. Parkes remarks, "And how much greater the joy 'if you could still see what you had once seen and hear what you had once heard there'—which could happen if those moments came around again, if they recurred."[61] In the sage's unwavering fidelity to "one body" through changing "every day together with all things," even death can be affirmed as yet another transformation of *qi* energies, and consequently, even if events do not eternally recur in exactly the same way, ancestors live through us in blood and soul, culture, and kinship with all things.[62] This ancestral web invites a sense of death true to the earth and that "nourishes life" (*yang sheng* 養生), instead of the drive toward conquering the earth leading us and millions of species to extinction.

In becoming future ancestors, it is possible to not just become assistants to nature, but apprentices to its great creative powers, akin to Nietzsche's alchemist—the true benefactor of humanity who can create something valuable out of what is unwanted or even terrible. As Parkes describes in "Nietzsche's Care for Stone: The Dead, Dance, and Flying," Zarathustra must learn (as Nietzsche did) "the trick of the alchemist: of turning the base metal of experience into gold by willing its eternal return."[63] Parkes shows how Zarathustra overcomes the Spirit of Heaviness—"his greatest therapeutic task"—through stone: "the philosopher's stone and the stone of death, and the rock on which he sits and dances, and from which he ultimately takes off and flies."[64] In the long history from ancient alchemy to modern depth psychology, the most powerful philosopher's stone is *anima mundi*. Hillman reminds us that alchemy does not just seek the transformation of the human soul, but also the soul of the world and, in doing so, reconnects our efforts to the ecological world.[65] As well, the practice of *fengshui* shares roots (conceptual, historical, and shamanistic) with Chinese internal alchemy or *neidan* 內丹. *Fengshui* attends to the dynamic *qi* energies in the environment, while *neidan* works with the *qi* energies in the human body to "reverse" or "trace to the source" (*dian dao* 颠倒) from separation back to the one *ti* body of all things; in both cases, flourishing and realization are cultivated through resonance with the world as flux. Among the practices mentioned here, none of them seek the secrets of nature in a calculative manner or impose human laws on it; rather, they awaken the creative processes of the heavens and earth in ourselves that are needed to both unhumanize nature and naturalize humanity.

Siren Songs

Alchemy, it turns out, is important to *fengshui*, *anima mundi*, and *homo natura*, and the question of whether the arcane arts can be separated from the more commonsense aspects of *fengshui* has become a more general question about the tension between common sense and alchemy. We should heed Nietzsche's insight that "The believer in magic and miracles reflects on how to *impose a law on nature*—: and, in brief, the religious cult is the outcome of this reflection," while also recognizing that alchemy, at least the versions of it discussed here, does not impose laws on nature.[66] Instead, it accords with and emulates the reciprocal, transforming energies of nature: heaviness and lightness, *yin* and *yang*, life and death, and will to power as "a play of constantly self-renewing interpretive forces."[67] Mystification and charlatanry are more likely when we fail to undertake the hard work of overcoming egoic, purpose-driven, and anthropocentric perspectives—when we fall short of perspectives beyond the human and stand lead-footed, imposing our desires on nature and unattuned to the music of the world. But we are also vulnerable to missing the wandering dance if we stick too closely to common sense, which is often heavy with sedimented metaphysical beliefs and historical errors, presumptions of the current times, and the kind of common meanings that led Nietzsche to say, "*In comparison to music* all communication through *words* is shameless … The word makes what is uncommon common."[68] Or Zarathustra to declare bird-wisdom: "Are all words not made for those who are heavy? Do all words not lie for one who is light! Sing! speak no more!"[69] Whatever the status of other arcane arts, music—such a common and yet uncommon art—remains inseparable from the alchemical routes followed here.

Parkes has done invaluable work to help us become musical listeners who can hear Nietzsche's writing with a *"third* ear" and a sense of life as music.[70] In honoring this, learning the text of nature involves hearing it as music just as becoming *homo natura* needs us to become like music. In his discussion of translating the human back into nature in *Beyond Good and Evil* §230, Nietzsche describes the future human as standing before nature with "intrepid Oedipus eyes and sealed Odysseus ears, deaf to the siren songs of old metaphysical bird catchers, who have been piping at him all too long, 'you are more, you are higher, you are of a different origin!'"[71] For future ancestors, the siren song of metaphysics gives way to Zarathustra's song—a siren song too, but like the songs of the heavens and earth heard in the winds and waters of *fengshui* or the song of the world heard in *anima mundi*, they draw us to a different kind of death—not extinction or the failed hero's journey of humanity—but the end of perspectives that prevent loving the world, affirming the eternal recurrence, or becoming one body with all things. They sing of right living on the earth, honoring the dead, reversing separation, and creating new worlds with the same extravagance as nature.

Following Parkes along some of the distant routes he has traveled, and made, shows how *feng shui, anima mundi,* and *homo natura* converge in the vital world of *qi* energies, in a place where we can consider human-driven ecological ruin from different perspectives than those that currently dominate. A Daoist poem called "Yuan You" 遠遊 or "Far Roaming" captures this kind of journey well. The wanderer of the poem opens with the sorrows of the world, travels by chariot across the earth and into the heavens, finally arriving at the Great Beginning (*tai chu*):

> Grieving at the pressing constraints of the age's vulgarity …
> What is inward I examine, indeed, with discipline most firm–
> And seek that which is the source of truest Vitality.
> Silently, by attenuation and stillness, I find pleasure and contentment …
> Facing into the wind at morning, I unloose my feelings.
> Kao-yang is now remote, far away—
> How shall *I* take to his route? …
> I comply with the triumphal wind, to follow its roamings …
> Suddenly glancing down, discerned my homeland of old.
> My coachman grew wistful, my own heart was grieved—
> *Luan*-birds wafted loftily, hovering in flight.
> The strains of music spread everywhere, boundless, endless—
> Going beyond doing nothing, and into utmost clarity,
> Sharing in the Great Primordium, I now became its neighbor.[72]

Wherever one lands among the diverse perspectives on nature, Parkes is right to suggest that the devastation of the earth is too urgent to indulge in fruitless speculations about truth or to assume a quietist stance or nihilistic reading of the eternal recurrence.[73] Power matters, in Parkes's description, "it all comes down to a question of will to power, conflicts between competing interpretations and worldviews" and the ability to affect human views about nature. All the more reason to encourage our fellow humans to reunite with the indefatigable, transformative power (*xuan de*) available in each moment's Great Beginning and, as Parkes says, generate "love for this radically ephemeral life that eternally recreates itself at every moment."[74]

Notes

1   Graham Parkes, "Human/Nature in Nietzsche and Taoism," in *Nature in Asian Traditions of Thought: Essays in Environmental Philosophy*, edited by J. Baird Callicott and Roger T. Ames (Albany: State University of New York Press, 1989), 79.
2   Graham Parkes, "Winds, Waters, and Earth Energies: *Fengshui* and Sense of Place," in *Nature across Cultures: Views of Nature and the Environment in Non-Western Cultures*, edited by Helaine Selin (London: Kluwer Academic Publishers, 2003), 185.
3   Heraclitus's φύσις κρύπτεσθαι φιλεῖ ("nature loves to hide") and the Daoist description of *dao* 道 as *xuan* 玄 ("dark," "mysterious," and "abstruse") come to mind here.

4   Parkes, "Winds, Waters, and Earth Energies," 185.
5   Parkes writes that people in the United States and Europe "are paying fengshui 'experts' large sums of money to align their expensive coffee tables with their even more expensive sofas, in the hope of bringing more wealth, and perhaps some happiness, into their already affluent households. This seems a gross perversion of the basic spirit of fengshui—which would say that happiness, and certainly some wealth, would come more easily if these people simply sold off all the furniture and other clutter that's obstructing their contact with their natural surroundings" (Parkes, "Winds, Waters, and Earth Energies," 186). In fact, nowadays there is no shortage of this version of *fengshui* among the wealthy in China too. In its beginnings, *fengshui* was used in service of abundance and happiness, but in ways that aligned more strongly with humility toward nature rather than with superficial symbols of capitalist success. Even still, early on "*Fengshui* went from being an environmental knowledge system based on observation to a method of site selection for good luck" (Michael John Paton, *Five Classics of Fengshui: Chinese Spiritual Geography in Historical and Environmental Perspective* (Leiden: Koninklijke Brill NV, 2013), 107).
6   Parkes, "Winds, Waters, and Earth Energies," 186.
7   Parkes, "Winds, Waters, and Earth Energies," 205.
8   *Di li* 地理 is an ancient term associated with *fengshui* that means "geography" in modern Mandarin. Along with earth principles and patterns, a *fengshui* practitioner learns the principles and patterns of the heavens (*tian li* 天理) and humans (*ren li* 人理).
9   Parkes, "Winds, Waters, and Earth Energies," 193.
10 Parkes, "Winds, Waters, and Earth Energies," 200.
11 Parkes, "Winds, Waters, and Earth Energies," 204.
12 Paton, *Five Classics of Fengshui*, 172.
13 Michael John Paton, *Five Classics of Fengshui: Chinese Spiritual Goegraphyin Historical and Environmental Perspectives* (Leiden: Koninkjike, 2003), 204.
14 In "Thinking Rocks, Living Stones: Reflections on Chinese Lithophilia," Parkes notes that presuppositions about the physical world and the invisible "impede acceptance of the idea of energies flowing through the earth along 'lifelines' that cannot be directly seen but can be intuitively discerned by the well-trained practitioner" despite the fact that the Stoics and Epicureans offered a precedent in the West for a "physics associated with a transformation in our experience through practice" and contemporary physics accepts the existence of the earth's magnetic field with its invisible, flowing energy lines (Parkes, "Winds, Waters, and Earth Energies," 81–2).
15 Parkes's sustained attention to the vitality of stone has done much to illuminate the limits of biophilia. In *How to Think about the Climate Crisis: A Philosophical Guide to Saner Ways of Living*, he writes, "While Western advocates of biophilia and deep ecology extend their concern to all living things, they tend to get stuck at the state of 'biocentrism', failing to go all the way and include the mineral realm of rocks and mountains—and from there also things of use and other artefacts" (Graham Parkes, *How to Think about the Climate Crisis: A Philosophical Guide to Saner Ways of Living* (London: Bloomsbury Publishing, 2021), 172).
16 *Sensus communis* has multiple complex meanings besides "community sense," many of which have relevance here. For example: a cognitive function or sense that unifies and integrates the other bodily senses; the sensory power by which we perceive we are perceiving; and the sensation of being alive.
17 In "Feng-shui: Ideology and Ecology," E. N. Anderson argues that *fengshui* does more than set guidelines; it also motivates communities to follow the guidelines. "*Feng-shui*

sold good planning very effectively. All peasants knew that their own future good fortune depended on following its rules. If a peasant was tempted to disregard possible future welfare and choose his own present self-interest, his neighbors would immediately raise the alarm: he was wrecking the community's *feng-shui*" (1996, 26). In this regard, *fengshui* is a counterexample to the problematic yet popular theory of the "tragedy of the commons," which presumes that self-interested individuals will competitively exhaust common resources rather than sustain them through collective cooperation.

18 Tu Wei-Ming, "The Continuity of Being: Chinese Visions of Nature," in *Nature in Asian Traditions of Thought: Essays in Environmental Philosophy*, edited by J. Baird Callicott and Roger T. Ames (Albany: State University of New York Press, 1989), 78
19 Parkes, "Human/Nature in Nietzsche and Taoism," 78.
20 Parkes, "Winds, Waters, and Earth Energies," 204–5.
21 *Kanyu* 堪輿 or "chariot of earth/canopy of heaven" is one of the traditional names for *fengshui*.
22 Parkes, *How to Think about the Climate Crisis*, 172.
23 Parkes uses Brook Ziporyn's translation here *Zhuangzi: The Essential Writings*. Translated by Brook Ziporyn (Indianapolis/Cambridge: Hackett Publishing, 2009), 69.
24 This is Charles Le Blanc's translation of Chapter 6 (sections 11/15b) of the *Huainanzi*. I have changed "corresponding" notes to "correlated." This example of sympathetic resonance between two guqins occurs in several other texts, such as the *Zhuangzi* (24/158), *Chu Ci* (13/156), *Lüshi Chunqiu* (13:2/127), and *Chunqiu Fanlu* (57/4b–5a).
25 Graham Parkes, "Befriending the Things around Us, Respecting the Soul of the World." *NotaBene* Issue 46. Edited by Antoaneta Nikolova, 2019, 6.
26 Parkes, "Befriending the Things around Us, Respecting the Soul of the World," 7.
27 Hillman writes, "The depression we're all trying to avoid could very well be a prolonged chronic reaction to what we've been doing to the world, a mourning and grieving for what we're doing to nature and to cities and to whole peoples— the destruction of a lot of our world. We may be depressed partly because this is the soul's reaction to the mourning and grieving that we're not consciously doing. The grief over neighborhoods destroyed where I grew up, the loss of agricultural land that I knew as a kid" (James Hillman and Michael Ventura, *We've Had a Hundred Years of Psychotherapy—And the World's Getting Worse* (New York: HarperCollins Publishers, 1992), 45).
28 James Hillman, *The Thought of the Heart and the Soul of the World* (Thompson, Connecticut: Spring Publications, 2021), 61.
29 Hillman, *The Thought of the Heart and the Soul of the World*, 60.
30 Hillman, *The Thought of the Heart and the Soul of the World*, 183.
31 Hillman and Ventura, *We've Had a Hundred Years of Psychotherapy—And the World's Getting Worse*, 39.
32 Everyday tasks can become "ways." Parkes writes, "He's talking about what the Japanese call *dō* (Chinese *dao*): *ways* of living, artistically; practices that derive from, and in turn encourage, a long-standing culture of attention to aesthetic detail" (Parkes, *How to Think about the Climate Crisis*, 183). Maurice Merleau-Ponty's work on the "flesh" as the "natal bond" between the mutually embodied subject, known through *aesthesis* or living sense perception, is helpful to Hillman and Parkes's depictions of *anima mundi* here and to understanding the possibilities of *sensus communis*. Parkes, *How to Think about the Climate Crisis*, 183.

33 Hillman, *The Thought of the Heart and the Soul of the World*, 64.
34 Hillman, *The Thought of the Heart and the Soul of the World*, 36.
35 Hillman, *The Thought of the Heart and the Soul of the World*, 36.
36 Graham Parkes, "Awe and Humility in the Face of Things: Somatic Practice in East-Asian Philosophies," *European Journal for Philosophy of Religion* 4 (3) (Autumn 2012): 70.
37 Parkes, "Awe and Humility in the Face of Things," 77.
38 Graham Parkes, "In the Light of Heaven before Sunrise: Zhuangzi and Nietzsche on Transperspectival Experience," in *Daoist Encounters with Phenomenology: Thinking Interculturally about Human Existence*, edited by David Chai (London/New York: Bloomsbury Publishing, 2020), 77.
39 *Zhuangzi Yinde*, 4.26–28. Here I have chosen to translate *xu* 虛 as "empty," but note that alternate translations such as "space," "tenuous," and "insubstantial" have their merits as well. The character depicts a gap between two mountains.
40 Quoted by Parkes from *Zhuangzi* chapter 6, 42. Translation by Ziporyn (2009). Parkes, "Awe and Humility in the Face of Things," 77.
41 *Zhuangzi Yinde*, 2.1–7.
42 Quoted by Parkes from *Zhuangzi* chapter 3, 22. Translation by Ziporyn (2009). Parkes, "Awe and Humility in the Face of Things," 88.
43 Parkes, "Awe and Humility in the Face of Things," 78.
44 Many philosophers have compared understanding music to understanding a face. See, for example, Ji Kang's 嵇康 (223–62 CE) "Music Has Neither Joy Nor Sorrow" (*Sheng Wu Ai Le Lun* 聲無哀樂論) for a tremendous debate about what can be known from music and faces. Much later, Wittgenstein echoed Ji Kang in arguing that despite the apt comparison between music and faces, we should not be fooled into searching for the real meaning behind the face of music, rather "Music conveys to us *itself*!" (1964, 178). This helps avoid mistakenly projecting dualistic views about meaning onto music, and more to the point here, the music of the earth.
45 Parkes, "In the Light of Heaven before Sunrise," 80.
46 Parkes, "In the Light of Heaven before Sunrise," 79.
47 Parkes, "In the Light of Heaven before Sunrise," 79.
48 Friedrich Nietzsche, *Human, All Too Human: A Book for Free Spirits*. Translated by R. J. Hollingdale (Cambridge: Cambridge University, 1996), 197.
49 Nietzsche's shares the image of musical distance from the self in the same passage in which he famously declares, "Life without music is simply an error, exhausting, an exile" (Friedrich Nietzsche, "Letter to Köselitz, January 15, 1888," in *Twilight of the Idols* (Indianapolis: Hackett Publishing Company, 1997), vii).
50 Parkes emphasizes there are many examples in the *Zhuangzi* that also complicate the quietist reading of the sage, especially those depicting skilled virtuosity in the world: "a cicada catcher, a ferryman, a maker of bell-stands, a charioteer, and an artisan" (Parkes, "In the Light of Heaven before Sunrise," 79).
51 Projecting ourselves onto nature can be especially wily when we imagine we are living in accord with nature. In *Beyond Good and Evil* §9, Nietzsche challenges the Stoics on this issue: "Imagine a being like nature—extravagant without limit, indifferent without limit, without purposes and consideration, without pity and justice, simultaneously fruitful, desolate, and unknown—imagine this indifference itself as a power—how *could* you live in accordance with this indifference?" In "Zhuangzi and Nietzsche on Human and Nature," Parkes rightly notes that Daoists, who also aspire to live in accordance with nature, are also vulnerable to Nietzsche's

provocation here. Perhaps this is why chapter 5 of the *Laozi* 老子 can be unsettling: "Heaven and earth are not humane, they regard all things as straw dogs. The sage is not humane, they regard all things as straw dogs."

52 Friedrich Nietzsche, *The Gay Science*. Translated by Walter Kaufmann (New York: Random House, Inc, 1974), 168.

53 Friedrich Nietzsche, *Thus Spoke Zarathustra*. Translated by Graham Parkes (Oxford: Oxford University Press, 2005), 283.

54 Parkes quotes Nietzsche here: "The highest human being is to be conceived in the image of nature" (2005, xix). Nietzsche, *Thus Spoke Zarathustra*, xviii.

55 Parkes Graham, "The Wandering Dance: Chuang Tzu and Zarathustra," *Philosophy East and West,* 33 (3) (Jul., 1983): 237.

56 Parkes, "In the Light of Heaven before Sunrise," 80.

57 *Zhuangzi: The Essential Writings*, 150.

58 Zhuangzi: The Essential Writings, 78. "In the great beginning, there is nothing: without something and without name. Out from where one emerges, there is one not yet formed. When things grasp it as the means by which they are generated, this is called virtuosic power (*de*). In what is not yet formed, there is a division; as it moves it is boundless. This is called destiny. Out of stillness and movement, things are generated; when they are ready, patterns are generated; this is called form. The embodied form shelters spirit, each with its own conditions and principles; this is called natural endowment. In cultivating our natural endowment we return to our virtuosic power. When this power is at its utmost, we accord with the great beginning. Being in accord with the great beginning, we are emptied. Emptied, we become great. Unifying with the opening and closing of beaks, the opening and closing of beaks are unified. Together with the heavens and earth this is unity. This unity is mindless, as if foolish, as if turbid. This is called hidden virtuosity (*xuan de* 玄德) and harmonizes with the great accord." *Zhuangzi Yinde* 37.14.13–30. *Xuan de* 玄德 also occurs in the *Laozi* (Chapters 10, 51, 65). The term defies easy translation. Alternate translations could include "arcane or dark virtuosity" or "profound potency or power." This passage could be easily read as the emergence of *anima*, and perhaps Parkes would hear will to power in these dynamics.

59 *Zhuangzi: The Essential Writings*, 150.

60 Nietzsche, *The Gay Science*, 168.

61 Parkes, "In the Light of Heaven before Sunrise," 76.

62 Notably, the "one body" (*yi ti* 一體) formed by the sage with all things is a *ti* body, which is the most corporeal and somatic of the many kinds of bodies referenced in Chinese philosophy, medicine, and alchemy. This is the sensing body of *aisthesis*; it can extend into and receive other *ti* bodies, and is the site of "embodiment." The Chinese character *ti* is closely related to "ritual" (*li* 禮), and therefore music, as echoed in the *Liji* 禮記 ("Book of Ritual") where the "great body" of ritual is said to embody heaven and earth, model the seasons, follow *yin* and *yang*, and accord with human emotions.

63 Graham Parkes, "Nietzsche's Care for Stone: The Dead, Dance, and Flying," in *Nietzsche's Therapeutic Teaching*, edited by Horst Hutter and Eli Friedland (London: Bloomsbury, 2013), 185.

64 Parkes, "Nietzsche's Care for Stone," 175.

65 In *Alchemical Psychology*, Hillman writes, "Alchemy is cosmological work; to follow an alchemical psychology at once leads to working with the world" (2010, 55).

66 Nietzsche, *Human, All Too Human*, 64.

67 Parkes, "Nietzsche's Care for Stone," 9.
68 Quoted from Georges Liébert's *Nietzsche and Music*, translated by Parkes and David Pellauer. The original passage is from KSA 12: 493. Georges Liébert, *Nietzsche and Music*. Translated by David Pellauer and Graham Parkes (Chicago: The University of Chicago Press, 2004), 3.
69 Liébert, *Nietzsche and Music*, 203.
70 This is most obvious in his painstaking analysis of the symphonic structure of *Zarathustra*, but one of the great contributions of Parkes's work on Nietzsche is that he faithfully reads him musically, as Nietzsche wanted, and passes on at least some of his hearing to us. See his introduction to *Zarathustra* (2005) as well as "The Symphonic Structure of *Thus Spoke Zarathustra*: A Preliminary Outline" (2008).
71 Friedrich Nietzsche, *Beyond Good and Evil: Prelude to a Philosophy of the Future*. Translated by Walter Kaufmann (New York: Random House, Inc, 1966), 161.
72 "Yuan You" is part of the famous anthology of songs known as *Chuci* or *Lyrics of Chu*, dating from the early third century BCE to early second century CE. Daoist cultivation practices are referenced throughout the poem's imagery. This translation is Paul Kroll's ("On 'Far Roaming," *Journal of the American Oriental Society* 116 (4) (Oct.—Dec. 1996): 660-3). An alternative translation of *tai chu* is "Great Beginning," which I have been using throughout this chapter. Reproduced with permission.
73 Graham Parkes, "Staying Loyal to the Earth: Nietzsche as an Ecological Thinker," in *Nietzsche's Futures* (London: Palgrave Macmillan, 1999), 185.
74 Graham Parkes, "The Symphonic Structure of Thus Spoke Zarathustra: A Preliminary Outline," in *Nietzsche's Thus Spoke Zarathustra: Before Sunrise*, edited by James Luchte (London: Continuum, 2008), 27.

References

Anderson, Eugene N. 1996. "Feng-shui: Ideology and Ecology." In *Ecologies of the Heart*, 15–27. Oxford: Oxford University Press.
Hillman, James. 2010. *Alchemical Psychology*. Putnam, CT: Spring Publications, Inc.
Hillman, James. 2021. *The Thought of the Heart and the Soul of the World*. Thompson, CT: Spring Publications.
Hillman, James and Ventura, Michael. 1992. *We've Had a Hundred Years of Psychotherapy—And the World's Getting Worse*. New York: HarperCollins Publishers.
Kroll, Paul. 1996. "On 'Far Roaming." *Journal of the American Oriental Society*. 116 (4) (October–December 1996): 653–69.
Le Blanc, Charles. 1985. *Huai-nan Tzu: Philosophical Synthesis in Early Han Thought: The Idea of Resonance (Kan-Ying) with a Translation and Analysis of Chapter Six*. Hong Kong: Hong Kong University Press.
Liébert, Georges. 2004. *Nietzsche and Music*. Translated by David Pellauer and Graham Parkes. Chicago: The University of Chicago Press.
Nietzsche, Friedrich. 1966. *Beyond Good and Evil: Prelude to a Philosophy of the Future*. Translated by Walter Kaufmann. New York: Random House, Inc.
Nietzsche, Friedrich. 1968. *The Will to Power*. Translated by Walter Kaufmann and R. J. Hollingdale. New York: Random House.
Nietzsche, Friedrich. 1974. *The Gay Science*. Translated by Walter Kaufmann. New York: Random House, Inc.

Nietzsche, Friedrich. 1996. *Human, All Too Human: A Book for Free Spirits*. Translated by R. J. Hollingdale. Cambridge: Cambridge University Press.
Nietzsche, Friedrich. 1997. "Letter to Köselitz, January 15, 1888." In *Twilight of the Idols*, trans. Richard Polt, intro. Tracy Strong, vii. Indianapolis: Hackett Publishing Company.
Nietzsche, Friedrich. 2005. *Thus Spoke Zarathustra*. Translated by Graham Parkes. Oxford: Oxford University Press.
Parkes, Graham. 1983. "The Wandering Dance: Chuang Tzu and Zarathustra." *Philosophy East and West* 33 (3) (July 1983): 235–50.
Parkes, Graham. 1989. "Human/Nature in Nietzsche and Taoism." In *Nature in Asian Traditions of Thought: Essays in Environmental Philosophy*, edited by James Baird Callicott and Roger T. Ames, 79–97. Albany: State University of New York Press.
Parkes, Graham. 1999. "Staying Loyal to the Earth: Nietzsche as an Ecological Thinker." In *Nietzsche's Futures*, edited by John Lippitt, 167–88. London: Palgrave Macmillan.
Parkes, Graham. 2003. "Winds, Waters, and Earth Energies: *Fengshui* and Sense of Place." In *Nature across Cultures: Views of Nature and the Environment in Non-Western Cultures*, edited by Helaine Selin, 185–209. London: Kluwer Academic Publishers.
Parkes, Graham. 2008. "The Symphonic Structure of *Thus Spoke Zarathustra*: A Preliminary Outline." In *Nietzsche's Thus Spoke Zarathustra: Before Sunrise*, edited by James Luchte, 9–28. London: Continuum.
Parkes, Graham. 2012. "Awe and Humility in the Face of Things: Somatic Practice in East-Asian Philosophies." *European Journal for Philosophy of Religion* 4 (3) (Autumn 2012): 69–88.
Parkes, Graham. 2013. "Nietzsche's Care for Stone: The Dead, Dance, and Flying." In *Nietzsche's Therapeutic Teaching*, edited by Horst Hutter and Eli Friedland, 175–190. London: Bloomsbury.
Parkes, Graham. 2019. "Befriending the Things around Us, Respecting the Soul of the World." *NotaBene* (46): 1–22. Edited by Antoaneta Nikolova.
Parkes, Graham. 2020. "In the Light of Heaven before Sunrise: Zhuangzi and Nietzsche on Transperspectival Experience." In *Daoist Encounters with Phenomenology: Thinking Interculturally about Human Existence*, edited by David Chai, 61–84. London/New York: Bloomsbury Publishing.
Parkes, Graham. 2021. *How to Think about the Climate Crisis: A Philosophical Guide to Saner Ways of Living*. London: Bloomsbury Publishing.
Paton, Michael John. 2013. *Five Classics of Fengshui: Chinese Spiritual Geography in Historical and Environmental Perspective*. Leiden: Koninklijke Brill NV.
Wei-Ming, Tu. 1989. "The Continuity of Being: Chinese Visions of Nature." In *Nature in Asian Traditions of Thought: Essays in Environmental Philosophy*, edited by James Baird Callicott and Roger T. Ames, 67–78. Albany: State University of New York Press.
Wittgenstein, Ludwig. 1964. *Blue and Brown Books*. Oxford: Blackwell.
Zhuangzi: The Essential Writings. 2009. Translated by Brook Ziporyn. Indianapolis/Cambridge: Hackett Publishing.
Zhuangzi Yinde [*A Concordance to Chuang Tzu*]. 1956. 洪業主編《莊子引得》哈佛燕京學社引得特刊第 20. Harvard-Yenching Institute Sinological Index Series, Supplement no. 20. Cambridge, MA: Harvard University Press.

4

The Role of Aesthetics in Nietzsche's Philosophy and Japanese Culture

Yuriko Saito

The philosophical inquiry into Modern Western aesthetics that emerged during the eighteenth century is characterized by two features.[1] First, it is primarily concerned with the aesthetic experience of the viewers, audience, and readers. Second, with its emphasis on disinterestedness as a characteristic mark of an aesthetic attitude and experience, the aesthetic is considered distinct from other areas of human concerns, such as moral, political, religious, scientific, and practical. Thus, the model for the aesthetic discourse is a spectator deriving an aesthetic experience from an object for its own sake. This model still dominates contemporary aesthetics.

Friedrich Nietzsche is one of the philosophers who recognized these characteristics of Western aesthetics. He observes that Western aesthetics "formulated the experiences of what is beautiful, from the point of view of the *receivers* in art" and takes Kant as representative of this stance that "considered art and the beautiful purely from that of the 'spectator.'"[2]

Nietzsche's own aesthetics instead focuses on issues generated from the creator's viewpoint. However, his notion of creators encompasses human beings in general as artists of their selves and lives. For him, a man who fancies that "he is a *spectator* and *listener* who has been placed before the great visual and acoustic spectacle that is life … overlooks that he himself is really the poet who keeps creating this life."[3] As such, aesthetics is inseparable from all life concerns.

Paralleling Nietzsche's emphasis on the creative dimension of aesthetics, the Japanese aesthetic tradition is dominated by writings by the *practitioners* of various arts, such as painting, tea ceremony, flower arrangement, haiku, linked verse, Noh theatre, and martial arts. While discussion of the aesthetic experience of the *receiver* of art is not absent, as one commentator of Japanese aesthetics observes, "Japanese aestheticians … have generally very little to say about the relationship between the work and the audience, or about the nature of literary and art criticism."[4] Further, what may at first appear to be training manuals for artistic practices are ultimately treatises on what constitutes a good life and how to live such a life. There is no separation between and among the aesthetic (via artistic activities), the moral, the spiritual, and the practical.

In what follows, the role of aesthetics in Nietzsche's philosophy of life and the main themes of traditional (pre-modern) Japanese aesthetics is explored. I find a number of commonalities, as well as differences, between these approaches to be illuminating, particularly in view of the fact that both are concerned with how aesthetics can aid in living a good life. This thesis is developed in the following discussion in light of Graham Parkes's works on Japanese aesthetics, continental philosophy, and comparative study, as I find them to be particularly pertinent and helpful.

Aesthetic Overcoming of Life's Difficulties

Both Nietzsche's aesthetics and Japanese aesthetics can be interpreted as providing a strategy to cope with life's difficulties. For Nietzsche, the primary cause of difficulties in life has to do with "accident" and "chance." Life never proceeds the way we want or plan; curve balls are thrown at us all the time. Nietzsche depicts Zarathustra as "fight[ing] step for step with the giant Chance" and trying to become a "redeemer of coincidence" by "cook[ing] every *chance* event" in his pot.[5] This cooking metaphor continues:

> And only verily, many a chance event came to me imperiously: but even more imperiously did my *will* speak back to it—then it went down imploring on its knees—Imploring that it might find shelter and heart with me, and urging me flatteringly: "But see, O Zarathustra, how only a friend comes to a friend!"[6]

The challenge posed by life's contingencies is not simply that unexpected things happen to us, but even more significantly there is no rhyme or reason for such accidents. Attribution of such accidents to the will of God or any other philosophical or theological scheme that takes us away from this world, according to Nietzsche, is a form of bad faith and a sign of weakness. An accident in life is an accident insofar as we believe it lacks justification or necessity, thinking that it could have been missing or it could have been otherwise. The only strategy to turn life's accidents into necessity that is acceptable to Nietzsche is to devise an *aesthetic* justification for them.

Nietzsche subscribes to the classical notion of a work of art as an organic unity consisting of completeness, parsimony, and necessity of relationship among parts. His model seems to be music that is composed of an overall harmonious arrangement of consonance and dissonance. He explains the value of tragedy as an artistic genre by pointing out "the joy aroused by the tragic myth has the same origin as the joyous sensation of dissonance in music" and declares "only music, placed beside the world, can give us an idea of what is meant by the justification of the world as an aesthetic phenomenon."[7] A dissonant sound taken out of context may not sound pleasant, but in Western classical music (Bach, Mozart, Beethoven, and Wagner are mentioned in *The Birth of Tragedy*), it has its raison d'être, leading to a resolution to constitute a harmonious whole. Hence, in both music and tragedy, "even the ugly and disharmonic are part of an artistic game."[8] They cannot be missing. Thus, pain and misery in life can also be rendered necessary elements of life through aesthetic justification. In short, "it is only as an *aesthetic phenomenon* that existence and the world are eternally justified"[9] and "existence and the world seem justified only as an *aesthetic phenomenon.*"[10] If successful, every contingency in life will assume inevitability:

> whatever it is, bad weather or good, the loss of a friend, sickness, slander, the failure of some letter to arrive, the spraining of an ankle, a glance into a shop, a counter-argument, the opening of a book, a dream, a fraud—either immediately or every soon after it proves to be something that *"must not be missing."*[11]

By fashioning one's life as an aesthetic phenomenon, a work of art, we can embrace life with all its contingencies and challenges: "*as an aesthetic phenomenon* existence is still bearable for us, and art furnishes us with eyes and hands all the good conscience to be able to turn ourselves into such a phenomenon."[12]

However, to render our life an aesthetic whole is not to remain a spectator by *viewing* life *as if* it were a work of art. It requires, for Nietzsche, a more active and creative engagement. Specifically, through our *willing* we make each part of life indispensable. Accidents distress us and we resent their occurrence as long as they *happen to us*, leaving us powerless. However, if we *will* each accident to happen in the way it does, we can make it a *necessary* element of life. In other words, I accept each accident by saying "yes" to it; moreover, I make it *mine* by saying that "Thus I willed it" or "Thus do I will it" or "Thus shall I will it."[13] If successful, Zarathustra states:

> The time has flowed past when accidents could still befall me; and what *could* still fall to me now that would not already be my own! It simply comes back, it finally comes home to me—my own self, and what of myself has long been abroad and scattered among all things and accidents.[14]

Nietzsche's figure of the *Übermensch* can be characterized as a creator of an aesthetic phenomenon: life. For those who can construct a work of art from their lives by saying "yes" to every part, including those parts that are difficult to accept and appreciate, thereby render them indispensable for constituting an aesthetic whole. This artistic project would have to include the end of life as well. Zarathustra praises "free death" which "comes to me because *I* will it," while he despises those who are "like the ropemakers ... [who] drag out their threads at length and so are themselves always walking backwards."[15] If one conceives of life as one long piece of music, the coda should not be needlessly prolonged. As a fitting finale to the whole piece, we can "patiently accept death with the feeling, 'now we are ripe for it.'"[16] Zarathustra teaches us to "die at the right time," particularly because "many die too late, and some die too early."[17] Although this claim is ironic in light of the last few years of Nietzsche's own life, we can interpret him to be advocating suicide in certain cases if the aesthetic considerations demand it.

In this regard, Nietzsche may well have praised Mishima Yukio's 三島由紀夫 death, if not its manner, then at least its timing. One of the most celebrated writers of twentieth-century Japan, Mishima committed *seppuku* in 1970, the old-fashioned way for a warrior to commit suicide by disemboweling himself with a sword, followed by beheading by his "assistant." Mishima's act was carried out after a public appearance at the Self Defense Army Headquarter in Tokyo on the day he sent the last volume of his four-part *tour de force* to his publisher. While the pretext of his suicide was to protest the national indifference and disrespect toward the traditional Japanese values embodied in the Emperor, it is more commonly interpreted as the intentional act of completing his life as an aesthetic whole. Graham Parkes points out that "there is a sense that Mishima had reached the height of his powers, that he knew it, and that he had no wish to go on to live a life of both physical and artistic decline. His suicide would thus be his ultimate *aesthetic* act."[18]

Facilitating one's timely death as an aesthetic act is in keeping with the way of life extolled by Nitobe Inazō 新渡戸稲造 in his *Bushidō: The Soul of Japan* (1899). The so-called way of the Samurai was established by a number of writings during the feudal period in Japan, the most representative and popular one is *Hagakure* 葉隠 by Yamamoto Tsunetomo 山本常朝 written in the eighteenth century.[19] The influence of this work on Mishima is indicated by his annotated modern translation, *Hagakure Nyūmon* 葉隠入門 (*Introduction to Hagakure*) published in 1967. Although Nitobe's celebration of *Bushidō* can be criticized today for fueling a nationalistic sentiment during the period of rapid Westernization that led Japan to engage in its own imperialistic agenda, for our purpose what is important is the aesthetization of the timely death through the symbol of cherry blossoms. Nitobe begins by likening the warrior's determination to die at the right moment to the cherry blossoms falling from the trees by declaring that "chivalry is a flower no less indigenous to the soil of Japan than its emblem, the cherry blossoms."[20] Particularly in comparison with a rose which, according to Nitobe, is the symbol of beauty for the Europeans characterized by its "tenacity with which she clings to life," cherry blossoms' aesthetic asset resides in being "ever ready to depart life at the call of nature." Clinging to the branches after their prime is "unsightly."[21]

In the Japanese tradition, this concern with providing an aesthetic justification for something difficult to accept is not limited to dealing with death. Just as Nietzsche is concerned with all kinds of contingencies in life, the Japanese aesthetic tradition also develops a strategy to cope with various challenges in life.[22] Mostly Zen priests or disciples, Japanese artists pursue their practice ultimately as a way of dealing with those conditions in life that are not satisfactory, whether they be imperfection, insufficiency, or disappointment. Perhaps the most explicit and well-known pronouncement of this aesthetic sensibility is found in the following passages from *Tsurezuregusa* 徒然草 (*Essays in Idleness*) by Yoshida Kenkō 吉田兼好 (c. 1283–c. 1350), a retired Buddhist monk.

> Are we to look at cherry blossoms only in full bloom, the moon only when it is cloudless? To long for the moon while looking on the rain, to lower the blinds and be unaware of the passing of the spring—these are even more deeply moving. Branches about to blossom or gardens strewn with faded flowers are worthier of our admiration.[23]

In an earlier passage, Kenkō laments:

> It is only after the silk wrapper has frayed at top and bottom, and the mother-of-pearl has fallen from the roller that a scroll looks beautiful. I was impressed to hear

the Abbot Kōyū say, "It is typical of the unintelligent man to insist on assembling complete sets of everything. Imperfect sets are better." In everything, no matter what it may be, uniformity and completeness are undesirable.²⁴

Implied in these passages is the assumption that people will generally only gravitate toward cherry blossoms in full bloom, a clear view of the moon, objects in good conditions, and a complete set of things. Indeed, the first passage above continues as follows:

> People *commonly* regret that the cherry blossoms scatter or that the moon sinks in the sky, and this is *natural*; but only an exceptionally insensitive man would say, "This branch and that branch have lost their blossoms. There is nothing worth seeing now."²⁵

What Kenkō is advocating is an aesthetic capacity that challenges this all-too-common propensity.

Many Japanese art forms in fact specifically highlight difficult-to-appreciate qualities. Art forms achieve this condition by celebrating minimalism, imperfection, and indirect communication; in a way, they issue an aesthetic challenge. For example, the classical writings on poetry-making often invoke the imagery of an obscured moon or misty landscape as the ideal beauty for their art. Consider the following passage by Kamo no Chōmei 鴨長明 (1153–1216), another retired monk well-known for *Hōjōki* 方丈記 (*An Account of My Hut*). He claims that "expressing the whole of one's heart in words, rather like describing the moon without clouds or praising flowers for being beautiful, is not at all difficult." However,

> when looking at autumn mountains through mist, the view may be indistinct yet have great depth. Although few autumn leaves may be visible through the mist, it is alluring. The limitless vista created in imagination far surpasses anything one can see more clearly.²⁶

By far the most prominent art form that challenges people to embrace imperfection and insufficiency is the tea ceremony. From its entrenched minimalism (such as a sparse interior and the small size of a tea hut; meager amounts of snacks served; economized bodily movement; simplicity of utensils) to its incorporation, and sometimes intentional creation, of imperfection (such as a cracked tea bowl, water jug, and vase), the tea ceremony elevates those qualities that are normally depreciated to an aesthetic height, creating a different kind of aesthetics: *wabi* aesthetics. As Graham Parkes points out, "implements with minor imperfections are often valued more highly, on the *wabi* aesthetic, than ones that are ostensibly perfect; and broken or cracked utensils, as long as they have been well repaired, more highly than unbroken."²⁷

This obsession with celebrating the otherwise unappreciable qualities is not simply to encourage a different aesthetic taste. Similar to Nietzsche's use of aesthetics for the same purpose, the Japanese cultivation of such a taste is further motivated by making the acceptance of difficulties in life easier. For example, after characterizing *wabi* as "lacking things, having things run entirely contrary to our desires, being frustrated in our wishes,"[28] the author of *Zen-cha-Roku* 禅茶録 (*Zen Tea Record*, 1828) explains that *wabi* aesthetics is a strategy to turn such difficulties of life into positive assets.

> Always bear in mind that *wabi* involves not regarding incapacities as incapacitating, not feeling that lacking something is deprivation, not thinking that what is not provided is deficiency. To regard incapacity as incapacitating, to feel that lack is deprivation, or to believe that not being provided for is poverty is not *wabi* but rather the spirit of a pauper.[29]

Furthermore, the cultivation of a positive attitude toward deficiency and imperfection can assume a political dimension by urging people's acceptance of status quo, no matter how much hardship is involved. For example, Ii Naosuke 井伊直弼, a powerful nineteenth-century statesman, writes in his *An Essay on the Tea Cult as an Aid to Government* (茶道の政道の助けとなるべきを論へる文):

> If pleasure is not gratification accompanied by a sense of contentment, it is not real pleasure … if each individual is satisfied with his lot and is not envious, he will enjoy life because he knows contentment and will be contented because of enjoying his lot … if the art of drinking tea were widely practiced throughout the country … both high and low would be content with their lots, would enjoy but not grieve, and would do no wrong … The country would become peaceful and tranquil spontaneously.[30]

This appeal to the *wabi* aesthetics of tea ceremony in cultivating the attitude of contentment with the status quo is not unique to the feudal regime which Ii was defending. A post-war writing in the twentieth century also invokes a similar idea by claiming that "the tea cult stresses 'accordance with one's lot' and 'knowing contentment and resting upon complacency'" and "in the state of finding beauty in imperfect things, modesty begins to function and a love of privation and simplicity develop."[31]

Thus, *wabi* aesthetics that stresses the positive aesthetic values of difficult-to-appreciate qualities such as insufficiency and imperfection is not simply a different kind of aesthetics from what is normally embraced, for it has existential and even political dimensions that encourage people to accept and cherish what otherwise may not be appreciable. Just as Nietzsche was concerned with affirming the contingencies of life that cause pain and suffering, so were the advocates of *wabi* aesthetics concerned with privation and defects in life. Both proposed an attempt to cope with life's difficulties and what is interesting is that their strategies were directed by aesthetics.

There is a difference between their aesthetic strategies, however. For Nietzsche, by conceiving one's life as a work of art with death as its finale, strictly speaking, we won't know the aesthetic role of various difficult-to-justify contingencies until the very end. The contribution of each part has to be determined in the context of the whole, but we won't know what the whole is like until life is completed. Perhaps there can be more diminutive works of art throughout one's life which can be appreciated after we know their respective completion. This reminds us of Roquentin's question in Jean-Paul

Sartre's *Nausea*: whether or not we can have an "adventure" while living. Sartre's conclusion is that only in "recounting" can one have an adventure because in narrating what happened we can give a structure to the plot, rendering each part indispensable and relationship among them necessary, thereby composing an organic whole. But if we are supposed to regard our entire life as a work of art like a piece of music, as Nietzsche seems to suggest, then it is not clear whether we can ever be in a position to grasp, let alone appreciate, the way in which the organic whole is structured.

Japanese *wabi* aesthetics, in contrast, does not seem to place imperfection and defect in a larger context in order to justify its existence and value. It is true, however, that sometimes those qualities are discussed as a *contrast* or *complement* to the opposite qualities, such as opulence and brilliance. For example, a well-known tea master, Sen no Rikyū 千利休 (1522–91), does state that "a plain tea bowl of present-day porcelain should be combined with an exquisite antique piece of Chinese tea-caddy."[32] Similarly, Rikyū's predecessor, Murata Jukō 村田珠光 (d. 1502) expresses the importance of this contrast by the verse: "A prize horse looks best hitched to a thatched hut."[33] However, unlike Nietzsche who relies heavily on the metaphor of music that exists in time, the contrast between the plain, imperfect, deficient and the opulent, perfect, abundant invoked in explaining the appeal of *wabi* qualities relies on spatial analogies, making it possible for one to justify the raison d'etre of imperfect, deficient qualities without having to wait until some future time. Those difficult-to-justify qualities in one's life, such as poverty and misfortune, will be justified, accepted, and appreciated as they occur.

Despite this difference, Nietzsche's aesthetic justification of life and *wabi* aesthetics share one problem in common. As the political ramification of *wabi* indicates, justifying every aspect of life aesthetically runs the risk of accepting and being complicit with societal conditions that give rise to misery in one's life. For example, if I suffer from poverty and discrimination because of social injustice, is it advisable for me to justify those difficulties through aestheticization? Perhaps that may be the only possible and pragmatic way of coping with something so considerable that I cannot change. If it is clear that one cannot fight city hall, as the expression goes, it may be prudent to try to turn life's lemons into lemonade. However, if everyone adopted this attitude, we would not have had Rosa Parks, Martin Luther King Jr., Mahatma Gandhi, or Nelson Mandela. Granted they are exceptional human beings, but we are indeed fortunate they did not sit content with their lot in life and solely developed an aesthetic appreciation for the suffering caused by social injustice. Thus, there is a danger in both aesthetic strategies of coping with life insofar as they seem to recommend *indiscriminate* aestheticization of every part of life. They may provide valuable and prudent advice regarding those conditions over which we have no control, such as literal accidents or natural disasters that destroy our health or possessions. However, if life's difficulties are caused by social ills, it is problematic to adopt the same accepting and aestheticizing attitude.

Aesthetic Improvement of Self

There is another role aesthetics plays in people's lives in both Nietzsche's philosophy and the Japanese tradition. Nietzsche was concerned not only with the challenge posed by life's contingencies that happen to people, such as illness and bad fortune, but also with the defects and weaknesses in one's personality. None of us are perfect; we all inevitably suffer from some character flaws. Just as in dealing with life's contingencies, Nietzsche's strategy in coping with one's weaknesses was to mold a work of art from one's personal qualities. Rather than trying to render ugly parts beautiful, the construction of self aims at making the whole beautiful. The task here is "to 'give style' to one's character," which he calls "a great and rare art" that is "practiced by one who surveys everything his nature offers in the way of weaknesses and strengths, and then fits it into an *artistic plan* until each element appears as artistic and reasonable and even the weaknesses delight the eye."[34] As such, "in man *creature* and *creator* are united: in man there is material, fragment, excess, clay, dirt, nonsense, chaos; but in man there is also creator, form-giver, hammer hardness, spectator divinity, and seventh day."[35] Moreover, a man who cannot fashion himself in this way poses a threat to others, for "whoever is dissatisfied with himself is continually ready for revenge, and we others will be his victims, if only by having to endure his ugly sight."[36]

The strategy here once again reminds us of his characterization of music where a harmonious music piece consists of the arrangement of consonance and dissonance. Likewise, it is also reminiscent of Plato's discussion in *The Republic* of an ideal society that invokes aesthetic analogies. Plato claims that we cannot make a sculptural piece beautiful by applying gorgeous colors to various parts; rather, he recommends to "look to see whether by dealing with each part appropriately we are making the *whole statue* beautiful."[37] Plato's criticism of democracy also uses an aesthetic analogy by pointing out that it is like "a cloak embroidered with every kind of ornament."[38] It is not clear whether Nietzsche proposed a specific formula in the same way as did Plato with respect to one's soul and an ideal society. However, it is instructive that Nietzsche's strategy of creating one's own self seeks the overall beauty.

The artistic creation of one's self, therefore, does not mean we have to transform all weaknesses into strengths, because doing so would compromise the integrity of the self, which would be similar to how we would destroy the overall beauty of a cloak or a sculptural piece by adorning every inch of its surface with beautiful ornaments or colors. Parkes points out that "rather than eradicate the tendencies in the soul that are branded as vices, one can work with them in a wide variety of ways, such that they ultimately become not only aesthetically pleasing but also fruitful."[39]

It is noteworthy that this strategy to work with what is given, rather than forcing every part to change into what is deemed best, perfect, or most beautiful, can imply an environmental consequence. Consider Parkes's reference to Nietzsche's imagery of garden cultivation. If we only have "a paltry flow of water," that is what we will have to work with; we may devise the direction of water to flow effectively or plant vegetables and flowers that do not need much water, and so on.[40] The strategy of working *with* what is given by nature, whether regarding ourselves or the natural environment, is a particularly effective alternative to the marching order issued by the Western scientific revolution and Enlightenment agenda. According to this directive best articulated in Francis Bacon's utopia in *New Atlantis*, the good life is one that is free from limitations placed by nature, and nature is a raw material that needs to be "improved" by humans. While some improvements have served humans well, we now know that the problems created by this agenda have become quite overwhelming—the most serious of which is the environmental harm that threatens life not only today but also the future sustainability of life on the earth.

In this regard, Nietzsche's notion of the all-too-human perspective to which we are tethered can be interpreted as a form of anthropocentrism, which is considered to be a major factor underlying today's environmental problems. Nietzsche encourages us to overcome anthropocentrism in a passage highlighted by Parkes: "If we could communicate with the gnat, we would realize that it flies through the air with a similar seriousness and feels within itself the flying center of the world."[41] Parkes continues by suggesting that "the possibility of 'perspectiveless' seeing … involves broadening our perspective beyond restricted human viewpoints, in such a way that we attain a genuinely 'trans-anthropocentric' way of looking at the world."[42] Parkes points out, however, that Nietzsche's non-anthropocentrism goes beyond biocentrism insofar as he includes inorganic objects, such as rocks, as something through which we have to be willing to experience the world. As Parkes cites Nietzsche, "*How one is to turn to stone.*—Slowly, slowly to become hard like a precious stone—and finally to lie there still and to the joy of eternity."[43] We will see shortly that this radical non-anthropocentrism is also encouraged in the Japanese tradition.

As we have seen with Nietzsche's philosophy and in the Japanese tradition as well, aesthetics plays an important role for self-improvement. Selfless devotion to the chosen artistic medium and constant practice is not only the means to excelling in the art but it is also a way of experiencing enlightenment. For example, Rikyū is recorded as defining "the art of tea" as "the way through which one attains spiritual awakening."[44] The Japanese philosophical and aesthetic tradition does not subscribe to the Western distinction between theory and practice, or mind and body. Philosophy in the sense familiar to the West, as a theory, did not appear in Japan until Westernization began after 1868; in fact, the term *tetsugaku* 哲学 was coined at the time to refer to philosophy as an academic discourse. This does not mean, however, that philosophical thinking did not exist prior to the application of the term. Rather, philosophical thinking was developed as a guideline for practices, such as artistic practices predominantly guided by Zen Buddhism. Parkes observes that "the Western dichotomy between theory and practice, or reflection and action … fails to hold" in the Japanese tradition, because "Zen is not a set of dogmas but rather a practice that transforms one's relationships to the world," and recommends that in pursuing Japanese ways of thinking one studies a variety of practices, such as literature, art and architecture, films, martial arts, theater, and food.[45] The Great Books of Japanese aesthetics consists of teachings by art masters and we can find a common theme among their instructions. They all emphasize the importance of transcending one's ego and devoting oneself to a particular art simply for its own sake, rather than trying to attain fame, please others, or even realize self-satisfaction.

Arguably the most important work of Japanese Zen Buddhism is the *Shōbōgenzō* 正法眼蔵 (*The Storehouse of True Knowledge*), written by Dōgen 道元 in the early thirteenth century. Dōgen defines enlightenment in this way: "Acting on and witnessing myriad things *with the burden of oneself* is 'delusion.' Acting on and witnessing oneself in the advent of myriad things is enlightenment."[46] Enlightenment starts with realizing that one's world is dependent upon one's limited perspective, which is constituted by one's identity as well as species membership. This realization and the practice of expanding one's horizon requires "overcoming," "transcending," or "forgetting" oneself.

> Studying the Buddha Way is studying oneself. Studying oneself is forgetting oneself. Forgetting oneself is being enlightened by all things. Being enlightened by all things is causing the body-mind of oneself and the body-mind of others to be shed. There is ceasing the traces of enlightenment, which causes one to forever leave the traces of enlightenment which is cessation.[47]

Specifically, in his *Gakudō Yōjinshū* 学道用心集 (*Guidelines for Studying the Way*), Dōgen repeats the importance of selfless devotion: "Proceed with the mind which neither grasps nor rejects, the mind unconcerned with name or gain. Do not practice buddha dharma with the thought that it is to benefit others."[48] In short, "just *forget* yourself for now and practice inwardly" and "practice buddha-dharma *solely for the sake of* buddha-dharma."[49]

While disinterested self-devotion toward enlightenment can be realized through engaging in any kind of craft or activity, it is in the training of artists where we find the most explicit and specific application. The recurrent theme is that it is only when one transcends one's ego and intention to excel in art that a truly superb result will emerge; as long as one has not overcome one's intention, no matter how skillful and talented, one will not be able to create a masterful work or produce a masterful performance.[50] Furthermore, there is no stopping point.[51] One cannot rest on one's laurels or past glory; enlightenment has to be sustained through continuous and selfless devotion. No matter what particular artistic training one engages in, excellence emerges from unceasing practice with an attitude of selflessness and disinterested devotion.

In those arts that represent, imitate, or give expression to objects and characters, when one attains a masterful stage through successfully shedding of one's ego, or to borrow the seventeenth-century master haiku poet, Bashō's 芭蕉 expression, through "slenderness of mind," one succeeds in having the object or character spontaneously express itself. If we want to represent a bamboo or a pine tree through painting or poetry, we first have to "become" a bamboo or a pine tree; then the artistic expression will emerge on its own.[52] The same point is made by the Noh Theatre master Zeami 世阿弥 (c. 1363–c. 1443) who states that "if the essence is a flower, the performance is its fragrance. Or they may be compared to the moon and the light which it sheds. When the essence has been thoroughly understood, the performance develops of itself."[53]

These teachings regarding artistic activities essentially encourage transcending anthropocentrism. This repudiation of a human-centered viewpoint is stressed in a number of passages in the *Shōbōgenzō*. By pointing out various viewpoints held by fish, birds, ghosts, and dragons, in addition to humans, it may appear that Dōgen is advocating biocentrism.[54] However, like Nietzsche, Dōgen goes further than biocentrism. In the fascicle *Busshō* 仏性 (Buddha Nature) in the *Shōbōgenzō*, Dōgen presents the most important passage of Mahayana Buddhism that is usually read as: "All sentient beings without exception have the Buddha-nature."[55] However, as Abe Masao explains, Dōgen risks committing a grammatical mistake by reinterpreting this passage composed of Chinese characters as: "All are sentient beings, all beings are ... the Buddha-nature," thereby denying a privileged position the normal (and more grammatically correct) reading gives to sentient beings.[56] Indeed, Dōgen elsewhere claims:

> If we take grasses and trees, tiles and stones to be inanimate we lack sufficient study. ... Even common things like grasses or trees that are seen by human beings are not easy to categorize with ordinary distinctions. ... In general, we divide beings into animate and inanimate, but actually some grasses or trees are human beings and animals—there is no clear distinction between animate and inanimate.[57]

This egalitarian view of Buddha nature of everything, sentient or non-sentient, underlies his notion of *mujō seppō* 無情説法 (teachings by non-sentient beings) that acknowledges Buddhist teachings include not only birds and fish, but also plants, the sound of a stream, and the color of a mountain (*keisei sanshoku* 渓声山色).[58] It is also revealing that the objects most commonly focused upon in Zen meditative practice are rocks in dry landscape gardens.[59]

The Japanese artistic training is thus more a means to becoming a certain kind of certain person and experiencing enlightenment rather than merely attaining artistic skills. The ultimate purpose of artistic training is existential: how to live one's life. Thus, the Japanese aesthetic tradition and Nietzsche's aesthetics are both offering a strategy to practice a good life. In Nietzsche's case practicing a good life is through an artistic project of creating self by making use of all the materials including weaknesses and in the Japanese case it is through rigorous training and selfless devotion.

One crucial difference may at first appear to be that Nietzsche's strategy is guided by asserting one's self and one's will through and through, while the Japanese tradition urges overcoming, transcending, and "forgetting" one's self and conscious intent. However, this difference may not be as marked as it may at first appear. First, although Zen training emphasizes a series of "forgetting," including forgetting the act of forgetting itself, such a process of constant transcendence is not possible without a strong will power. Although the whole point is to overcome one's self, Zen training and Zen-inspired artistic practice require strict self-discipline, which is similar to the exercise of will extolled by Nietzsche.

Second, if one could characterize Nietzsche's program of self-control (through accepting and justifying one's naturally endowed weaknesses as a part of an aesthetic whole) as engaging in an unnatural act (insofar as we are not simply letting all the natural endowments take their own course), then we could also characterize the Japanese Zen and artistic disciplines as engaging in a rather unnatural training. That is, we have to practice transcending "natural" inclination to want to become a good artist, to attain enlightenment, to enjoy fame, or even to please others. This paradox of the "natural" and the "unnatural" is astutely identified by Parkes:

> We can also appreciate the further paradox that the Zen emphasis on *natural* action, when it re-emerges as a consequence of intense physical and mental discipline in something like the spontaneous activity of the consummate swordsman, coincides

now with a way of being that is quite *unnatural*. This is not meant to suggest that it is artificial, but rather that the way to realize one's full humanity lies in going against what is naturally given so that one may sublimate, as it were, one's human nature.[60]

Of course, the definition of what counts as a natural inclination or an unnatural practice is debatable and it is doubtful whether a clear distinction can be made. For example, one could argue that wanting to improve oneself or achieve enlightenment is a natural urge in us; hence, the whole enterprise of self-overcoming or selfless devotion is a natural human act. However, for the present purpose, what interests me is the common view on aesthetics shared by both Nietzsche and the Japanese tradition as a strategy, whether natural or unnatural, to achieve a good life.

In conclusion, both Nietzsche's aesthetics and the Japanese aesthetic tradition are concerned with issues beyond what is normally associated with aesthetics, such as art and beauty. They address how to live a good life, how to improve one's self, and how best for humans to work with nature. As such, the value of their views extends beyond their specific historical and cultural contexts.

Notes

1. I would like to thank David Jones, the editor of this volume, for his helpful suggestions and meticulous editing work on my initial draft of this chapter.
2. Friedrich Nietzsche, *The Will to Power*. Translated by Walter Kaufmann and R. J. Hollingdale, edited by Walter Kaufmann (New York: Vintage Books, 1968), 429; emphasis added; Friedrich Nietzsche, "On the Genealogy of Morals," in *Basic Writings of Nietzsche*, translated and edited by Walter Kaufmann (New York: The Modern Library, 1968), 539.
3. Friedrich Nietzsche, *The Gay Science*. Translated by Walter Kaufmann (New York: Vintage Books, 1974), 241.
4. Makoto Ueda, *Literary and Art Theories in Japan* (Cleveland: Press of Case Western Reserve University, 1967), 226.
5. Friedrich Nietzsche, *Thus Spoke Zarathustra: A Book for Everyone and Nobody*, E-book. Translated by Graham Parkes (Oxford: Oxford University Press, 2009), 66–7; 172; 148; emphases added.
6. Nietzsche, *Thus Spoke Zarathustra*, 148.
7. Friedrich Nietzsche, "The Birth of Tragedy," in *Basic Writings of Nietzsche*, translated and edited by Walter Kaufmann (New York: The Modern Library, 1968), 141.
8. Nietzsche, *The Birth of Tragedy*, 141.
9. Nietzsche, *The Birth of Tragedy*, 52.
10. Nietzsche, *The Birth of Tragedy*, 141; emphasis added.
11. Nietzsche, *The Gay Science*, 224; emphasis added.
12. Nietzsche, *The Gay Science*, 164; emphasis added.
13. Nietzsche, *Thus Spoke Zarathustra*, 122.
14. Nietzsche's aesthetic justification of life by invoking music anticipates Jean Paul Sartre's use of music for the same purpose. See his *Nausea* (Jean-Paul Sartre, *Nausea*. Translated by Lloyd Alexander [New York: New Direction, 1969], 22, 133); Nietzsche, *Thus Spoke Zarathustra*, 131.

15   Nietzsche, *Thus Spoke Zarathustra*, 62 and 63.
16   Nietzsche, *The Gay Science*, 337.
17   A similar sentiment was expressed by a fourteenth-century Japanese Buddhist monk Yoshida Kenkō: "If man were never to fade away like the dews of Adashino, never to vanish like the smoke over Toribeyama, but lingered on forever in the world, how things would lose their power to move us! … We cannot live forever in this world; why should we wait for ugliness to overtake us?" (Kenkō Yoshida [referred to as Kenkō], *Essays in Idleness*. Translated by Donald Keene [New York: Columbia University Press, 1967], 7–8). I will follow the Japanese order and place the family name before the given name. Thus, Yoshida is the family name. However, it is customary in the Japanese tradition to refer to him with his given name, Kenkō. Nietzsche, *Thus Spoke Zarathustra*, 62.
18   Graham Parkes, "Ways of Japanese Thinking," in *Japanese Aesthetics and Culture: A Reader*, edited by Nancy Hume (Albany: State University of New York Press, 1995), 103.
19   The title is translated as *The Book of the Samurai* by William Scott Wilson (Tsunetomo Yamamoto, *Hagakure: The Book of the Samurai*. Translated by William Scott Wilson [Tokyo: Kodansha International, 1979]).
20   The aesthetics of the transient beauty of cherry blossoms and its symbolization of a moral virtue, ultimately utilized by the Japanese military leading to WW II, are the subject matters of Emiko Ohnuki-Tierney's *Kamikaze, Cherry Blossoms, and Nationalisms: The Militarization of Aesthetics in Japanese History* (Emiko Ohnuki-Tierney, *Kamikaze, Cherry Blossoms, and Nationalisms: The Militarization of Aesthetics in Japanese History* [Chicago: The University of Chicago Press, 2002]). Inazō Nitobe, *Bushidō: The Soul of Japan* (Rutland: C. E. Tuttle, 1988), 1.
21   Nitobe, *Bushidō*, 165.
22   I give a more extensive discussion on this issue in Yuriko Saito, "The Japanese Aesthetics of Imperfection and Insufficiency," *The Journal of Aesthetics and Art Criticism* 55 (4) (1997): 377–85: "The Aesthetics of Weather," in *The Aesthetics of Everyday Life*, edited by Andrew Light and Jonathan M. Smith (New York: Columbia University Press, 2005), 156–76; *Everyday Aesthetics* (Oxford: Oxford University Press, 2007).
23   Kenkō, *Essays in Idleness*, 115.
24   Keene's translation has "uniformity" instead of "uniformity and completeness" at the end of this passage. I added "completeness" here to capture the entire meaning of the original term: "*koto no totonōritaru.*" Kenkō, *Essays in Idleness*, 70: emphasis added.
25   Kenkō, *Essays in Idleness*, 115; emphasis added.
26   Cited by Kōshirō Haga, "The *Wabi* Aesthetic," translated by Martin Collcutt in *Tea in Japan: Essays on the History of Chanoyu*, edited by Paul Varley and Kumakura Isao (Honolulu: University of Hawaii Press, 1989), 204.
27   I give some examples in Saito, "The Japanese Aesthetics of Imperfection and Insufficiency," 377–85; Graham Parkes, 2005a. "Japanese Aesthetics," in *Stanford Encyclopedia of Philosophy*. Accessed March 19, 2017. https://plato.stanford.edu/.
28   Cited by Haga, "The *Wabi* Aesthetic," 195.
29   Cited by Haga, "The *Wabi* Aesthetic," 196.
30   Cited by Hiroshi Minami, *Psychology of the Japanese People*. Translated by Albert R. Ikoma (Toronto: University of Toronto Press, 1971), 88.
31   Cited by Minami, *Psychology of the Japanese People*, 88–9.

32 Sōkei Nanbō, "A Record of Nanbō," in *The Theory of Beauty in the Classical Aesthetics of Japan*, translated and edited by Toshihiko and Toyo Izutsu (The Hague: Martinus Nijhoff Publishers, 1981), 146.
33 Cited by Haga, "The *Wabi* Aesthetic," 196.
34 Nietzsche, *The Gay Science*, 232; emphasis added.
35 Friedrich Nietzsche, "Beyond Good and Evil," in *Basic Writings of Nietzsche*, translated and edited by Walter Kaufmann (New York: The Modern Library, 1968), 344.
36 Nietzsche, *The Gay Science*, 233.
37 We must remember that ancient Greek sculpture used to be colored. Plato, *Plato's Republic*. Translated by G. M. A. Grube (Indianapolis: Hackett Publishing Company, 1974), 86–7; emphasis added.
38 Plato, *Plato's Republic*, 206–7.
39 Graham Parkes, *Composing the Soul: Reaches of Nietzsche's Psychology* (Chicago: The University of Chicago Press, 1994), 169.
40 The idea here comes close to the notion of "design with nature," regarded as one of the most important principles of sustainable design. I explore this notion in Yuriko Saito, "Ecological Design: Promises and Challenges," *Environmental Ethics* 24 (3) (2002): 243–61; Parkes, *Composing the Soul: Reaches of Nietzsche's Psychology*, 160–1.
41 Cited by Graham Parkes, "'Floods of Life' around 'Granite of Fate': Emerson and Nietzsche as Thinkers of Nature," *ESQ A Journal of the American Renaissance* 43 (1997): 217.
42 Parkes, "'Floods of Life' around 'Granite of Fate,'" 231.
43 Cited by Graham Parkes, "Staying Loyal to the Earth: Nietzsche as an Ecological Thinker," in *Nietzsche's Futures*, edited by John Lippitt (Houndsmills: MacMillan Press Ltd, 1999), 182.
44 Nanbō, "A Record of Nanbō," 155.
45 Parkes, "Ways of Japanese Thinking," 89 and 96.
46 Dōgen, *Shōbōgenzō: Zen Essays by Dōgen*. Translated by Thomas Cleary (Honolulu: University of Hawaii Press, 1986), 32; emphasis added.
47 Dōgen, *Shōbōgenzō: Zen Essays by Dōgen*, 32.
48 Dōgen, "Guidelines for Studying the Way," in *Moon in a Dewdrop: Writings of Zen Master Dōgen*, translated and edited by Kazuaki Tanahashi (New York: North Point Press, 1995), 34.
49 Dōgen, "Guidelines for Studying the Way," 35 and 32; emphasis added.
50 I discuss this aspect of Japanese aesthetics in Yuriko Saito, "Representing the Essence of Objects: Art in the Japanese Aesthetic Tradition," in *Art and Essence*, edited by Stephen Davies and Ananta Ch Sukla (Westport: Praeger, 2003), 125–41.
51 See Dōgen, "Guidelines for Studying the Way," 41: "In mind and body there is no abiding, no attaching, no standing still, and no stagnating" and "You who study the way will come to awakening in the course of study. Even when you complete the way, you should not stop."
52 One of the well-known statements attributed to Bashō is "Of the pine-tree learn from the pine tree. Of the bamboo learn from the bamboo." Reported by Hattori Dohō in Dohō Hattori (referred to as Dohō), "The Red Booklet," in *The Theory of Beauty in the Classical Aesthetics of Japan*, translated and edited by Toshihiko and Toyo Izutsu (The Hague: Martinus Nijhoff Publishers, 1981), 162. (It is customary to refer to him with his given name, Dohō.)

53 Seami, "The Book of the Way of the Highest Flower," in *Japanese Aesthetics and Culture: A Reader*, edited by Nancy Hume (Albany: State University of New York Press, 1995), 68.
54 The best fascicle for this point is "*Sansuikyō*" ("The Scriptures of Mountains and Waters").
55 Dōgen, *A Complete English Translation of Dōgen Zenji's Shōbōgenzō (The Eye and Treasury of the True Law)*. Translated by Kōsen Nishiyama, et al., Vol. IV (Tokyo: Nakayama Shobō, 1983), 27.
56 Masao Abe, *Zen and Western Thought* (Honolulu: University of Hawai'i Press, 1985), 27.
57 The title of the fascicle where this passage appears is "*Mujō Seppō*" ("The Proclamation of the Law by Inanimate Beings"). Dōgen, *A Complete English Translation of Dōgen Zenji's Shōbōgenzō (The Eye and Treasury of the True Law)*, Vol. IV, 70.
58 Both *mujō seppō* and *keisei sanshoku* receive discussion in the fascicles with these titles.
59 And this is the main theme of Parkes's discussion on "The Role of Rock in the Japanese Dry Landscape Garden" (Graham Parkes, "The Role of Rock in the Japanese Dry Landscape Garden," in François Berthier's *The Japanese Dry Landscape Garden*, translated by Graham Parkes [Chicago: The University of Chicago Press, 2000], 85–155, preface and x–xi). Also see his "Nietzsche's Environmental Philosophy: A Trans-European Perspective" (Graham Parkes, "Japanese Aesthetics," in *Stanford Encyclopedia of Philosophy*. 2005. Accessed March 19, 2017. https://plato.stanford.edu/).
60 Parkes, "Ways of Japanese Thinking," 88–9.

References

Abe, Masao. 1985. *Zen and Western Thought*. Honolulu: University of Hawai'i Press.
Dōgen. 1983. *A Complete English Translation of Dōgen Zenji's Shōbōgenzō (The Eye and Treasury of the True Law)*. Translated by Kōsen Nishiyama, et al., Vol. IV. Tokyo: Nakayama Shobō.
Dōgen. 1986. *Shōbōgenzō: Zen Essays by Dōgen*. Translated by Thomas Cleary. Honolulu: University of Hawaii Press.
Dōgen. 1995. "Guidelines for Studying the Way." In *Moon in a Dewdrop: Writings of Zen Master Dōgen*, translated and edited by Kazuaki Tanahashi, 31–43. New York: North Point Press.
Haga, Kōshirō. 1989. "The *Wabi* Aesthetic." Translated by Martin Collcutt. In *Tea in Japan: Essays on the History of Chanoyu*, edited by Paul Varley and Kumakura Isao, 195–230. Honolulu: University of Hawaii Press.
Hattori, Dohō (referred to as Dohō). 1981. "The Red Booklet." In *The Theory of Beauty in the Classical Aesthetics of Japan*, translated and edited by Toshihiko and Toyo Izutsu, 159–67. The Hague: Martinus Nijhoff Publishers.
Minami, Hiroshi. 1971. *Psychology of the Japanese People*. Translated by Albert R. Ikoma. Toronto: University of Toronto Press.
Nanbō, Sōkei. 1981. "A Record of Nanbō." In *The Theory of Beauty in the Classical Aesthetics of Japan*, translated and edited by Toshihiko and Toyo Izutsu, 135–58. The Hague: Martinus Nijhoff Publishers.

Nietzsche, Friedrich. 1968a. "Beyond Good and Evil." In *Basic Writings of Nietzsche*, translated and edited by Walter Kaufmann, 191–435. New York: The Modern Library.
Nietzsche, Friedrich. 1968b. "The Birth of Tragedy." In *Basic Writings of Nietzsche*, translated and edited by Walter Kaufmann, 15–144. New York: The Modern Library.
Nietzsche, Friedrich. 1968c. "On the Genealogy of Morals." In *Basic Writings of Nietzsche*, translated and edited by Walter Kaufmann, 449–599. New York: The Modern Library.
Nietzsche, Friedrich. 1968d. *The Will to Power*. Translated by Walter Kaufmann and R.J. Hollingdale, edited by Walter Kaufmann. New York: Vintage Books.
Nietzsche, Friedrich. 1974. *The Gay Science*. Translated by Walter Kaufmann. New York: Vintage Books.
Nietzsche, Friedrich. 2009. *Thus Spoke Zarathustra: A Book for Everyone and Nobody*. E-book. Translated by Graham Parkes. Oxford: Oxford University Press.
Nitobe, Inazō. 1988. *Bushidō: The Soul of Japan*. Rutland: C. E. Tuttle.
Ohnuki-Tierney, Emiko. 2002. *Kamikaze, Cherry Blossoms, and Nationalisms: The Militarization of Aesthetics in Japanese History*. Chicago: The University of Chicago Press.
Parkes, Graham. 1994. *Composing the Soul: Reaches of Nietzsche's Psychology*. Chicago: The University of Chicago Press.
Parkes, Graham. 1995. "Ways of Japanese Thinking." In *Japanese Aesthetics and Culture: A Reader*, edited by Nancy Hume, 77–108. Albany: State University of New York Press.
Parkes, Graham. 1997. "'Floods of Life' around 'Granite of Fate': Emerson and Nietzsche as Thinkers of Nature." *ESQ A Journal of the American Renaissance* 43: 207–40.
Parkes, Graham. 1999. "Staying Loyal to the Earth: Nietzsche as an Ecological Thinker." In *Nietzsche's Futures*, edited by John Lippitt, 167–88. Houndsmills: MacMillan Press Ltd.
Parkes, Graham. 2000. "The Role of Rock in the Japanese Dry Landscape Garden." In François Berthier's *The Japanese Dry Landscape Garden*, translated by Graham Parkes, 85–155. Chicago: The University of Chicago Press.
Parkes, Graham. 2005a. "Japanese Aesthetics." *Stanford Encyclopedia of Philosophy*. Accessed March 19, 2017. https://plato.stanford.edu/.
Parkes, Graham. 2005b. "Nietzsche's Environmental Philosophy: A Trans-European Perspective." *Environmental Ethics* 27 (1): 77–91.
Plato. 1974. *Plato's Republic*. Translated by G. M. A. Grube. Indianapolis: Hackett Publishing Company.
Saito, Yuriko. 1997. "The Japanese Aesthetics of Imperfection and Insufficiency." *The Journal of Aesthetics and Art Criticism* 55 (4): 377–85.
Saito, Yuriko. 2002. "Ecological Design: Promises and Challenges." *Environmental Ethics* 24 (3): 243–61.
Saito, Yuriko. 2003. "Representing the Essence of Objects: Art in the Japanese Aesthetic Tradition." In *Art and Essence*, edited by Stephen Davies and Ananta Ch Sukla, 125–41. Westport: Praeger.
Saito, Yuriko. 2005. "The Aesthetics of Weather." In *The Aesthetics of Everyday Life*, edited by Andrew Light and Jonathan M. Smith, 156–76. New York: Columbia University Press.
Saito, Yuriko. 2007. *Everyday Aesthetics*. Oxford: Oxford University Press.
Sartre, Jean-Paul. 1969. *Nausea*. Translated by Lloyd Alexander. New York: New Direction.
Seami. 1995. "The Book of the Way of the Highest Flower." In *Japanese Aesthetics and Culture: A Reader*, edited by Nancy Hume, 64–9. Albany: State University of New York Press.

Ueda, Makoto. 1967. *Literary and Art Theories in Japan*. Cleveland: Press of Case Western Reserve University.
Yamamoto, Tsunetomo. 1979. *Hagakure: The Book of the Samurai*. Translated by William Scott Wilson. Tokyo: Kodansha International.
Yoshida, Kenkō (referred to as Kenkō). 1967. *Essays in Idleness*. Translated by Donald Keene. New York: Columbia University Press.

5

Heidegger's Evasion of Music

Kathleen Higgins

Graham Parkes is a multifaceted thinker. His work in East Asian (particularly Japanese) philosophy is complemented by work on Heidegger and Nietzsche, and often he brings figures from East and West together, discussing Nietzsche and Zhuangzi, for example, or Heidegger and the Kyoto School. Parkes's work in aesthetics draws on figures of two "continental" traditions (those of Europe and of Asia) and is informed by his activities as an artistic practitioner in theater and film. He has done translations from several languages, including an English language translation (along with David Pellauer) of Georges Liébert's book *Nietzsche and Music*. The topic I will be considering, Heidegger's relationship to music, thus engages with several facets of Parkes's work.

This chapter pursues a perplexity I have long experienced in connection with two of Heidegger's influential works, *Being and Time* and "The Origin of the Work of Art." My question is why Heidegger largely ignores the topic of music, given points at which these two works seem to set up at least some discussion.[1] One possibility is that Heidegger just was not very much interested in music or felt that he knew too little about it. Parkes reports on a conversation he had with Hans-Georg Gadamer on why Heidegger so rarely mentions Daoist thought despite his evident interest in it, noting Gadamer's comment, "You have to understand that a scholar of the generation to which Heidegger belongs would be very reluctant to say anything in print about a philosophy if he were himself unable to read and understand the relevant texts in the original language."[2] Perhaps Heidegger had a similar reticence with respect to music because of what he viewed as limitations in his own knowledge about music. He does not feel so restrained in commenting on artworks accessed visually, but reticence about music as opposed to visual arts would not set him apart among philosophers, even across generations.[3]

Of course, one can only speculate about a deceased thinker's conscious motives for ignoring a topic. My purpose here will not be to establish what Heidegger's actual motives were, but to indicate various features of his project that might have given him philosophical motive or rationale for not discussing music. My point will be that despite the ways in which discussion of music might have been useful for communicating some of his ideas, Heidegger's failure to provide such discussion would be consistent with certain of his goals, and these goals may have prompted this omission.

I am not alone in noting that Heidegger gives short shrift to music. Andrew Bowie describes Heidegger's avoidance of music as "repression" of the topic, noting many points at which he veers away from mentioning music despite its relevance to what he is discussing.[4] Julian Young raises "the question of the absence, in Heidegger, of any sustained discussion of music," and suggests that "to a degree" his "musical deafness diminishes his thinking about art."[5] I will consider some of Bowie's and Young's ideas about what led Heidegger to underemphasize music. I will begin, however, by considering themes in *Being and Time* and "The Origin of the Work of Art" that might easily have been elaborated with reference to music.

Openings for Musical Discussion

Both *Being and Time* and "The Origin of the Work of Art" bring up ideas that might have been helpfully amplified by discussion of music. The idea in *Being and Time* that I have in mind is the notion of *Stimmung*, or mood. By the point in the work where Heidegger introduces it, he has already established that he will be considering the existential structures of Dasein, literally "being-there," the term he uses for the kind of being that each of us is. Heidegger's point is that we do not first of all appear in the world as subjects trying to gain knowledge of the world and its contents as objects. From the beginning, we are always already engaged in our world, relating to it as actors with projects. In Section 29 he introduces the idea that Dasein is at every point characterized by some *Stimmung*. In other words, a *Stimmung* is operative in every experience that Dasein has. Heidegger observes,

> Dasein always has some mood … the possibilities of disclosure which belong to cognition reach far too short a way compared with the primordial disclosure belonging to moods … A mood makes manifest "how one is, and how one is faring" … The Being of the "there" is disclosed moodwise in its "that-it-is."[6]

Mood, according to Heidegger, offers a primordial disclosure of the world in which Dasein always already finds itself. Heidegger goes on to use this notion of mood to characterize the revelatory power of experiences of anxiety, which he diagnoses as betraying our own concern for our continued existence.

It would be hard to overestimate the importance of mood in Heidegger's account, for mood fills out the character of our role as active agents who are inherently "Being-in-the-world." Our typical stance is to act within the world and to relate to objects in the context of our projects, taking cognitive distance from items within this context only in unusual circumstances, such as the case in which one of our tools breaks. Mood is decisive in each situation for *how* we are "in-the-world" and how we relate to everything that surrounds us. To attend to anything within the world involves taking some mood toward it, even if the mood is affectively rather dull. "Only because the 'there' has already been disclosed in a state-of-mind can immanent reflection come across 'Experiences' at all," says Heidegger. He goes on to observe: "The mood has already disclosed, in every case, Being-in-the-world as a whole, and makes it possible first of all to direct oneself toward something."[7]

Reference to music would seem a natural way of expanding on the idea of *Stimmung*, for the term refers to "tuning" (as in the tuning of a musical instrument) and "intonation" (in musical performance) as well as "mood." Bowie points out that the German word *Stimmung* makes the connection between mood and musical mode and "attunement" obvious in a way that is no longer evident in English.[8] John Macquarrie and Edward Robinson note that the original meaning of *Stimmung* was the musical sense, and that the term was only later applied to states of mind. They also note Heidegger's use of the related term *Gestimmtsein*, "attunement," another term that both in German and in English translation has a musical meaning, which may be heard as the primary sense of the word.[9]

Given that music is often utilized as a means for setting a mood, directing attention to the musical associations would not undermine the focus on mood that Heidegger intends. Indeed, he might have referred to music's power to affect our attitude in order to bring out, by comparison, the pervasiveness of mood's impact in priming certain orientations toward the world. One might imagine that the musical connotations of *Stimmung* were not especially salient to Heidegger, particularly since he was using the term to single out a non-musical denotation. However, in light of his own verbal play with variants of words built from common roots and his later preoccupation with poetry, it is hard to imagine him having a complete lack of interest in the verbal resonances of any of his central terms.[10]

Heidegger again fails to pick up on musical connotations of central terms in his later *Contributions to Philosophy*, "now considered by many commentators to be Heidegger's second greatest work" according to Michael Wheeler.[11] Joan Stambaugh comments that the work shows a "preference for terms taken from the realm of sound," which she takes to be in keeping with his interest in blocking the tendency (inherited from the Western metaphysical tradition) to objectivize and substantialize, typically making use of visual terms.[12] One of Heidegger's central terms in this context is *der Anklang*, which Richard Rojcewicz and Daniela Vallega-Neu translate as "resonating," and which Thomson interprets as "a kind of echo of reminiscence."[13] *Anklang* makes use of another root that references music (as "resonate" does in English), for the German word *Klang*, which means "sound," is often used in reference to specifically musical sound. Heidegger's decision not to pick up on the musical imagery associated with some of his key terms may not have been the result of a deliberate policy, but it does seem out of keeping with his tendency to meditate on the clusters of meanings associated with many of his central terms.

One reason why unpacking the associations of *Stimmung* with musical tuning and attunement might have been at least metaphorically useful to Heidegger is that the simultaneously inner and outer character of music would facilitate his efforts to undermine the models that emphasize a split between subject and object and fail to recognize the human being's interconnections with the world. Mood is not a condition of the subject that it projects onto the world; instead it characterizes the connection between Dasein and the world. "Mood discloses how one stands in the world," observes Karsten Harries. "But 'state of mind' is misleading in that it invites a too subjective interpretation."[14]

The fact that mood pertains to the way Dasein is plugged into its world according to Heidegger would make the musical condition an illustrative metaphor, for music undermines a sharp distinction between "inner" mind and "outer" world.[15] Music attunes the listener with the world; the listener and the environment are simultaneously made to resonate in accordance with the same music. At the same time, the listener often identifies with music, even though it originates in the external world, as though it were the direct expression of his or her affective life. The distinction between "inner" and "outer" seems largely inapplicable. Even when one hears it through earphones, music is a mode in which one acts as a participant within the world. Bowie draws attention to the work of musicologist Heinrich Besseler, a student of Heidegger, who emphasizes that in such social contexts as a dance, the person hearing music "behaves in an active-outpouring manner, without taking the music expressly as objectively present … It is not there in an objective form … for him."[16]

The Rift between World and Earth

Another context in which discussion of music might have been useful to Heidegger is his essay "The Origin of the Work of Art." There he claims that in order to fulfill art's proper function, an artwork must preserve the conflict between "world," the context of intelligibility that the work sets up, and "earth," the larger reality that resists intelligibility. The essay was originally drafted in 1936 and 1937, based on lectures given in the 1930s, but Heidegger revised the essay later and published it only in 1950. We can thus, I think, regard it as reflecting the "later" Heidegger's view of the world, even if much of the analysis came from an earlier time.

"The Origin of the Work of Art" considers the significance of the phenomenon of art taken as a whole. In this respect Heidegger follows Hegel, who considered art's fundamental purpose to have been to make humanity aware of itself as spirit. According to Hegel, art does this by embodying spiritual content in keeping with the degree of comprehension available at a given historical moment. Shifts in artistic tendencies from one period to another are the consequence of humanity's development of an increasingly comprehensive spiritual vision, which at the high point of classical art was successfully embodied in the artwork. But this vision eventually becomes so replete that no artwork can fully represent it, and thus art starts to gesture beyond itself toward meanings that can only be grasped through reflection. Humanity's aspiration is truth, which is achieved when the human spirit gains a comprehensive vision of reality and its own nature. This is ultimately attainable only through reflective thought, according to Hegel. Thus, historically a point is reached at which art, with its commitment to sensuous media, can no longer lead the way for Spirit (the collective consciousness of humanity) but can only direct attention to matters that must be addressed in reflective thought.

Heidegger also sees art as having a spiritual mission, and he also understands its aim in terms of truth. But truth for him is not a matter of the rational intelligence attaining transparent clarity about all that is, as it is for Hegel.[17] Instead, truth involves our becoming aware of aspects of reality that have been obscured. These aspects come into focus, or are "disclosed" to us, and Heidegger considers such moments of disclosure as events in themselves. Thus, he describes truth as a happening (*Ereignis*).[18] Art is a means through which truth happens, for art is effective in disclosing features of our world of which we have previously been unaware.

That Heidegger is responding to Hegel is clear. He directly describes Hegel's *Vorlesungen über die Asthetik* as "the most comprehensive reflection on the nature of art the West possesses" in the epilogue to the essay.[19] Moreover, the primary examples that Heidegger employs are the temple, the statue of the god, and a painting by Van Gogh. These instances respectively correspond to the three ages of art history that Hegel discusses in his lectures on aesthetics. Hegel refers to the period of early art as "symbolic," claiming that at this stage humanity's conception of reality is underdeveloped. Art at this stage only obliquely indicates spiritual content because its limited vision is not yet complete enough to provide the formative principle that shapes sensuous material. The external shape of the artwork is abstractly related to the spiritual content, but the work does not straightforwardly present it. Hegel links the temple to this era, for it does not directly embody spiritual content, but instead makes a place for "the god."

By contrast, in the era of "classical" art, humanity's conception of reality has developed sufficiently that it can provide the shaping principle behind the artwork, so that the optimal balance of form and content, between embodiment and spiritual content, is attained. In this connection, Hegel refers to sculpture that expresses the human Spirit in its freedom by presenting the embodied human being, in whom the life of the soul is active.

Romantic art develops when the spiritual content has become too capacious for any sensuous embodiment to do justice to it. Romantic art, in order to represent spiritual content at all, has to point beyond what is evident on its surface toward the human being's inner life. But that life is ever-changing. The media that Hegel sees as best able to indicate life in its continual transformation inherently involve a multiplicity of elements, which the perceiver experiences in consecutive moments: painting, music, and poetry. The impossibility of the artwork's giving full embodiment to the spiritual content is made evident through the fugitive way in which different aspects of the work present themselves to the perceiver. They are therefore capable of conveying the transient subjective dimension of human life, but by means of suggestion, not literal statement.

Heidegger's example of the Van Gogh painting of a pair of shoes utilizes one of these optimally "romantic" media. In keeping with Hegel's account of romantic art, Heidegger's emphasizes the work's pointing to something it does not present, namely the world of the person who habitually wore the shoes.[20] Despite this example from painting, poetry nevertheless ends up being the premier art, while music is not considered except in the sense that it is included when Heidegger mentions "all art."[21] The only point at which music is explicitly considered in the essay is in a list of the "thingly character" of works in various media, which is "sonorous in a musical composition."[22] Another sentence, elaborating on this point, again lists music in a series of kinds of works.[23] Otherwise, the presence of music in the essay is mainly implicit or unremarked, for example, in referring to beings "ringing out" when a work of poetry brings them into "the Open."[24]

Heidegger considers all the arts to involve poetry in the broad sense of being that which brings forth.

> Truth, as the clearing and concealing of what is, happens in being composed, as a poet composes a poem. *All art,* as the letting happen of the advent of the truth of what is, is, as such, *essentially poetry* ... What poetry, as illuminating projection, unfolds of unconcealedness and projects ahead into the design of the figure, is the Open which poetry lets happen, and indeed in such a way that only now, in the midst of beings, the Open brings beings to shine and ring out.[25]

Why does Heidegger take poetry to be pre-eminent among the arts? A partial explanation is that he uses the term poetry comprehensively to include all of the arts. Heidegger goes on to point out that poesy, poetry in the restricted sense of the making of poems, "is only one mode of the lighting projection of truth, i.e., of poetic composition in this wider sense." But what immediately follows is the privileging of poetry in exactly this narrow sense: "Nevertheless, the linguistic work, the poem in the narrower sense, has a privileged position in the domain of the arts." This is so because "language alone brings what is, as something that is, into the Open for the first time."[26]

Nevertheless, Heidegger does not think that poetic language reveals everything. It conceals even while it unveils. And this is the context in which he points to the distinction between world and earth. The world is what stands forth, and he sees the artwork as setting up the world. That is, Heidegger considers the artwork's role as making reality and what is important in it clear to a human community. The artwork articulates the community's "implicit sense" of these things, making it visible for them.[27] What it articulates is the community's particular "world," that context of intelligibility in which individuals' activities are understood as having meaning. At the same time, the artwork preserves "the earth," the larger, nebulous context out of which what is made intelligible is brought forth. Every item that is disclosed has an aspect that is veiled from our view. It never becomes completely accessible through our symbolic means of appropriating it. What is drawn into the clearing can be explicitly named, while the earth is that aspect of what is encountered that remains ineffable.[28] In this respect, disclosure has two faces, along the lines of yin and yang.

In order to do the work of letting truth "happen," an artwork must preserve the conflict, the "rift," between earth and world.

> In the creation of the work, the conflict, as rift, must be set back into the earth, and the earth itself must be set forth and used as the self-closing factor. This use, however, does not use up or misuse the earth as matter, but rather sets it free to be nothing but itself.[29]

For Heidegger, the particular value of poetic language, as opposed to language in general, is that it makes evident that not everything is clearly revealed for us. This is because poetic language is polysemous, demanding interpretation but in such a way that no particular interpretation can exhaust it. This forces us to recognize that not everything has been made intelligible.

But why is this not a job for music, or perhaps music in conjunction with texts? Does not music make it clear that not everything is transparent to our intelligence? Music is said to be ineffable, not exhaustible by any particular construal or interpretation. Julian Young, in fact, uses this feature of music to defend Heidegger's account of the "bringing-forth" of what is implicit. Young compares Heidegger's model to the idea (articulated by Michelangelo) that the sculptor releases through his work the figure that is trapped within the stone. Young anticipates the objection that there are multiple particular figures that could have been "brought forth," arguing that while "no unique figure is determined by the given," the given determines a limited range of possibilities. He compares the situation to that of a performed symphony: "Though there is no definitive performance … the work nonetheless imposes limits outside which the performance is no longer a performance of that work but rather its violation."[30] Following on this image, we may say that musical works, like works of poetry, direct the view of the perceiver, not toward something that could just as well be explicated literally, but toward a specific enough spectrum that the perceiver can experience his or her own moment of disclosure. Leonard Bernstein makes a related point when he directly compares music with poetic language in its evocative character, claiming that both arts bypass literal denotation in favor of what is more figurative or metaphoric.[31]

One way in which Heidegger takes music insufficiently into account is by insisting that poetic language is needed to reveal that not everything is intelligible despite the fact that music could also fulfill this role. Another is that he neglects the social role of music, a point that Bowie emphasizes.[32] The consequence is that Heidegger seems to exclude music when he describes the artwork's role in making clear the shared vision of a people, thereby "founding" a people as a people. He contends that art enables members of the community to recognize their vision as shared, and it can accordingly renew members' common sensibility. Heidegger does not consider music in connection with this conception of art's political work, but music is a potent means for unifying communities. Music's power to create solidarity is evident whenever people gather and music is performed. Literally, music entrains individuals with each other, prompting people to coordinate their efforts to the point that they walk lockstep. Given that Heidegger was active in the Nazi era and himself a defender of the Nazi effort, it is hard to imagine that he was not aware of music's ability to unite a crowd.

Perhaps Heidegger saw music's ability to forge solidarity as a matter of regularized rhythm in sound working on the neurophysiology of the human being. This may be the implication of Heidegger's allusion in "The Nature of Language" to "the danger of understanding melody and rhythm also from the perspective of physiology and physics, that is, technologically, calculatingly in the widest sense," a comment that Bowie describes as "pretty odd."[33] To the extent that Heidegger considered music as a mere stimulus, he may not have seen it as capable of bringing forth a vision that members of a society could share. But because he cannot have been naïve about the use of music to mobilize political enthusiasm for a cause, he must have known that music could at least reinforce such a vision. This leads directly to a possible reason behind Heidegger's failure to discuss music at any length. I will proceed to consider some candidates for an explanation of this omission.

Possible Explanations

I will consider three conceivable explanations for Heidegger's dearth of discussion of music, though this list does not exhaust the possibilities. The first, argued by Julian Young and already noted, is the idea that music does not "bring forth" a world. Young argues that that the work Heidegger intends for art to do requires representational capacity and that he does not see music as having this potential. Language, by contrast, is a medium for representation, the very one that we use to pick out anything from the experiential flux. Consequently, music is significantly different from language in that it cannot (on its own) assume the role Heidegger expects art to play if it is to have its full cultural significance.

Young also observes that Heidegger seems to project an "outer"/"inner" dichotomy onto music, a point that Bowie also suggests when he attributes Heidegger's avoidance of music to his "suspicion of the 'subjective' nature of the emotions" and his association of music with immersion in feeling.[34] Both Young and Bowie reference Heidegger's criticism of Richard Wagner for pursuing the dominance of music among the art forms within his music dramas, which Heidegger sees as amounting to "the dominance of the pure state of feeling."[35] Young construes this as an indication that Heidegger took music, particularly instrumental music, to represent an "inner world" of feeling that is incapable of setting forth a world in his "rich, onto-ethical sense."[36] Young also notes that while the later Heidegger expresses admiration for certain musical works, none of these is an instance of absolute (that is, purely instrumental) music, which suggests to Young that "insofar as he thought about music at all, he continued to insist that in 'valid' musical artworks, music must always be subordinate to a linguistic text, subordinate ... to 'poetry.'"[37] Bowie also considers implausible Heidegger's "exclusion of feeling from world disclosure error."[38]

The suggestion that Heidegger dismissed music because it did not "represent" in the manner of language strikes me as convincing, as does Young's suggestion that Heidegger missed something important in denying music the potential to set forth a world: "Music ... can itself give 'birth' to drama, to action, to a world ... [M]usic possesses, in fact to a consummate degree, the power to be an *Ereignis*-experience. Heidegger's discounting of absolute music is thus ... in his own terms, a serious error." Despite this criticism, Young suggests that "those who are hypersensitive to one art form are typically afflicted by a compensatory blindness to another," and that Heidegger's misjudging of music is "the price we pay" for his profound understanding of poetry.[39] This strikes me as working rather hard to be charitable to Heidegger on this matter. Young's claim is also open to challenge, since it seems that sensitivity in discerning details within a particular form of art should give a person some advantage for acquiring skill in discerning nuances in other arts.

Had he not ignored or been oblivious to some of music's capabilities, Heidegger might have found that reference to musical experience could have furthered his philosophical goals. For music can provoke awareness of the fact that Being occurs in time, that it flows. Indeed, this is an aspect of music that Hegel emphasizes when he considers it a paradigmatically romantic art (though Hegel emphasizes the subjective dimension of experience, which reinforces the grip of the "inner/outer" interpretation of self and world that Heidegger wished to avoid). The fact that instrumental music brings the flowing character of Being into view without connection to a specific language makes it especially serviceable for drawing our attention to the mystery that Heidegger wants us always to keep in view, the amazing fact that there is a world, rather than nothing at all.[40] In this sense, it would encourage his goal of advocating a "recollection of Being." Texted music, on the other hand, might serve as an ideal model for earth and world in relation, with the words calling particular things into the open while the music suggests a more obscure matrix on which these called forth features of the world depend. Thomson also points to music as embodying "the earth/world tension."[41]

Another possible explanation for Heidegger's relative silence regarding music depends on the supposition that Heidegger was well aware of music's capacity to forge solidarity. Perhaps he was more concerned with the historical situatedness of art than its potential to make us aware of Being. Nietzsche emphasized music's "Dionysian" power to obliterate awareness of social boundaries, conjoining people regardless of their political positioning.[42] Heidegger's requirement that the artwork serve the role of solidifying situated historical communities may be achievable in other arts, but on Nietzsche's account, music (in so far as it is Dionysian, not subordinated to words that constrain it) can impress people with their connection whether or not they are members of the same historical community. Its power is not restricted to revealing a particular society's implicit vision. Indeed, music can draw attention to what people qua human beings have in common, or more aptly, it can make us *feel* this connection. No particular identification with a given community is required.[43]

On this view, Heidegger might have de-emphasized music because it does not conform to his model of the artwork founding and preserving situated communities. Coupled with his probable conviction that music on its own does not "bring forth" into the open, Heidegger would have had a double motive for keeping music in the margins of his discussion. To the extent that he was moved by the problem that music may be insubordinate to the aim of furthering a given society's view of itself and a sense of priorities, he deflects attention from Being as a whole, though recollection of Being is a high priority in *Being and Time*. This would also be in dissonance with the Daoistic vision of all being part of the same Dao, and thus a limitation to the "pre-established harmony" between Heidegger and the Daoist perspective that Parkes observes.

A third possible explanation for Heidegger's reticence regarding music is linked to the other two. Perhaps Heidegger actually subsumed musical qualities under language, despite his claim that we will almost certainly go astray in connection with the musical features of language because of the "danger" of understanding these in terms of physics and physiology, as noted previously.[44] The very solidarity-generating characteristics of music that put it potentially at odds with efforts to secure clear boundaries between communities (particularly rhythm) are available within language itself. The power of rhythm is utilized precisely in the poetic language that he admires, and its role is not to occasion disclosure but to enchant us. And if the goal is to persuade auditors to one's way of thinking, all the better if this musical impact on hearers is not recognized.

This account is perhaps most convincing if one assigns dubious motivations to Heidegger. Heidegger, while emphasizing art's capacity to bring features of the world into the clearing, is himself using rhetorical ploys that rely on rhythm. His own constructions making use of variant terms deriving from the same etymological root in consecutive sentences often give his language an incantational character. On this view, though Heidegger speaks in opposition to the "bewitchery" of our era and its worship of technological progress, his rhetoric depends on a traditional kind of bewitching, that accomplished by rhythm and rhyme.[45] This is in keeping with the view of Heidegger's rhetoric defended by Markus Weidler. According to Weidler, Heidegger exploited philosophical rhetoric in the service of promoting a national language cult and resisting Schelling's more egalitarian and theological vision. As opposed to Schelling, who saw the meaning of the symbol of Christ for the Christian community to be one that could and should evolve, Heidegger encouraged a view of an un-situated "demand" coming from Being itself. In the actual political context, this encouraged what Weidler terms "political idolatry," in which "the purported demand (*Anspruch*) of Being calls upon a community ('the Germans,' in this case) not only to revere their cultural heritage but to revere their own reverence for their cultural heritage, as well."[46]

Supposing that Heidegger's political agenda had an impact on his assessment of the relative powers of music and language, he may have taken a greater interest in music had he been aware of ways that even absolute music can address a particular historical society and its political concerns. Music might have struck him as a phenomenon that could be adduced to serve his goals. However, he still might have had reservations because the power of music to render political lines irrelevant would still threaten to undermine significant premises of his project. In any case, Heidegger makes use of musical characteristics in language and does not expose the impact of its more propulsive and less disclosive aspects. These are perhaps particularly effective in "poetic" language that is not straightforward about the details of what is brought into view. In that rhetorical appeals are most irresistible when subliminal, bringing these musical dimensions of his language into view would not have served his rhetorical aims.

The three possible explanations for Heidegger's avoidance of discussion of music that I have considered are not mutually exclusive. He could have thought music both too subjective to be disclosing and too Dionysian to unify particular communities to serve the role he thought art should ideally serve. This is compatible with his making use of certain compelling features exhibited by music, in particular rhythm and repetition, in the non-standard use of language in which he engages.[47] Whether or not any or all of these motives influenced him, they indicate ways in which discussion of music would have been inconsistent with certain of his apparent goals.

Heidegger's interest in Daoism notwithstanding, with respect to music, he sides more with the Confucians than with the Daoists. The Confucians focus on music's role in promoting harmony within the community and inspiring cooperation in that context. The Daoists, by contrast, emphasize both the subjective and the Dionysian side of music. They do the former by romanticizing the intimate (even solitary) context of playing the *qin* (an unfretted lute). They do the latter by stressing music's role in manifesting the sympathetic resonances and interconnections among beings and everything within the Dao. Parkes is surely right in claiming that Heidegger amply demonstrates affinities for Daoism, but where music is concerned, Heidegger is far from a Daoist.

Notes

1. Compare Andrew Bowie, *Music, Philosophy, and Modernity* (Cambridge: Cambridge University Press, 2007), 74.
2. One wonders, if this speculation is accurate, what to make of Heidegger's insistence that great philosophy could only be done in Greek and German. Graham Parkes, *Heidegger and Asian Thought* (Honolulu: University of Hawai'i Press, 1987), 7.
3. Compare Bowie, *Music, Philosophy, and Modernity*, 290n14.
4. Bowie, *Music, Philosophy, and Modernity*, 74.
5. Julian Young, *Heidegger's Philosophy of Art* (Cambridge: Cambridge University Press, 2001), 168, 170.
6. For readers unfamiliar with the pagination of *Being and Time*, "H" refers to later German editions, which are shown in the margins of the Macquarrie and Robinson translation. The page number of their translation follows. Martin Heidegger, *Being and Time*. Translated by John Macquarrie and Edward Robinson (New York: Harper & Row., 1962), H. 134–5, 173.
7. Heidegger, *Being and Time*, H. 136–7, 175–6.
8. Bowie, 68.
9. Michael Inwood (*A Heidegger Dictionary* [Oxford: Blackwell, 1999], 130) has a somewhat different origin story, claiming that *Stimmung* derives from *Stimme*, voice, and that the related *stimmen*, one of whose meanings is "to make harmonious," was originally used in reference to making the human mind harmonious, though it is now used with the meaning of "to tune a musical instrument." Heidegger, *Being and Time*, H. 134n, 172n.
10. For consideration of Heidegger's word play, see, for example, Karsten Harries's discussion of the verbal overtones of variants of *Riss* as it is used in "The Origin of the Work of Art," in Karsten Harries, *Art Matters: A Critical Commentary on Heidegger's "The Origin of the Work of Art"* (Dordrecht: Springer, 2009), 156.
11. Michael Wheeler, "Martin Heidegger," *The Stanford Encyclopedia of Philosophy*. (Fall 2020 Edition), ed. Edward N. Zalta. https://plato.stanford.edu/archives/fall2020/entries/heidegger/.
12. This passage is cited by Iain Thomson, *Heidegger, Art, and Phenomenology* (New York: Cambridge University Press, 2011), 181. Stambaugh translates *Anklang* as "assonance" or "sounding." Joan Stambaugh, *The Finitude of Being* (New York: State University of New York Press, 1992), 113.

13 Martin Heidegger, *Contributions to Philosophy (Of the Event)*. Translated by Richard Rojcewicz and Daniela Vallega-Neu (Bloomington and Indianapolis: Indiana University Press, 2012), 85ff; Thomson, *Heidegger, Art, and Phenomenology*, 176n. Thomson comments that this "ringing echo" is literally a "reminiscence" triggered by sounds received through the ear.
14 Karsten Harries, *Heidegger's "Being and Time,"* 2014. Accessed May 25, 2021. https://cpb-us-w2.wpmucdn.com/campuspress.yale.edu/dist/8/1250/files/2016/01/Heidegger-Being-and-Time-1qtvjcq.pdf.
15 Compare Bowie, *Music, Philosophy, and Modernity*, 306.
16 Heinrich Besseler, *Aufsätze zur Musikästhetik und Musikgeschichte* (Leipzig: Reclam, 1978), 33; translation in Bowie, *Music, Philosophy, and Modernity*, 294.
17 Compare Harries, *Art Matters*, 11.
18 Harries characterizes an *Ereignis* as "a happening to which I belong and that belongs and therefore matters to me." Harries, *Art Matters*, 111.
19 Martin Heidegger, "The Origin of the Work of Art," in *Poetry, Language, Thought*. Translated by Alert Hofstadter (New York: Harper & Row, 1971), 79.
20 Heidegger envisions the world of a peasant woman who wore the shoes, but art historian Meyer Schapiro points out that the painting that Heidegger seems to indicate depicts a pair of shoes that most likely belonged to Van Gogh himself (Meyer Schapiro, "The Still Life as a Personal Object: A Note on Heidegger and Van Gogh," in *The Reach of the Mind*, edited by M. L. Simmel (New York: Springer, 1968), 206).
21 Heidegger, "The Origin of the Work of Art," 73.
22 Heidegger, "The Origin of the Work of Art," 19.
23 Heidegger, "The Origin of the Work of Art," 19.
24 Heidegger, "The Origin of the Work of Art," 72.
25 This notion of beings "ringing out" as well as "shining" is a straightforwardly musical metaphor, but not one that Heidegger explicates as such (Heidegger, "The Origin of the Work of Art," 72).
26 Heidegger, "The Origin of the Work of Art," 73.
27 Iain Thomson, "Heidegger's Aesthetics." *Stanford Encyclopedia of Philosophy*. 2015. Accessed June 22, 2018. https://plato.stanford.edu/entries/heidegger-aesthetics/.
28 Heidegger, "The Origin of the Work of Art," 64.
29 Harries, *Art Matters*, 156.
30 Young elsewhere, however, suggests that Heidegger does not see music on its own as bringing anything into the open and that this explains the asymmetry he sees between music and poetic language. I will take up this point in what follows. Julian Young, *Heidegger's Later Philosophy* (Cambridge: Cambridge University Press, 2002), 40.
31 Compare Leonard Bernstein, *The Unanswered Question: Six Talks at Harvard* (Cambridge, MA: Harvard University Press, 1976), chapter 3.
32 Bowie, *Music, Philosophy, and Modernity*, 295.
33 Martin Heidegger, *On the Way to Language*. Translated by Peter D. Hertz (New York: Harper & Row, 1971), 98; Bowie, *Music, Philosophy, and Modernity*, 302.
34 Bowie, *Music, Philosophy, and Modernity*, 76.
35 See Martin Heidegger, *Nietzsche*. Vol. 1. Translated by David F. Krell (San Francisco: Harper-Collins, 1979), 86.
36 Young contends that most theorists of music buy into this duality of inner and outer, and unless they are formalists, they tend to see music as concerned with

the inner world of affect instead of the outer world. This does not acknowledge the many theorists of music who think that music reflects patterns within social life and those who indicate many ways in which listeners relate instrumental music to events and experiences within the social world and political world. For a range of approaches to this idea, see Laurence Berman, "*The Musical Image: A Theory of Content* (Westport, Connecticut: Greenwood Press, 1993); Steven Feld, "Sound Structure as Social Structure," *Ethnomusicology* 28 (1984b): 383–409; David Hesmondahalgh, *Why Music Matters* (Oxford: Wiley Blackwell, 2013); Gregory Karl and Jenefer Robinson, "Shostakovich's Tenth Symphony and the Musical Expression of Cognitively Complex Emotions," in *Music and Meaning*, edited by Jenefer Robinson (Ithaca Cornell University Press, 1997), 154–78. See also Steven Feld, "Communication, Music, and Speech about Music," *Yearbook for Traditional Music* 16 (1984a): 74–113. Feld points to the "interpretive moves" made by listeners, who relate what they hear to matters of relevance to them in their lives (8).
37   Young, *Heidegger's Philosophy of Art,* 168-9.
38   Bowie, *Music, Philosophy, and Modernity*, 303.
39   Young, *Heidegger's Philosophy of Art,* 170.
40   Martin Heidegger, *An Introduction to Metaphysics*. Translated by Ralph Manheim (New Haven: Yale University Press, 1959), 7–8.
41   Thomson, *Heidegger, Art, and Phenomenology,* 128n.
42   Friedrich Nietzsche, *The Birth of Tragedy and The Case of Wagner*. Translated by Walter Kaufmann (New York: Random House, 1967), 64.
43   Compare Thomson, *Heidegger, Art, and Phenomenology,* 128n. Thomson observes that in song we sometimes have "an experience that transports us beyond the mundane realm of differentiated human meaning in an ecstatic feeling of transpersonal mystical union with the All." He concurs with Schopenhauer and the early Nietzsche that "the medium of *music* more easily facilitates such a mystical experience than any of the other arts."
44   Heidegger, *On the Way to Language,* 98.
45   Compare Heidegger, *Contributions to Philosophy* (*Of the Event*), 98; Friedrich Nietzsche, *The Gay Science*. Translated and edited by Walter Kaufmann (New York: Random House, 1974), §84, 140 and §175, 202.
46   Markus Weidler, "Heidegger's Theft of Faith: A Campaign to Suspend Radical Theology," PhD diss., The University of Texas at Austin, 2005, 14.
47   Bowie also points out that Heidegger's concern with poetry, which overcomes the "language of metaphysics" by "re-ordering words in new configurations," brings him "close to the irreducibility of the order of the elements of a particular piece or performance to any other in music." Bowie, *Music, Philosophy, and Modernity*, 75.

References

Berman, Laurence. 1993. *The Musical Image: A Theory of Content*. Westport, CT: Greenwood Press.
Bernstein, Leonard. 1976. *The Unanswered Question: Six Talks at Harvard*. Cambridge, MA: Harvard University Press.
Bessler, Heinrich. 1978. *Aufsätze zur Musikästhetik und Musikgeschichte*. Leipzig: Reclam.

Bowie, Andrew. 2007. *Music, Philosophy, and Modernity*. Cambridge: Cambridge University Press.
Feld, Steven. 1984a. "Communication, Music, and Speech about Music." *Yearbook for Traditional Music* 16: 74–113.
Feld, Steven. 1984b. "Sound Structure as Social Structure." *Ethnomusicology* 28: 383–409.
Harries, Karsten. 2009. *Art Matters: A Critical Commentary on Heidegger's "The Origin of the Work of Art."* Dordrecht: Springer.
Harries, Karsten. 2014. *Heidegger's "Being and Time."* Accessed May 25, 2021. https://cpb-us-w2.wpmucdn.com/campuspress.yale.edu/dist/8/1250/files/2016/01/Heidegger-Being-and-Time-1qtvjcq.pdf
Heidegger, Martin. 1959. *An Introduction to Metaphysics*. Translated by Ralph Manheim. New Haven: Yale University Press.
Heidegger, Martin. 1962. *Being and Time*. Translated by John Macquarrie and Edward Robinson. New York: Harper & Row.
Heidegger, Martin. 1971a. "The Origin of the Work of Art." In *Poetry, Language, Thought*. Translated by Alert Hofstadter, 17–87. New York: Harper & Row.
Heidegger, Martin. 1971b. *On the Way to Language*. Translated by Peter D. Hertz. New York: Harper & Row.
Heidegger, Martin. 1979. *Nietzsche*, Vol. 1. Translated by David F. Krell. San Francisco: Harper-Collins.
Heidegger, Martin. 2012. *Contributions to Philosophy (Of the Event)*. Translated by Richard Rojcewicz and Daniela Vallega-Neu. Bloomington and Indianapolis: Indiana University Press.
Hesmondahalgh, David. 2013. *Why Music Matters*. Oxford: Wiley Blackwell.
Inwood, Michael. 1999. *A Heidegger Dictionary*. Oxford: Blackwell.
Karl, Gregory, and Robinson, Jenefer. 1997. "Shostakovich's Tenth Symphony and the Musical Expression of Cognitively Complex Emotions." In *Music and Meaning*, edited by Jenefer Robinson, 154–78. Ithaca: Cornell University Press.
Nietzsche, Friedrich. 1967. *The Birth of Tragedy and the Case of Wagner*. Translated by Walter Kaufmann. New York: Random House.
Nietzsche, Friedrich. 1974. *The Gay Science*. Translated and edited by Walter Kaufmann. New York: Random House.
Parkes, Graham. 1987. *Heidegger and Asian Thought*. Honolulu: University of Hawai'i Press.
Schapiro, Meyer. 1968. "The Still Life as a Personal Object: A Note on Heidegger and Van Gogh." In *The Reach of the Mind*, edited by Marianne L. Simmel, 203–9. New York: Springer.
Stambaugh, Joan. 1992. *The Finitude of Being*. New York: State University of New York Press.
Thomson, Iain. 2011. *Heidegger, Art, and Phenomenology*. New York: Cambridge University Press.
Thomson, Iain. 2015. "Heidegger's Aesthetics." *Stanford Encyclopedia of Philosophy*. Accessed June 22, 2018. https://plato.stanford.edu/entries/heidegger-aesthetics/.
Weidler, Markus. 2005. "Heidegger's Theft of Faith: A Campaign to Suspend Radical Theology." PhD diss. The University of Texas at Austin.
Wheeler, Michael. 2011. "Martin Heidegger." *Stanford Encyclopedia of Philosophy*. Accessed May 25, 2021. https://plato.stanford.edu/entries/heidegger/.
Young, Julian. 2001. *Heidegger's Philosophy of Art*. Cambridge: Cambridge University Press.
Young, Julian. 2002. *Heidegger's Later Philosophy*. Cambridge: Cambridge University Press.

6

Improvising with Soul: Philosophizing in an Interpretative Key

Peter D. Hershock

Philosophizing is an embodied practice. In the context of what is often referred to as comparative, global, or intercultural philosophy, this is not a controversial claim. But until relatively recently, in European and American philosophy departments and their intellectual clones, philosophizing was epitomized by Descartes's armchair meditations and the weighty repose of Rodin's *Le Penseur*.[1] Philosophers might have had bodies, but their work did not depend intimately on them. Philosophizing was presumed to be clear and clean mental labor.

Graham Parkes (1983) introduced the image of philosophizing as a "wandering dance" in one of his early efforts to interpret the philosophies of Zhuangzi and Nietzsche. The image of wandering dance neatly encapsulates his career-defining commitment to explore the intercultural textures of philosophizing as a physically instantiated and environmentally situated performative practice.[2] For Parkes, all philosophies *of* action properly emerge from and resolve in relationally manifest philosophies *in* action. Bodies and their environs matter philosophically.

Yet, Parkes's most recent and most activist work, *How to Think about the Climate Crisis: A Philosophical Guide to Saner Ways of Living* (2021), is no fluid wander, cloud hidden, whereabouts unknown. It is a trenchantly impassioned "tango" aimed at inspiring the kinds and qualities of global collaboration needed to heal the relational harms of anthropogenic climate disruption. As such, it raises a question that runs implicitly throughout all of Parkes's work: who do we need to be *present as* to engage skillfully (and, ideally, virtuosically) in the embodied, activist performance of philosophy that makes a difference in the world?

One answer is intimated by Parkes's characterization of Nietzschean psychology as an emotionally charged and somatically situated practice of "composing the soul."[3] This characterization entailed reading Nietzsche both decisively against the grain of authoritarian appropriations of will to power, and in ways that foregrounded a radical embrace of the polycentric, intersectional, and normatively open nature of human presence. This embrace of human complexity aptly supported Parkes's ongoing efforts to mediate Nietzsche's philosophical iconoclasm and the relational conceptions and ideals of personhood forwarded in East Asian Confucian, Daoist, and Buddhist thought and practice.

Yet, although conceiving of the self as "put together" through the situated *incorporation* of manifold positions and perspectives serves the good purpose of stressing the intimacy of psyche and society, it has limited utility in elaborating what is involved in being present as needed to engage in the kind of philosophizing—the kind of intercultural and international "tango"—that is needed practically to resolve global predicaments like climate change.

Composition is a process that takes place "offline" in a deferred performative present. While composers may work out ideas by playing an instrument like the piano, composing music is not *performing* music. Musical passages are run together and decomposed at will, played in repetitive bursts with small variations until fluent, and eventually sedimented safely in place—committed to paper or computer file. Composers *make up* music in an abstract present, without venue, without time of day or season, and without the moody indeterminacies of audience reception. This is nothing like performing precariously in organically real, irreversible time.

As a metaphor for enacting changes in human presence of the kinds and depths needed to practice activist philosophy and transform imminent climate catastrophe into a catalyst for more humane human-world relations, composing the soul will not do. The climate crisis cannot be taken offline and resolving the values predicaments that inform its dynamics cannot be resolved by concatenating measures composed in a stop and start fashion in closed rooms by negotiators insulated temporally from the precarities of climate uncertainty. It may well be that internationally binding climate accords will need to be composed. But that compositional endeavor will be effective only if it occurs as a way of capturing and formalizing the improvised performances of embodied and precariously embedded, predicament-resolving virtuosos.

In what follows, I want to develop an alternative to the compositional metaphor—an alternative passage, cantilevered out over the familiar, to becoming present as we must to engage in international, intercultural, and intergenerational predicament resolution. The approach I will take is significantly anticipated by Parkes in his relational reading of Nietzschean will to power as interpretation, and in his epistolary "open letter" exchanges with colleagues that are dynamic kin to the practice of "trading fours" and interweaving soloing explorations in improvisational jazz and blues. To be present as needed for embodied participation in resolving the values conflicts made manifest by climate change, persistent global hunger, and the new global attention economy and surveillance capitalism, composing the soul must yield to improvising with soul.

Vital Presence as Meaning

No matter how much one might wish otherwise, there are things that cannot be unsaid. That is arguably true in the case of Nietzsche's characterization of the vitality of all existents as "will to power"—a locution that implies singular determination and a forceful drive to be free from any compulsion (and perhaps even any inclination) to defer. That does not seem to have been Nietzsche's understanding of the phrase, which he employed to intensify the inexhaustible procreative will of life invoked by Schopenhauer. But his use of the phrase was sufficiently indefinite that readings of it continue to multiply. Although Parkes is not able to unsay "will to power" for Nietzsche, he has charted itineraries through the terrain of Nietzsche's oeuvre that support an emancipatory, East Asian Buddhist reading of will to power and Nietzschean perspectivism—and one that bears significantly and positively on who we need to be present as to engage in global predicament resolution.

One of the conceptual commitments common to the major East Asian Buddhist traditions—Tiantai (J: Tendai), Huayan (J: Kegon), Chan (J: Zen), and Zhenyan (J: Shingon)—is subverting either/or reasoning and affirming the nonduality of all things. In the more esoteric leaning traditions, this led to seeing the "six elements" (earth, water, fire, wind, space, and consciousness) as the Dharmakāya or "reality-body" of the Buddha, and more generally to affirming that all things have Buddha-nature. In keeping with both Indian Buddhist injunctions to see all things as impermanent and Confucian and Daoist conceptions of the cosmos as intrinsically dynamic and immanently ordered, the East Asian Buddhist conception of Buddha-nature was not that of a fixed essence contained in all things—a conception that would have imposed a foreign appearance/reality distinction and allied nonduality with uniformity.

Instead, Buddha-nature was understood as consisting in a disposition for or being intent upon realizing enlightening or liberating relational dynamics. Thus, all things are inclined, like the historical Buddha, to dynamically and relationally express both the means to and meaning of liberation from conflict, trouble, and suffering—the means to and meaning of uncompelled presence. In Kūkai's (774–835) memorable phrasing: "Soaring mountains are brushes; vast oceans, ink; heaven and earth, the box preserving the sutra. Yet contained in every stroke of its letters is everything in the cosmos. From cover to cover, all the pages of the sutra are brimming with the objects of the six senses, in all their manifestations."[4]

Granted the validity of an appearance/reality distinction, seeing all things as part of the reality/teaching body of the Buddha might be presumed to entail seeing the world around us—a world that we do not normally experience as a sutra or medium of inspiration and instruction—as a world in need of deciphering. But this would misconstrue Buddhist nonduality and strip Nietzsche's perspectivism of all but epistemic value, reducing his identification of will to power with interpretation to an inauthentic and megalomaniacal translation project of egocentrically determining what all things mean. Parkes (2000) insists otherwise, invoking as an interpretative medium the Huayan Buddhist conception of nonduality.

As elaborated by the Chinese philosopher-monk, Fazang (643–712), the nonduality of all things is realized through the practices of seeing all things as impermanent (*anitya*), as mutually conditioning (*pratītyasamutpāda*), and as characterized by emptiness (*śūnyatā*) or the absence of any fixed essence or position. Doing so reveals that the interdependence of all things entails their interpenetration. Interdependence is an internal and constitutive relation, not an external and contingent one. The cosmos is a meaning-generating matrix within which each thing or situation (*shi* 事) consists in simultaneously causing and being caused by the totality—a cosmos in which each thing *is* what it contributes functionally to the dynamically patterned articulation (*li* 理) of that totality. In short, realizing Buddhist nonduality is not an erasure of difference. It is realizing all things are the same only insofar as each differs meaningfully from and for all others. In short, each thing *is* what it *means* to and for others.

Interpretation as Generative Presence

Given the dynamic and difference conserving nature of nonduality, the world as sutra or the reality/teaching body of the Buddha is not an object in need of deciphering. Each thing in the cosmos is a dynamically evolving interpretation of all other things. Interpreting in this sense is not fundamentally a process of translating or establishing semantic equivalence. As implied by its Latin (*interpres*) and proto-Indo-European (*per*) roots, interpreting consists in "distributing meaning among" things or "trafficking in significance." Thus, with Buddhist assistance, Parkes can ably argue that when Nietzsche claims the will to power is interpretation, he is doing so with the understanding that the world "includes within itself an infinity of interpretations" (*The Joyful Wisdom* no. 374). The will to power is not being determinately intent on semantic narrowing. On the contrary, will to power expresses vital and fecund openness to engendering new meanings. As humanly enacted, it is openness to exceeding how one has been present up until now, not in pursuit of some preexisting truth or some fixed conception of ultimate reality, but in virtuosic expansion of ways of being/becoming meaningful.

There is a sense, then, in which all things consist in the will to power, not as a drive toward transcendence, but as a vital urge toward semantically liberating somatic immanence—a restoration of unbounded creative presence within the horizonless relational matrix of the cosmos. Parkes is thus free to *interpret* the cool harmonization advocated by Zhuangzi and the iconoclastic heat of Zarathustra's will to power as performative cousins, each intent to "dissolve the atrophying self back into the network of the world."[5] Rather than a drive toward assertions of power over others in the maintenance of a central core of egoic sameness in the face of a world of differences, the will to power is relinquishing the self/other horizon not as a negation of self, but as an affirmation of generative presence and creative genius.

This Buddhist-inspired view of Nietzsche's will to power as interpretation can be contested. Indeed, one of Parkes's epistolary interlocutors, Bret Davis (2015), has done just that, and with considerable care. The point here, however, is not whether Parkes is right or wrong about being able to separate Nietzsche's will to power as interpretation from egoist assertions of power over others. It may be, as Davis suggests and Parkes agrees, that Nietzsche is best understood as a "zebra" with both anti-Buddha and bodhisattva stripes. The point is whether Parkes's interpretation offers guidance in leaving behind both the "composing the soul" metaphor and a Buddhist-inspired "decomposing the soul" approach to articulate a vision of how we must be present to engage in the ethically creative labor of resolving global predicaments. I think that it does, and that it carries us usefully back to the conception of philosophy as embodied practice.

From Composing the Soul to Improvising with Soul

I have argued elsewhere that resolving global predicaments like that of climate change or that of intelligent technology cannot be accomplished from within any existing ethical system—each of which has notable blind spots—or through the construction of a generically universal or global ethics. Instead, it will require the emergence of a global ethical ecosystem in which ethical differences are engaged as opportunities for mutual contribution to engendering truly shared—and not presumptively common— values.[6] In short, resolving global predicaments will depend on fostering a transition from ethical plurality or variety to ethical diversity through the sustained practice of ethical improvisation. The performative philosophizing that Parkes enjoins us to undertake demands nothing less than improvisationally embodying ethical virtuosity. What does this entail in practice?

In reflecting broadly on the requirements for philosophizing as embodied practice, drawing on Confucian valorizations of ritual conduct, Parkes notes that by means of "honing the body's movements in relation to other people and things, one becomes more open and responsive to the world," but that this is not "simply a matter of getting the prescribed movements right: the activity must also be infused with soul." As he insists, "your heart has to be in it, or else the performance will fall flat."[7] As a practice, performing philosophy is impossible without the body-mind unity of total investment.

In music-making, this whole-hearted investment is characterized as playing with "feel" or "soul." It is not just body-mind unity, but an affectively charged body-mind-music unity through which a musician goes beyond playing music to non-subjectively inhabiting the music and then to being the vitally beating heartmind *of* the music. To improvise in this wholly invested manner is to leave behind playing variations on a theme or charting a novel pathway through a set of memorized harmonic changes. We can call that musical *innovation*—a way (*dao*) of making music that can be computationally realized and exemplifies *closed creativity*: a capacity for opening new routes to already-anticipated ends or goals. In contrast, wholly invested musical *improvisation* is the embodied enactment of *open creativity*: a capacity for indeterminacy embracing and yet evaluatively guided and resolutely non-instrumental exploration. "Improvising is a process of working out from within existing norms and circumstances in the direction of what is both unprecedented and qualitatively enriching"—a process of "continually expanding horizons of anticipation."[8]

To build a bridge between musical and ethical improvisation, consider the *presence* involved when two guitarists engage in virtuosic musical improvisation. If they are embarking on free improvisation—not standards-based or harmonically specified innovation—they do not begin with a known/knowable environment. Beginning in shared silence, they embark on realizing previously inexistent and thus unknown musical realms. As they begin improvising in epistemic emptiness, they are sustained by deep listening as each sonically enacts a readiness to affect and be affected by the other in shared appreciation of musical possibility. This affectively intent and relationally attuned readiness, embedded in histories of all past music-making, is extended temporally in a growing present as fingers interact with strings to open qualitatively distinct spaces of sonic opportunity. Embodying shared investment in expanding the horizons of musical anticipation, music-improvising minds are distinctively intent on performing differently *from* each other to make appreciable differences *for* one another.

What emerges with freely improvised music is a sonic/aesthetic ecology: the emergence out of emptiness of a *jointly evoked, wholly shared* world: an immediate coordination of values or modalities of appreciation in the absence of representation and reflection. This achievement of improvisational practice demonstrates that, with the right kind and quality of affective/responsive presence, interests and intentions can be productively cantilevered out into the unknown in the joint realization of creative presence. Improvising with soul is not just a matter of playing patterns of notes that have never been played before. It is sonically conjuring an affectively enriching time-space—a shared realization of the relational fecundity of meaningful inexistence.[9]

Ethics as Moral Music and the Need for Moral Virtuosos

Inviting someone to engage in ethical improvisation may sound a bit like asking someone to do some creative accounting. Getting creative ethically is a nice way, perhaps, of saying that one is using ethics instrumentally: using it as means to other-than-ethical ends. In this sense, the ethical improviser is a non-virtuous magician performing moral sleight-of-hand tricks. But this interpretation arguably rests on presuppositions that ethics involves using predetermined—that is, rationally derived—standards to determine the moral merit of individual moral agents, their actions, and how their actions impact relevant moral patients. Ethics of this kind is a process of putting things in order—setting out the terms and conditions of being a good person, leading a good life, and contributing to the workings of a good society. Practicing ethics in this way is thus something like moral *design*.

One of the difficulties in placing East Asian ethical traditions into one of the three mainstream Western approaches to normative ethics based on presuming the primacy of virtues, duties, and consequences is that Confucian, Daoist, and Buddhist approaches to evaluating persons, actions, and the consequences of actions have focused dynamically on relational quality and have (at their best) been concerned with the realization of relational virtuosity. One implication of this is that ethics takes on aesthetic qualities and becomes radically historical. In much the same way that musical virtuosity consists in exceeding current standards of musical performance in ways that the musical community affirms as establishing new standards of performance, ethical virtuosity is not about living up to existing ethical norms; it is about exceeding them. The practice of ethics is not just doing good and avoiding bad. It is going beyond all that is currently deemed good, evil, and mediocre to realize superlative or exemplary relational dynamics. Ethics in this sense is the necessarily improvisational art or way (*dao*) of virtuosic human course correction.

This association of ethics with exceeding standards is one way of reading Nietzsche's conviction—expressed in *The Joyful Wisdom* and elsewhere—that philosophizing should not devolve into a search for fixed truths or the codified wisdom of the ancients. It should consist in continuously fashioning what has until now never existed: contributing with superlative freedom to an eternally growing world of values, perspectives, affirmations, and negations. Parkes's East Asian Buddhist reading of this characterization of philosophical performance aligns it with the Chan/Zen iconoclasm that depicts reading the words of past exemplars as an exercise in dung eating and regurgitation and advises killing the Buddha if one happens to meet him on the road. The point here is to vigorously refrain from accepting standards of responsive virtuosity and then falling prey to a Zeno's paradox while trying to traverse the distance between being a mere sentient being and a fully enlightened and liberated being. Responsive virtuosity is not achieved *through* Buddhist practice. It is not a goal to be reached. It is a qualitatively distinct achievement *of* practice, with this very body, and in this very moment, in situationally apt exemplifications of its interpretative infinity.

The embodied philosophical practice of ethics should not be about designing the good life. It should not be about zeroing in on reality or cutting the world at its joints or engaging in any of the other transcendentally performed mysteries aimed at *fixing* things. Ethics is not the disciplining mediation of intention or will and the world. It is the aesthetic adventure of human creativity—an art of shared relational appreciation and enhancement.

The question running implicitly throughout Parkes's career of philosophical performance has been: Who do we need to be present as to venture aesthetically into the unknown and draw forth from the treasury of inexistence enlightening resources for enacting the celebratory mantra closing the Heart Sutra—going beyond, beyond and the even beyond the beyond in consummate appreciation of the inexhaustible meanings of enlightenment? The climate predicament, the predicament of global hunger, the predicament of intelligent technology, and the many other predicaments that could be enunciated here are expressions of who we have been humanly present as. Resolving these predicaments is predicated on us improvising ways of being otherwise present and interpretatively enabling things to be differently interdependent.

Indeed, in thinking through the implications of Nietzsche's psychology, Parkes advocates throwing over the Western "monarchy of I and the tyranny of univocal reason" to imagine the psyche as analogous to a theater company that may initially benefit from rotating directorships—but the ideal would be to come to a point where there would be no need for directors and the company would become "fully improvisational."[10] Given an irreducibly relational conception of personhood, this ideal can be anticipated to scale up to serve as an ideal of the kind of persons we need to be to succeed as ethical agents who are committed to realizing moral communities conducive to the emergence of ethical diversity—agents who are analogously comfortable engaging creatively in director-less ethical improvisation.

Philosophizing should matter. But it will do so only to the degree that our wandering dance settles intensively into a transformative tango in which apparent opposites enact creatively shared appreciations of opportunity for being present otherwise. In resolving global predicaments, it is only ethical virtuosos who can improvisationally intimate the way.

Notes

1. In *The New York Times*'s philosophy blog, *The Stone*, a recent, and sensible, argument has been made by Bryan Van Norden and Jay Garfield in a 2016 piece titled "If Philosophy Won't Diversify, Let's Call It What It Really Is" suggests that what is taught in most philosophy departments around the world is really some version of regional philosophizing—typically a generic "Western philosophy" or some combination of "Anglo-Analytic," "Euro-Continental," or "American-Pragmatic" philosophy—and should be labeled as such. The broader term "Philosophy" should be used as an all-encompassing approach that embraces African, Asian, Indigenous, and all other positional approaches to the discipline.
2. Graham Parkes, "Awe and Humility in the Face of Things: Somatic Practice in East Asian Philosophies," *European Journal for the Philosophy of Religion* 4 (3) (Autumn 2012): 69–88.
3. Graham Parkes, *Composing the Soul: Reaches of Nietzsche's Psychology* (Chicago: University of Chicago Press, 1994).
4. Adapted from Abe, 288.
5. Graham Parkes, "The Wandering Dance: Chuang Tzu and Zarathustra," *Philosophy East & West* 33 (3) (1983): 247.
6. The distinction of the shared and common that is invoked here is that of advance by Jean-Luc Nancy (*Being Singular Plural*. Translated by Robert Richardson and Anne O'Bryne [Stanford, CA: Stanford University Press, 2000]). Peter D. Hershock, *Buddhism and Intelligent Technology: Toward a More Humane Future* (London: Bloomsbury Academic, 2021), 135–41.
7. Arno Böhler, Loughnane, Adam, and Parkes, Graham, "Performing Philosophy in Asian Traditions," *Performance Philosophy* 1 (1) (2015): 141.
8. Hershock 2021, 173–4.
9. Inexistence is used here to refer to the nondual remainder of dissolving the categories of existence and nonexistence—the logically ineffable subject of Nagarjuna's tetralemma of that which can only be said to be: not x; not not-x; not both x and not-x; and not neither x nor not-x.
10. Parkes, *Composing the Soul*, 369.

References

Abé, Ryūichi. 1999. *The Weaving of Mantra: Kūkai and the Construction of Esoteric Buddhist Discourse*. New York: Columbia University Press.

Böhler, Arno, Loughnane, Adam, and Parkes, Graham. 2015. "Performing Philosophy in Asian Traditions." *Performance Philosophy* 1 (1): 133–47.

Davis, Bret W. 2015. "Reply to Graham Parkes: Nietzsche as Zebra: With Both Egoistic Antibuddha and Nonegoistic Bodhisattva Stripes." *Journal of Nietzsche Studies* 46 (1): 62–81.

Nancy, Jean-Luc. 2000. *Being Singular Plural*. Translated by Robert Richardson and Anne O'Bryne. Stanford, CA: Stanford University Press.

Nietzsche, Friedrich. 2016. *The Joyful Wisdom*. Edited by Oscar Levy and translated by Thomas Common, Project Gutenberg EBook #52881, Release Date: August 23.

Parkes, Graham. 1983. "The Wandering Dance: Chuang Tzu and Zarathustra." *Philosophy East & West* 33 (3): 235–50.
Parkes, Graham. 1994. *Composing the Soul: Reaches of Nietzsche's Psychology*. Chicago: University of Chicago Press.
Parkes, Graham. 2000. "Nature and the Human 'Redivinized': Mahâyâna Buddhist Themes in Thus Spoke Zarathustra." In *Nietzsche and the Divine*, edited by John Lippitt and James Urpeth, 181–99. Manchester: Clinamen Press.
Parkes, Graham. 2012. "Awe and Humility in the Face of Things: Somatic Practice in East Asian Philosophies." *European Journal for the Philosophy of Religion* 4 (3) (Autumn 2012): 69–88.
Parkes, Graham. 2015. "Open Letter to Bret Davis: Letter on Egoism: Will to Power as Interpretation." *Journal of Nietzsche Studies* 46 (1): 42–61.
Parkes, Graham. 2021. *How to Think about the Climate Crisis: A Philosophical Guide to Saner Ways of Living*. London: Bloomsbury Academic.
Van Norden, Bryan and Garfield, Jay. 2016. "If Philosophy Won't Diversify, Let's Call It What It Really Is." *New York Times, The Stone Opinion*, May 11, 2016.

7

Image Thinking: Cinematographic Flâneurs and Twenty-First-Century Philosophy

Andrew K. Whitehead

The twenty-first century has already brought about significant changes in what constitutes legitimate contributions to academic scholarship, as well as the emergence of new and distinct programs and fields of inquiry. From the successful defense of a graphic novel dissertation in Education at Columbia University in 2014,[1] to the submission of rap album dissertations—both at Clemson University and at Harvard—in 2017,[2] technologies are being re-engaged and re-deployed to transcend or compliment the limitations of the written word. Understandably, these developments have attracted both supporters and detractors. What is most interesting is that the nature of these debates has not changed significantly from those debates concerning the role of film in research and pedagogy that began almost a century ago. My interest here concerns the possibility of using alternative mediums for philosophical thinking and expression, specifically film, and the extent to which philosophy is or is not able to be practiced and expressed in mediums other than academic writing.

In effect, what is at issue is the role of language in philosophy. This is a problem that has served as a frame for a significant number of skeptics concerning film's capacity to yield anything like a genuine philosophical contribution. Underneath the claim of philosophy as a necessarily linguistic practice lies a conviction so strong in many cases it is often taken to be self-evident, namely that thinking is necessarily linguistic. However, this conviction has found strong challenges in recent phenomenological research. Dieter Lohmar, for example, in uncovering notions of non-linguistic thinking in the writings of Edmund Husserl and his theory of meaning, has concluded that "language is one system of representation" and that "we can in principle conceive of other systems of representation with the same performance."[3] In fact, says Lohmar, "the basic way of thinking is non-linguistic."[4] On the basis of Husserl's theory of meaning, Lohmar's reflections instead reveal the function of, for example, "scenic phantasmata," that is, images or sketches of objects, persons, events, or situations in both dreams and daydreams. These scenic phantasmata, according to Lohmar,[5] "emerge like short video clips that give rise to feeling and co-feelings connecting short scenes in a kind of story." As such, they manifest not only a base form of thinking, but they can in fact accelerate thinking and enable a synthesis of different aspects more effectively than linguistic and linear thinking ever could. Insofar as this is the case, film could actually be a more suitable medium for the expression of certain thoughts or modes of thought.

This idea of "phantasmatic" thinking is fruitful for our interests here, namely the role of film with regard to its philosophical capacities, both in terms of enhancing classroom learning and in terms of distinct medium specific contributions that film can make to the practice of philosophy.

In what follows, I critically review the distinction between philosophy of film and film as philosophy. With an aim to constructing a meaningful dialogue and exchange, I work to compare and contrast the central theses developed by two philosophers of radically different opinions concerning the possibility of using film philosophically.[6] Specifically, I place the writings and cinematographic works of Graham Parkes into dialogue with a skeptical voice and less-than-sympathetic interlocutor—here personified by Paisley Livingston—concerning the issue of the potential of the medium of film for philosophical insight and pedagogy. I aspire to clarify Parkes's critique of the "breathtakingly narrow view of the discipline" that takes for granted that non-linguistic (and, by extension, non-argumentative) approaches to philosophy are invalid and nonsensical.[7] Against such a belief, and in line with Parkes, I argue that "while reasoned argument may be a preferred strategy for dogmatists convinced they are in possession of the truth," contemporary philosophers are in the unique position of being able to "innovate by extending the use of film and video beyond the dissemination of knowledge and understanding to new modes of knowledge production."[8] In so doing, I hope to provide an argument in favor of the possibility (or even more strongly, the inevitability) that film be used to make genuine contributions to the field of philosophy. To the extent that this is the case, I do not believe that the constructed exchange I am presenting between Livingston and Parkes is limited to a disagreement concerning film for I believe that what is really at stake is the question: "What is Philosophy?"

Philosophy and Film

In approaching the topic of philosophy and film, it is important to recognize that philosophers deal with film in two distinct ways: one can philosophize about the medium of film itself, that is, about the technology, the state of production, and the many factors that lend themselves to the development of film and one can philosophize about the message, that is, about the content, the experience, the (life-)world of a given film. Each of these distinct ways is also subject to the sub-disciplinary division of philosophy of film (here understood as a particular branch of aesthetics) and film as philosophy (the practice of philosophizing through and with film). These analytic divisions are not always exclusive, but they do provide a useful distinction with which to approach the topic of philosophy and film.

In considering the first of these ways, the medium of film itself, it becomes easy to see how there exists overlap between, on the one hand, philosophical discussions concerning film and, on the other hand, the unique and genuine philosophical contribution made possible exclusively with film. As an example, let us consider the technological evolution of projectors, leading up to the home theaters of the last decades. Traditionally, one would describe film technology in terms of frames per second (FPS). A standard classic film would have been recorded and reproduced at a rate of 24 FPS in light of the determination that at this speed the human eye cannot detect the distinct frames, but instead senses genuine motion. In between each frame, a blinder that is synchronized to the specific frame rate obscures the transition from one frame to another. The old-fashioned "flickering" of films is a symptom of having mis-calibrated this specific frame rate.

With the advent of home televisions, FPS became relative to the hertz (Hz) of the particular display unit. Hz is measured in terms of cycles per second. In the case of televisions, Hz refers to the refresh rate, that is, the complete reconstruction of the image projected on the screen. If one has a television unit that is 60 Hz, indicating 60 pulses of light per second, then the image on the screen would be flashed sixty times. Owing to the power limitations of the time, traditional NTSC[9] television units therefore operated based on 30 FPS, using 60 Hz. In other words, each frame is projected twice, up to 30 frames per second across sixty distinct projections. These distinct projections are made up of 525 individual lines (480 of which are visible), divided across two fields. In other words, 480 visible lines are projected sixty times a second, with each projection being projected twice. PAL[10] television units traditionally operate on 50 Hz, displaying 25 FPS. These distinct projections are made up of 625 individual lines (576 of which are visible).

This basic technology remains largely unaltered in terms of how moving images are displayed or projected until today. Newer digital technologies, while more sophisticated, still project series of lines based on Hz. However, newer technologies have also returned to the proper mathematical ratio of 24 FPS, working in 72 Hz (each frame repeated three times), 96 Hz (each frame repeated four times), or 120 Hz (each frame repeated five times). High-definition digital televisions also distinguish between interlaced and progressive scanning technologies. Interlaced scanning technology displays the 480 visible lines of the NTSC system by breaking up the odd and the even lines, and digitally "interlacing" them at very high speeds. With each frame being able to be repeated up to five times (120 Hz), this type of interlacing is hardly perceptible. Progressive scanning technology displays the 480 visible lines of the NTSC system by working through the lines sequentially. This technology was originally introduced for computers, owing to the unpleasant and visible flicker of text when using interlaced scanning technology. It is now used for almost all high-definition television units for similar reasons.

At this juncture, the reader may be wondering what any of this has to do with philosophy and further, one may be wondering how this somehow evidences the overlap between philosophical discussions concerning film and unique and genuine philosophical contributions made possible exclusively with film.

In examining the medium of film, and coming to understand and recognize the processes by which images are captured, recorded, and then reproduced, we come to encounter issues concerning the distinction between appearance and reality, how perception works, and the nature of objectivity in relation to our epistemic and conscious experience of the world in which we live. Despite being well aware of the fact that what we see is nothing more than a series of still photographs, or, worse still, distinct individual lines dissected from still photographs, they remain beyond our sense-capacities for experience. We see motion and streamed continuity.

The medium of film therefore allows for information to be processed and expressed in a way radically different from written text (potentially rendered in even more effective ways), and this is the case prior to even considering technique or style. The technology itself, entailing moving image overlaid with sound, represents a fundamentally other form for philosophical expression, and therefore constitutes very fertile soil for both philosophical investigation and articulation. At the same time, the experience of this medium is itself open as a philosophical experience. Watching a film yields different effects on its spectators than a text does on its readers (which can be philosophically revealing in its own right). This becomes much more evident if we consider the many Native Americans and Aborigines of Australia, as well as several groups in Latin America, who, even today, experience film (whether in still photographs or sequenced moving pictures) with *Angst* or discomfort, believing that their souls, spirits, or essences are in jeopardy of being captured and/or stolen.

Watching a film of oneself is equally able to trigger the Lacanian mirror stage that facilitates apperception, an experience only heightened in the case of video, and this even and especially for adults. This experience is further amplified in the case of adults through the dynamics of youth and old-age, life and death. As Bernard F. Dick once remarked:

> Movie making is the transformation of living beings into dead images that are then given life by being projected on a screen. Movie going is watching dead images coming out of a projector, ordinarily at the rate of twenty-four frames per second. Since the stars have "died" by giving up their image to celluloid, they can be immortal both in their lifetime and after their death.[11]

We have arrived at a technological moment in which we can existentially encounter and experience our youth frozen in motion—a youth that will ultimately outlive our old age. Film affords us a means of expressing certain philosophical ideas, for example, concerning identity, mortality, or earnestness, in a way that communicates something distinctly different, or something in distinctly different ways, from philosophy in its written form.

These issues alone should suffice to draw attention to the wealth of possibilities for philosophical engagements with the medium of film and film's capacity to contribute to philosophy. We can see that it is neither acceptable to dismiss either facet of the discussion for the sake of maintaining consistency in one's argument nor adequate to dismiss film's capacity to contribute to philosophy—and to do so uniquely and independently—without developing a worthy account of the medium itself. Nonetheless, it remains true that a number of philosophers do precisely that.

Livingston and the "Bold Thesis"

In his article "Theses on Cinema as Philosophy," Paisley Livingston notes the "practical difficulty of simultaneously pursuing what are two rather distinct ends."[12] These ends refer to a more general tendency to discuss film's exclusive capacities for developing original philosophical contributions equally in terms of the "how" (the stylistic) and as the "what" (the thematic). As Livingston finds, "it is at least rhetorically very difficult to pursue them simultaneously."[13]

It is important to note that this engagement with Livingston's article is not to treat it as a unique example, but instead to consider it as an archetypal form. Livingston is not alone in his rejection of film as philosophy, nor is he alone in holding an oppressively narrow conception of what constitutes legitimate philosophy. He is, however, a clear example of what such a philosopher contends, and this in specific relation to what philosophy is and can be in relation to the cinematic medium.

The major thrust of his position goes against what he and others have come to call the "bold thesis" concerning film as philosophy. Livingston articulates this thesis as follows: "Films can make creative contributions to philosophical knowledge, and this by means exclusive to the cinematic medium."[14] Against the "bold thesis," he contends that while it is true that films can be used heuristically for the sake of complementing genuine philosophical works—in the form of developed linguistic argumentation—they cannot make creative contributions that are in and of themselves exclusive to the medium of film. He writes, "Films can provide vivid and emotionally engaging illustrations of philosophical issues, and when sufficient background knowledge is in place, reflections about films can contribute to the exploration of specific theses and arguments, sometimes yielding enhanced philosophical understanding."[15] In other words, one can use films as concrete illustrations that work in the service of some philosophical position or other, provided that the viewer has sufficient preliminary understanding; and this in the hope of enhancing or reinforcing how one understands actual philosophy. However, film cannot make original contributions to philosophy that are exclusive to its own medium.

Livingston finds that there are two parts to the "bold thesis," namely, those capacities determined to be exclusive to the cinematic medium and the significance and independence of those capacities respectively. Supporters of the thesis, according to Livingston, hold that "the cinematic medium's exclusive capacities involve the possibility of providing an internally articulated, nonlinguistic, visual expression of content, as when some idea is indicated by means of the sequential juxtaposition of two or more visual displays or shots."[16] Working in favor of these capacities are various cinematographic and editorial techniques, including the direction of camera movements and foci, the use of montage, and the use of audio and soundtracks.

Livingston further develops his imagined interlocutor's position in reference to the significance and independence of film's contribution to philosophy, noting that supporters of the "bold thesis" contend that films actually do in fact come to "realize historically innovative philosophical contributions," and are able to "provide ... historically innovative contribution[s] to knowledge regarding some philosophical topic, doing so in a significantly independent or autonomous manner."[17] In other words, according to Livingston, the "bold thesis" involves an argument in favor of film's ability to express innovative contributions to the field of philosophy that cannot otherwise be expressed. And it is here that we are most likely to encounter disagreement pertaining to what constitutes legitimate forms of argumentation.

Faced with his imagined interlocutor's position, Livingston is quick to note the "ruinous dilemma" with which they are faced. He does so on the grounds that "a bold 'cinema as philosophy' thesis of this ilk is difficult to defend."[18] His approach takes the form of a two-horned argument that finds the conception of exclusivity requisite for the "bold thesis" is at once both too narrow and too broad. It is too narrow, according to Livingston, if by exclusivity one means that it is only with film, and with no other medium that the innovative and autonomous contribution can be made. Such a belief involves the postulation of an ephemeral qualia, or a certain *je ne sais quoi*, that marks and characterizes the discursive form of film as fundamentally irreconcilable with the standard linguistic forms that have heretofore served as the medium for philosophy. He writes:

> If it is contended that the exclusively cinematic insight cannot be paraphrased, reasonable doubt arises with regard to its very existence. If it is granted, on the other hand, that the cinematic contribution can and must be paraphrased, this contention is incompatible with arguments for a significantly independent, innovative, and purely "filmic" philosophical achievement, as linguistic mediation turns out to be constitutive of (our knowledge of) the epistemic contribution a film can make.[19]

This limitation in terms of ability for paraphrasing—and the "reasonable doubt" believed to arise in the face thereof—seems to be a strange stipulation that echoes certain views of the logical positivists. Conversely, the idea that if the cinematic contribution can be paraphrased, then it being no longer significantly independent seems equally strange, especially because Livingston has already limited paraphrasing to the linguistic medium that he so readily accepts as being genuinely philosophical. At no point does he provide a suitable definition of what does and what does not constitute language, which is unfortunate because it therefore remains unclear how he intends its meaning throughout his text.[20] It is clear, however, that he assumes a narrow, verbal conception of language. For he contends that if the exclusive cinematic contribution cannot be stated in words, then adherents to the "bold thesis" must stand in hope "that others may have a similar experience and come to agree that philosophical insight or understanding has been manifested in a film."[21]

At the same time, Livingston finds the exclusivity of the "bold thesis" too broad. If one focuses too much on the medium and the stylistic "how," choosing to forego discussion of the message and the thematic "what," then the exclusivity of the thesis becomes so broad as to negate itself. As an example, Livingston notes that one iteration of the broad conception of exclusivity might claim that "cinema can make an exclusive contribution to philosophy by providing vivid audio-visual representations of genial philosophical conversations and lectures."[22] He finds that this representationalist appeal forces the concession that the genuine and innovative contribution provided by film is in fact not exclusive to it, but is instead captured and represented by it.

Livingston concludes that both alternatives fail. According to him, linguistic interpretation, in either case, is required either for the genuine contribution in the first place (as in the case of recording a conversation or a lecture) or for the interpretation that proves necessary in acknowledging such a genuine contribution (being able to explain and come to agreement on what such a genuine contribution in fact consisted in). Again, Livingston is making use of a very narrow conception of language and, with it, is deploying an even narrower conception of philosophy and its interpretive context. Both of these restrictive views come to inform his refusal of capacities that are exclusive to film in the generation and production of innovative philosophical contributions. Thus, for Livingston, film can be philosophical only reflectively, having been viewed under the appropriate conditions and with the appropriate background, and from within a given interpretive context that is conducive both to appreciating and to paraphrasing the contribution into the linguistic argumentative form. To this end, he writes: "An interpretative context must be established in relation to which features of the film are shown to have some worthwhile philosophical resonance."[23] In other words, in particular instances aspects of films can be said to resonate with genuine philosophy.

With this in mind, Livingston proposes an alternative to the "bold thesis." Rather than maintaining a philosophically privileged position for film, he suggests instead highlighting the pedagogical purposes film can be made to serve. Films can help illustrate philosophical problems, such as ethical problems. They can help students of philosophy through unique visualizations, or juxtapositions of either images or images and sound, in ways that are in fact unique to film, but are not exclusive to it or philosophical in their own right. Film can be made to work in service to philosophy proper, even if it is not itself able to constitute philosophy on its own.

It is worth noting that Livingston's position is valid, given his premises. If we take his premises as true, including the premise that philosophy is limited to its academic form, and specifically to its academically written argumentative form, then the conclusion necessarily follows. It is therefore also worth noting that the major point of contention that can be raised against Livingston's position will remain irreconcilable with it: from the perspective of what follows, his premise of a narrow conception of philosophy will not be sufficient. With that being said, Livingston would no doubt raise a similar version of his argument against a great many recognized philosophers and philosophies, noting that their method of expression precludes them from being considered philosophical in their own right. We might here think of figures such as Nietzsche, the fictional works of Beauvoir and Sartre, or Zen Buddhist poetry as possible examples. These figures and works are, however, considered philosophical by many in the academy. Similarly, alternative modes of communicating and evidencing academic ideas are already being used.

Parkes's Thinking Images

There are scholars who believe that film is not only able in principle to make genuine contributions to philosophy, but it already does. Again, the central issue seems to revolve around the commitment, or lack of commitment, to the linguistic medium and specifically to argumentation as the only philosophical medium. Graham Parkes, for instance, notes that "while talking used to be in the days of Socrates central to the philosophical enterprise, thinking and writing tend to predominate in modern times, and these do not lend themselves well to visual portrayal."[24] Parkes concedes that most attempts at philosophy in film have relied on language and argumentation as being merely reproduced on the screen. However, he remains convinced of the potential of film for philosophy. This inevitably follows from Parkes's commitment to a pluralist conception of philosophy and from his ideas of what this should entail. It is important to keep in mind that Parkes is a specialist in European-Continental and East Asian philosophies who questions a philosophical tradition for whom "it has to have arguments or it's not philosophy."[25] Against that tradition, Parkes finds that the emphasis on the discursive, as opposed to intuitive aspects of philosophizing, is unable to offer a full picture of the practice and discipline. Instead, Parkes invokes irony, poetic allusion, correlation of ideas or types, hyperbolic assertion, and enigmatic aphorism, among others, as evidence for a mode of philosophical thinking that is not limited to reasoned argument. Language itself, as easily demonstrated in the case of film, is not even necessarily required to convey successfully any number of these elements, but this is precisely where the argument loses traction with most opponents, such as Livingston. We might again turn to the work of Lohmar, and what he calls the prejudice of the language paradigm, and by extension the prejudice of the propositional paradigm, "which understands thinking only in terms of propositions, which are in turn connected to each other by logical rules."[26] If we are to remain open to the possibility of film as philosophy, then it would seem that one of the first obstacles to overcome is these prejudices by way of examination and description of the various modes of possible communication.

In taking up Husserl's notion of "meaning-bestowing acts" (*bedeutungsgebende Akte*) and how this idea relates to categorical intuition as the source of meaning, Lohmar acknowledges that "this is already an important starting point, since it suggests that language by itself is not knowledge, and that knowledge does not have a linguistic character from the very beginning."[27] This, in turn, allows for alternative symbolic mediums of expression, for example, what Parkes calls "writing in visual images (videography and cinematography),"[28] which serves as a way of extending writing beyond the written word. What is required, if such an endeavor is to prove successful, is a fundamental re-thinking of what philosophy is and how it is practiced. In taking up an altogether new medium, philosophers will have to work toward "*rightness* ('*Richtigkeit*'), which means appropriateness between categorical intuition and expression," as "this norm demands that the right external expression would allow another person to *emptily* think exactly the same what I *intuitively* thought earlier."[29] Lohmar is quick to point out that this norm would hold for every medium of expression. The question then becomes: What techniques lend themselves to the elaboration of philosophical ideas in the case of the medium of film?

Parkes lists six significant techniques in his article:

1. Set up an *interplay* between *visual images and music*.
2. Provide *visual context* for a philosopher's thought.
3. Present philosophical ideas and *exemplify, amplify, illustrate* them.

4. Practice *irony*, especially through tension between past and present.
5. Exemplify visually and orally *modes of awareness*.
6. Provide *aesthetic pleasure* while doing some of the above.[30]

These significant techniques are briefly discussed in their respective order to sufficiently draw out Parkes's position.

Interplay between Images and Music

The first technique discussed by Parkes concerns the communication of the non-linguistic and the extra-linguistic. He draws upon the effect musical sound can have in conveying something concerning the context, ambiance, and meaning behind images. Parkes notes that "this already happens to an extent when philosophy is written in poetry or poetic prose, where some of the meaning is coming through tempo, rhythm, assonance, cadence, etc."[31] This interplay between images and music effectively serves as an external expression allowing for different individuals to think in common and adheres, therefore, to the norm of rightness discussed earlier.

Visual Context

Parkes emphasizes how visual context can play a part in establishing philosophical meaning. He finds that "video offers the opportunity to show the place directly rather than having to describe it using language." Parkes continues that "this wouldn't matter in the case of abstract ideas or with a theoretical philosophy that presumes in its universality to transcend time and space, but is helpful in the case of a thinker like Nietzsche who often thinks it important that a particular idea should have come to him in a particular place."[32] While it is true that this can be limited to the idea that visual context helps in illustrating a particular philosophical point, it is equally true that the image itself, of the particular place itself, can facilitate philosophical insight of varying degrees and types; that is, the visual context can, in principle, allow for a thinking-together-with a particular phenomenon. In working back to the categorical intuition, as a ground of the meaning shared between subjects, visual context can therefore be understood to adhere to the norm of rightness.

Exemplify, Amplify, or Illustrate a Philosophical Idea Visually

Film can also work in the service of pedagogy, as accepted by Livingston seen above. It can work to exemplify, amplify, or illustrate philosophical ideas visually. Parkes discusses several different visual techniques, such as montage, slow-motion, and the close-up, as particularly apt means by which to assist in the understanding of a philosophical idea. In the case of amplification, it can also yield original philosophical ideas as it employs images that have the capacity to allude to possible avenues of further inquiry or insight. Here too we find the non-linguistic conveying of meaning, a writing with images that is able to express an original intuition.

Irony

Parkes identifies a few ways in which irony is used in film. He focuses again on the interplay (albeit this time disharmonious) of images and soundtrack that can disrupt the reality conveyed by either, and in the process this serves to reveal something novel about both. In these moments of disharmony, "the ironic tension prompts us to ask which one is real."[33] For certain philosophies of questioning—let alone any number of epistemological, ontological, or hermeneutical philosophies—philosophical meaning can be expressed by means of the ironic disharmony of filmic images.

Exemplify Modes of Awareness through Audiovisual Techniques

The fifth technique entails expressing modes of awareness, such as temporality. Parkes concludes that "film and video are eminently suited for presenting the multi-temporal nature of awareness."[34] This technique refers to practices such as superimposition of images, slow-motion, and "trails." An experiential meaning, as it pertains to the temporality of the existential moment, can therefore be expressed more succinctly and accurately via film. It is possible to "induce an experience of temporal overlap in the viewer,"[35] that is, an experience which can be measured against the norm of the rightness of the meaning-bestowing act.

Aesthetic Pleasure

The last technique discussed by Parkes exploits aesthetic pleasure. Quite simply, Parkes believes that, as is the case with written words, the writing of philosophies in images "will work best when they not only instruct but also captivate by way of the aesthetic pleasure they provide."[36]

Looking to the possibility of film as philosophy in the wake of new media in academia through the works of two diametrically opposed philosophers, I remain convinced the disagreement between them rests firmly in their presuppositions regarding what constitutes philosophy and the role of language in that constitution. Having borrowed insight from the works of Husserl on non-linguistic thinking and possible modes of expressing such thinking non-linguistically, I am persuaded that it is possible to make a genuine philosophical contribution via the medium of film. It is unquestionably possible to think in images. Likewise, using filmic images to express an original intuition to another subject is an expression that is equally subject to the norm of rightness in assessing the meaning-bestowing act. Returning to the point made regarding phantasmata, not only is film able to make a genuine philosophical contribution, some thoughts or modes of thought are very likely to be more suitably expressed in film than in written academic prose. It seems it is on this point of written academic prose and its privileged position within academia as *the* form of expression that the debate between Parkes and Livingston comes to rest. It is possible Livingston would even go so far as to conclude that philosophy can only be genuine philosophy as an academic discipline, thus making it dependent on its participation in conventional philosophical discourse. This is another point upon which it's likely Parkes would disagree. However, even if we were to concede that philosophy is necessarily academic, the cases from Columbia, Clemson, and Harvard clearly reveal academia's "breathtakingly-narrow" view of what constitutes acceptable expression is already broadening. Philosophy is sure not to be too far behind.

Notes

1. Nick Sousanis defended his dissertation, titled *Unflattening: A Visual-Verbal Inquiry into Learning in Many Dimension*, at Teachers College, Columbia University in May 2014. The dissertation was published a year later by Harvard University Press under the title *Unflattening*.
2. A. D. Carson submitted his thirty-four-track album Owning My Masters: The Rhetorics of Rhymes and Revolutions as a part of his PhD in rhetorics, communication, and information design at Clemson University in 2017, and that same year Obasi Shaw submitted his ten-track album Liminal Minds as his senior thesis at Harvard.
3. Dieter Lohmar, "Language and Non-Linguistic Thinking," in *The Oxford Handbook of Contemporary Phenomenology*, edited by Dan Zahavi (Oxford: Oxford University Press, 2012), 377.
4. Dieter Lohmar, "Non-Linguistic Thinking—From a Phenomenological Point of View," *Tijdschrift voor Filosofie* 79 (2017): 31.
5. Lohmar, "Language and Non-Linguistic Thinking," 383.
6. I make use of Paisley Livingston's 2006 article "Theses on Cinema as Philosophy," and Graham Parkes's 2009 article "Thinking Images: Doing Philosophy in Film and Video."
7. Graham Parkes, "Thinking Images: Doing Philosophy in Film and Video," *Educational Perspectives* 42 (1 and 2) (2009): 36.
8. Parkes, "Thinking Images," 37.
9. NTSC refers to the National Television System Committee, which was developed in 1941 and was the analog system used in most of the Americas, as well as many East Asian nation-states.
10. PAL refers to Phase Alternating Line, which was used for most countries broadcasting analog television at 50 Hz.
11. Bernard F. Dick, *Billy Wilder* (Boston: Twayne Publishers, 1980), 150.
12. Paisley Livingston, "Theses on Cinema as Philosophy," *The Journal of Aesthetics and Art Criticism* 64 (1), Special Issue: Thinking through Cinema: Film as Philosophy (Winter) (2006): 16.
13. He does concede that "if the research topic pertains to aesthetics or to the philosophy of art (and especially the philosophy of film!), ruminations over the specific style and themes of a given film may yield insights with regard to some well-framed question under discussion in the field."
14. Livingston, "Theses on Cinema as Philosophy," 11.
15. Livingston, "Theses on Cinema as Philosophy," 11.
16. Livingston, "Theses on Cinema as Philosophy," 12.
17. Livingston, "Theses on Cinema as Philosophy," 11.
18. Livingston, "Theses on Cinema as Philosophy," 11.
19. Livingston, "Theses on Cinema as Philosophy," 12.

20 I make use of Paisley Livingston's 2006 article "Theses on Cinema as Philosophy" and Graham Parkes's 2009 article "Thinking Images: Doing Philosophy in Film and Video."
21 Livingston, "Theses on Cinema as Philosophy," 13.
22 Livingston, "Theses on Cinema as Philosophy," 13.
23 Livingston, "Theses on Cinema as Philosophy," 15.
24 Parkes, "Thinking Images," 36.
25 Parkes, "Thinking Images," 36.
26 Lohmar, "Non-Linguistic Thinking—From a Phenomenological Point of View," 32.
27 Lohmar, "Language and Non-Linguistic Thinking," 378.
28 Parkes, "Thinking Images," 45.
29 Lohmar, "Language and Non-Linguistic Thinking," 382.
30 Parkes, "Thinking Images," 37.
31 Parkes, "Thinking Images," 38.
32 Parkes, "Thinking Images," 39.
33 Parkes, "Thinking Images," 41.
34 Parkes, "Thinking Images," 44.
35 Parkes, "Thinking Images," 44.
36 Parkes, "Thinking Images," 44.

References

Dick, Bernard F. 1980. *Billy Wilder*. Boston: Twayne Publishers.

Livingston, Paisley. 2006. "Theses on Cinema as Philosophy." *The Journal of Aesthetics and Art Criticism* 64 (1), Special Issue: Thinking through Cinema: Film as Philosophy (Winter): 11–18.

Lohmar, Dieter. 2012. "Language and Non-Linguistic Thinking." In *The Oxford Handbook of Contemporary Phenomenology*, edited by Dan Zahavi, 377–96. Oxford: Oxford University Press.

Lohmar, Dieter. 2017. "Non-Linguistic Thinking—From a Phenomenological Point of View." *Tijdschrift voor Filosofie* 79: 31–56.

Parkes, Graham. 2009. "Thinking Images: Doing Philosophy in Film and Video." *Educational Perspectives* 42 (1 and 2): 36–46.

Part Three

The Practice of Dance

8

Rubble and the Philosopher's Stone: The Practice of Philosophy and the Philosophy of Practice

Jason M. Wirth

A stone woman gives birth to a child in the night.

—Furong Daokai 芙蓉道楷 (Jpn. Fuyō Dōkai) (1043–1118)
(Quoted from Dōgen 2010, 154)[1]

By embracing the work and legacy of Graham Parkes's attentiveness to the strange and almost incomprehensible voice of rocks and stones as intrinsically valuable, we accept an invitation to be more philosophical about what matters as philosophy and how it does so. This likely sounds eccentric, although I argue that such reactions impoverish our sense of the powers and value of philosophical discourse.

Before turning to a discussion of lithophilia, both in general and in the work of Parkes, I begin with two anecdotes, both of which, given the prevailing norms of academic philosophical culture, may also appear eccentric.

I

As I write these words, the human world has come undone. It had been largely tied together through mandatory participation in runaway capitalism, whose "means of production and exchange," as Marx and Engels famously articulated it in the first chapter of the *Communist Manifesto*, are like the "sorcerer who is no longer able to control the powers of the nether world whom he has called up by his spells."[2] The Covid-19 has brought the nether world to a screeching halt. Although I generally subscribe to the writerly convention of avoiding associations with current events, which mercilessly date an essay when it becomes subject to historical retrospection, this event is of a different order. It marks the opening to a new world, and it can also serve as a dry run for the coming ecological crises, including the climate emergency, which will also radically shatter our complacency with the status quo. Whether or not we embrace these new openings, they are manifesting all around us, and through us.

At the time of this writing, Seattle, the city where I live, has largely been shut down except for "essential" services. It speaks to the charm of Seattle's culture that "essential" includes our wine shops, microbreweries, and cannabis outlets, but such transactions must be conducted quickly and with precise social distancing. Most of the time we are in our houses, often glued to our computer screens as we grasp for the vestiges of the former world. We are encouraged to take solitary walks and for me this means a daily walking meditation in the nearby and quite remarkable Kubota Garden. Begun in 1927 by a poor Japanese immigrant, Kubota Fujitarō, on land which he could not legally own in his name, and developed over the years with little money, but much sweat and heart, it is my refuge. Its existence was often precarious, including its abandonment during the wartime incarceration of the Kubota family at Minidoka in the high desert of the Snake River Plain in southern Idaho, despite the fact that Kubota's two sons were serving in the military. When he returned, the garden had been lost. Legend has it that he wept, but then went back to work not only restoring it, but dramatically enhancing it. When he died in 1973, he left his Garden to a culture that had subjected he and his family to unrelenting racism and other forms of discrimination.

Yet the Garden still "speaks," now more than ever. Prominent among its many treasures are its use of well-placed granite rocks mostly sourced from the nearby Cascades. As Parkes notes of the Japanese deployment of rocks in gardens in the context of his translation of the wonderful text on Ryōanji by François Berthier, "we don't normally think of rocks as having language, nor of gardens devoid of vegetation as posing questions, nor of stone as something that can be read, and possessing surfaces reflecting something in the depths of human being."[3]

Following the Japanese tradition, the stones, or as we more customarily say in English, rocks, in a Garden, including Kubota Garden, are unhewn or "untouched." The *kanji* for rock is *iwa* 岩 and the *kanji* for stone is *ishi* 石. In the upper part of the *kanji* for rock we can see mountain (山), indicating that stones remaining in their original mountain location are rocks. Although the stones for a Japanese Garden obviously do not remain in the mountains, they retain the power of rock and are not hewn. For his part, Parkes argues for calling them rocks and not stones. "Although rocks are originally found in nature, unlike stones they remain rocks when moved into a garden. (We speak of 'rock gardens' rather than 'stone gardens')."[4] The relationship of the Germanic *stone* (German *Stein*, Dutch *steen*, Swedish *sten*, and so forth) to the Greek word for pebble, στία, also reinforces the more diminutive status of stones compared to rocks. However, given the ambiguity in the English language and in the sources that I will be considering in this essay between rocks and stones, I do not always adhere to Parkes's admirable strategy. Nonetheless, even when I refer to *stones*, I at least tacitly invoke both their majesty and their pedigree as *rocks* that surge forth from the earth in mountains (in any of their orogenic processes) but also in lakes, oceans, glacial fields, exposed mantle, and so forth. In all of these original manifestations, rocks are subsequently shaped through their interaction with waters (watersheds, lakes, oceans, and so forth).

The *Sakuteiki* (*The Garden Making Record*), the Heian period classic and perhaps the oldest extant book on gardening in the world, instructs the garden architect to learn to *listen* to the stones and follow their "desires." In a section called the "Secret

Teachings on Setting Stones," we are instructed to "choose a particularly splendid stone and set it as the Main Stone. Then, following the request of the first stone, set others accordingly."[5] Parkes is also attuned to this remarkable passage:

> The primary principle to be observed is exemplified by the frequent occurrences of the locution *kowan ni shitagau*, which means "following the request [of the rock]." It is used to encourage a responsiveness on the part of the garden maker to what we might call the "soul" of the stone" … [that is, the] *ishigokoro*, meaning the "heart," or "mind," of the rock. Rather than imposing a preconceived design on the site and the elements to be arranged there, the accomplished garden maker will be sensitive to what the particular rocks "want." If he listens carefully, they will tell him where they best belong.[6]

How does one hear a stone's request? Kubota Fujitarō was a regional pioneer in introducing the usage of stone and Kubota Garden is full of dramatic stone personalities—"spirit stones" as he sometimes called them. To a casual observer, stones epitomize the dull inertia of the earth. In part the vast temporal remoteness from their origin allows us to be oblivious to the violent drama of their birth. Their peaceful state, often millions or even billions of years old, hides the fact that they did not first come into being in peace. Yet they appear utterly still, obdurate, hopelessly opaque. In this way, they are good examples of what Japanese Buddhists call "form." Rocks and stones evolve so slowly that they are ready figures of the almost imperceptibly slow pace of change on a geological scale. Yet they move! Even rocks and stones, seeming masters of holding to their form, are "empty" (*kū* 空). They are not just self-standing rocks and stones, but rather implicated in the interdependent vitality of all beings. When the *Heart Sutra*, the pith of Mahāyāna Buddhist practice, claims that "form is emptiness and emptiness is form," it is counseling us to become aware of their mutual reliance. Emptiness is often associated with water which has no form of its own, but which can take the shape of any form. Listening to stone is the capacity to detect the mutual interdependence of its form and emptiness, much in the way that Dōgen was later to celebrate this in his remarkable *Sansuikyō* (1241).[7] *The blue mountains are always walking*.

Strolling meditatively through Kubota Garden in the time of the coronavirus feels more than ever like a refuge, but not in the sense that it is merely offering an escape or diversion from something unpleasant. The Garden is transformative and serves as a reminder, with the flourishing of the flora and fauna issuing from the streams and these ancient rocky earth sentinels, that a new earth is first and foremost rooted in the most ancient of forces.

II

One of my most prominent memories of Graham Parkes was a slide show that he used to perform in the days before PowerPoint and other such computer extravagances. Using the same kind of slide projector that my teachers had used when I was in grammar school, he would show slides of the great *kare sansui* rock gardens in Japan, including, of course, Ryōanji. At a particularly apt moment, he would try to capture a dramatic, almost psychedelic aspect of the Gardens by rapidly cupping his hand over the projector light to create a kind of strobe effect. I do not know what was more remarkable: these strange images or the fact that he had the chutzpah to do such things in the otherwise staid atmosphere of an academic conference. (It was especially noteworthy to watch him do this at a meeting of the American Philosophical Association!)

Parkes is a careful scholar, reader of philosophy, student of languages, and interdisciplinary pioneer. There is nothing slapdash about his work, but it performs something that Nietzsche was very good at demonstrating: although philosophy of necessity demands hard work and unrelenting diligence, this does not imply that we must take everything so seriously. We habitually confuse the important and the difficult with the weighty. Indeed, the general mood of philosophy in its many European manifestations is what Nietzsche called the spirit of gravity, as if we were all helplessly weighed down by the rocks of our history, current predicaments, and the density and difficulty of the philosophical enterprise. Indeed, in German, the word for heavy, *schwer*, also means difficult.

We are embarrassed if the fruits of philosophical labor seem to soar, or laugh, or dance, as if that would mean that we were not working hard enough, that lightness is just another form of frivolity, that the difficulty of thought must therefore manifest as the weightiness and solemn seriousness of thought. We are consequently increasingly micrological, as if asking questions of sufficient size to be relevant is already too light, too irresponsible. Of necessity we are hard workers, but we have consequently committed the non-sequitur of becoming what Nietzsche's Zarathustra called "the conscientious in spirit [*der Gewissenhafte des Geistes*]." Absorbed in thought, Zarathustra trod on a truth seeker working in the mud, studying leeches—but not the whole leech, mind you, for that is too daunting a task, but just the brain of the leech. "How long I have been chasing after this one single thing, the brain of the leech, so that the slippery truth no longer here slips away from me. Here is my realm! For this I have thrown everything away, for this everything else has become the same."[8]

On the precipice of a new world, we cannot be weighed down entirely by the encumbrance of the old world. This includes a dawning openness derived from a sense of the non-obviousness of what matters as philosophy. Philosophy is a protean gesture, not a skillset and tool kit that we can take for granted. Moreover, the weightiness of our habits of thought and philosophical commitments cannot fundamentally be ascribed to an intellectual failing. Even the most unrepentantly solemn enactment of philosophical micrology requires considerable discipline and intelligence. Our stony gravity operates tacitly, the subterranean inertia of values that elude all but the most determined exercises in self-examination. Yet these subterranean values continue to shape our philosophical sensibility, resulting in the naivete that philosophy is simply something that we bring under our control with hard work and volition, a technique that we can master and practice self-sufficiently.

Although the subterranean dimensions of thinking and living can weigh us down without awareness, coming to terms with them is a practice without which philosophy cannot emerge from its depths. It is consequently a practice of the self that extends the self beyond the ego's delusion of being in charge. But how do I practice what is not merely a task for me to execute? Much as Georges Bataille once lamented, "Woe to those who, to the very end, insist on regulating the movement that exceeds them with the narrow mind of a mechanic who changes a tire."⁹

What better way for philosophy to soar than for it to find the requisite lightness already within the heaviness of what weighs it down? How do rocks teach us to fly when they embody the weighty *par excellence*? It is not that lithophilia is potentially of interest to philosophy, but rather that it is a clue to liberating philosophy as a practice.

III

We now turn to an ignominious pile of little stones, bits of rubble, and broken rocks that command little respect.

Parkes's own lithophilia rendered him sensitive to the many liberatory lithic elements in Nietzsche's thought. Yet despite a lithophilia that Parkes traces to Nietzsche's youth, one of the striking images of stone in *Also Sprach Zarathustra* is Zarathustra's own drive to reduce to rubble the seemingly supratemporal petrification of humanity brought about by Christianity, the "Platonism for the people."¹⁰ Parkes highlights a passage in the second book of *Zarathustra*—"Now my hammer rages fiercely against the prison. Fragments fly from the stone: what is that to me?"—observing that the birth of the *Übermensch* emerges out of the rubble of humanity's eidetic petrification. "Zarathustra, the awakened one, having seen in the raw material of the human the *Vorbild* of the Overhuman, will awaken the image to life by demolishing the prison from the outside, liberating the figure crushed into the dense conformity of stone. There will be fragmentation, pulverizing, suffering in shards and fragments."¹¹ But how does one shatter the heavy and stony *imago* of oneself? How does one turn the hammer on oneself and who, or what, would be wielding it? From within the depth of the self, occluded by our petrified egoistic senses of self, one must find an ancient rock harder than the petrified stone of ourselves. The trick is to get the heaviest and strongest rock of all—the implacable and indestructible granite of fate—to become liberatory rather than mere inertia and dead weight. The granite depths of the psyche shatter the traumatic petrification of the self. "Nietzsche's acknowledgment of the 'anorganic minerals' in the human body is at the basis of his idea that we have within us, psychologically, the firmness of stone: indeed not only 'some granite of spiritual fate' but also 'something that explodes rock.'"¹²

The granite depths of the psyche suddenly and paradoxically appear to dance. As such, Nietzsche's thought, whether he could fully appreciate it or not, resonates with a vital strain of East Asian thought and practice. For example, the great Song Dynasty Caodong (Jpn. Sōtō) Chan Master, Furong Daokai 芙蓉道楷, much to Dōgen's benefit in the *Sansuikyō*, famously proclaimed that "the blue mountains are always walking." Mountains, seemingly implacably and obdurately trapped in their form, are actually empty (in the sense of *śūnyatā*). And in their own way and time, they are dynamic, constantly becoming themselves anew. Mountains are not stuck in themselves, but rather interact and move interdependently with the fluid forces they comprise and that reciprocally comprise them. Parkes also finds this force at play in the great lithophilic traditions of classical Chinese culture. For example, the "17th-century garden manual by Ji Cheng, *The Craft of Gardens* (*Yuanye*), recommends that the rocks used for the peaks of artificial mountains should be larger at the top than below, and fitted together so that 'they will have the appearance of being about to soar into the air.'"[13] Rocks and stone are also expressions of *qi* energy, manifesting the "earth's 'essential energy.'"[14]

What first appear to be the opposite of movement and vitality—rocks, stones, mountains—are shown in their own way to be part of vast ecologies of life, beyond the duality of the organic and the inorganic. "Once the dichotomy between the animate and inanimate is seen as somewhat arbitrary, along with the borders between the animal and vegetal and mineral realms, the well arranged garden can be experienced as a field in which the energies constituting the human body are harmonized with the *qi* of all other inhabitants of the place."[15]

Yet, how did we come to see the distorted limestone formations of the "Grand Lake, Tai Hu," the source of some of the most venerated lithophilic wonders, not as monstrosities, but as awakening *philia*? How do we come to see stones dancing, mountains walking, even the implacable granite depths of the psyche as free and capacious? The default mode, whether it be Zarathustra's heaviness and melancholy, the obduracy of the mountain, the stony silence of the lithic realm, or the implacability of fate, is gravitational—the force that renders being heavy, as the German words for gravity, *die Schwerkraft*, and melancholy, *Schwermut*, poetically attest.

In other words, how do we burst the tyrannical gravity of the petrified into rubble? It would be the height of arrogance and folly to think that this is something we can do on our own steam and volition that we can suddenly become so powerful that we soar forth out of the burdensome weight of our pain, stupidity, delusion, and ideology. Although we can clearly recognize that our current mode of being will soon render life on earth as we know it no longer possible, we nonetheless act with scandalous insufficiency in response. Such inadequacy on our parts demonstrates the weightiness of what we have become. Left to ourselves, we are no longer nimble enough even to survive, let alone soar. We are so stuck in our ways we project our petrification on to capitalism and our ecologically destructive lifeways. As is often repeated these days, we can more easily imagine the destruction of the earth as we know it as well as the demise of our species, and many of the species with whom we share the earth, than we can imagine an alternative to capitalism. Yet, as the coronavirus abruptly brought the netherworld of capitalism to a halt, what seemed necessary suddenly appeared up for grabs. It is an opportunity for the hammer to reduce our stubborn stoniness to rubble. Yet what can rubble do? Is it not just a great fall from our prior heights?

In his contemporary classic, *Shin Buddhism: Bits of Rubble Turn into Gold*, Taitetsu Unno traced the titular reference to a poem by the great Tang Dynasty defender of Pure Land Buddhism, Huiri 慧日 (680-748). Refusing any hierarchy of the worthy and the unworthy, and merely evoking the abundant proclamation of the *nianfo* (Jpn. *nembutsu*),[16] the Amida Buddha vows that "I can make bits of rubble turn into gold" (Unno, 12). Two things stand out from this edict.

First, was not Nietzsche right when he prefaced the second edition (1887) of *The Gay Science* with his admonition that we should "laugh at every master who does not first laugh at himself"?[17] We awaken to a sense of ourselves as what Jōdo Shinshū and other Pure Land traditions call *bonbu* 凡夫, foolish beings. This dawning wisdom of our foolishness includes an awareness of the arrogance of having given ourselves all of the credit for our awakening, even the wisdom of knowing ourselves to be a hapless *bonbu*. We are broken shards of rock, hapless rubble, and the arrival of the Pure Land is not our doing. Although Huiri was a vehement defender of Pure Land devotionalism against its dismissal by the Chan school (he thought they were arrogant), this did not mean that in rejecting Chan he rejected meditation as such.[18] In fact, he valued it. We can see that both Taitetsu Unno and his son, Mark Unno, cite Dōgen as well as Suzuki Shunryū, the legendary founder of the San Francisco Zen Center, especially, given our present discussion, the latter's emphasis on beginner's mind (*shoshin* 初心): "In the beginner's mind there are many possibilities, but in the expert's mind there are few."[19] This resonates with Henry David Thoreau's awakening to what he called the "infinite expectation of the dawn,"[20] the wakefulness that does not take the world for granted, but which encounters each minute of it as if it were the first minute.

Although Huiri had to defend the practice of the *nianfo* (ritualistic evocation of the Amitabha Buddha) from its dismissal by Chan schools, the recitation of the primal vow and the practice of meditation were not always mutually exclusive, like they later tended to be regarded in Japan. Pushed to its limits, Chan practice is not simply something that one can write off as the elective power of the Path of the Sage, and hence a manifestation of self-power, *jiriki*, while the path of the Pure Land, the way of the *bonbu*, is simply opposed to it as a matter of other power, *tariki*. Just as one needs to resolve to practice Zazen, one must also in Pure Land practice resolve oneself to iterate the *nianfo* (Ch.), or *nembutsu* (Jpn.), that is, reciting the name of the Amitabha Buddha, again and again. And just as one cannot be arrogant or otherwise self-pleased by one's capacity to adhere to the primal vow, Dōgen was clear from his first writing (*Fukanzazengi*) that to practice Zazen, one should have "no mind on becoming a Buddha." Both the Nembutsu and Zen are practices of self-overcoming. But we can also hear that for Nietzsche—no advocate of the existence of an atomistic self ("all is will to power")—the eternal return was its own kind of practice.[21] Nietzsche's practice of self-overcoming was not a technique by which an actor achieves a goal. It is to undergo an overcoming of oneself from within oneself.

This self-overcoming is like realizing a mountain is always walking, or that a rock stretches to the skies, no longer subject to the exclusive tyranny of gravity locking it within its form. One might even remember in this context that the *kanji* for emptiness or *śūnyatā*, *kū* 空, also means the heavens or skies. Emptiness is the obdurate and weighty not exclusively trapped in the weight of its form. It also entails form reaching for the heavens or standing out against the field or horizon of the heavens. It is to realize the weighty form of the rock in its emptiness and therefore in its living and dynamic ecologies of interdependence. The mountain is not only earthbound and self-bound, but simultaneously includes its heaven-originating and heaven-bound dimension.

Second, we once again, as Nietzsche and Parkes advocate, "attend to the images of stone that occur in our dreams and fantasies."[22] The stony realm of recalcitrance and gravitational inertia reveal their alchemical quality as the philosopher's stones of the earth. Parkes has done seminal work in alerting us to the Asian spiritual resonances between Nietzsche and East Asian Buddha Dharma,[23] indeed, with East Asian practices as such. Parkes perspicaciously traces the lithophilic resonances between the two, as obdurate stones lift to the heavens, mountains constantly walk, and little pieces of rubble turn to gold. It was, after all, "by a magnificent pyramidal rock on the shore of the lake of Silvaplana that Nietzsche was first struck by the thought of eternal recurrence in the summer of 1881."[24]

In his *Composing the Soul: Reaches of Nietzsche's Psychology*, Parkes takes up a powerful 1882 note from Nietzsche to Franz Overbeck in which he confesses how difficult his life has become and that "unless I discover the alchemists' trick of turning this filth into *gold*, I am lost. I have here the *perfect* opportunity to show that for me 'all experiences are useful, all days holy, and all humans divine'!!!"[25] By what philosophical alchemy do we transform from rubble and filth into gold? From the perspective of the latter, we also see that rubble all along was gold. It is here that we can hear the celebrated words of Yunmen Wenyan (Jpn. Ummon Bun'en) in case six of the *Blue Cliff Record* (Jpn. *Hekiganroku*): *every day is a good day*.[26]

I think it is fair to say that if the practice of philosophy is limited to the production of athletic exercises in discursivity—as valuable as that can be in a larger context—then Yunmen's words are inaccessible. No such arguments will penetrate the gravity of human life, including the somber and exhausted machinations of much academic philosophical culture. Yet by transforming the icons of density and heaviness—the lithic realm—we can detect an avenue to a new sense of philosophy as a practice of overcoming. Philosophy becomes simultaneously an explication of that practice and its enactment.

Furong Daokai also proclaimed that *a stone woman gives birth to a child in the night*. Mountains, like beings, are in some manner barren. They do not give birth to themselves. The illusion of their implacable solidity arises from the misconception that a mountain always stays a mountain, and a rock is always stuck being a rock. Yet examining this more closely, a mountain does not remain a mountain by giving birth to itself every minute. The newness of the mountain—that it gives birth to a child, a new mountain, in the night, that is, mysteriously and interdependently, not because it follows from itself—means that a mountain does not manifest trapped within its form. Its emptiness, its lack of self-being, is its evolving interactivity with all the causes, conditions, and other interdependent beings with which it dynamically shares its being. A mountain is a mountain through the "passageless-passage"[27] of a kind of self-overcoming. A mountain is not a mountain merely by continuing to be a mountain. A mountain is a mountain ever anew. Dōgen called this the *continuous practice* (*shugyō*) of mountains and in it we find vital clues to the *practice* of philosophy.

Notes

1. Credit: Zen Master Dogen, excerpt from *Treasury of the True Dharma Eye: Zen Master Dogen's* Shobo Genzo, edited by Kazuaki Tanahashi. Copyright © 2010 by the San Francisco Zen Center. Reprinted by arrangement with The Permissions Company, LLC on behalf of Shambhala Publications Inc., Boulder, Colorado. www.shambhala.com.
2. Marx, *The Communist Manifesto*, 10.
3. François Berthier, *Reading Zen in the Rocks: The Japanese Dry Landscape Garden*. Translated and with a Philosophical Essay by Graham Parkes (Chicago: University of Chicago Press, 2000), viii.
4. Berthier, viii.
5. Sakuteiki: *Visions of the Japanese Garden*, 184.
6. Berthier, 111–12.
7. A critical literature has developed that challenges the prevalent identification of Japanese gardening principles with Zen practice. Granting the general point that the former has been disproportionately associated with the latter, it remains true that, whether or not rock gardens (*kare sansui*) were developed according to Zen principles, they are nonetheless de facto of immense value and relevance to Zen practice. Nevertheless, the range and power of Japanese gardening culture clearly exceed an exclusive reliance on Zen, or Japanese Mahāyāna practice more broadly, even if they rightly find something valuable in all such gardens. Kubota Fujitarō, for example, was a devout practitioner of a relatively new sectarian form of Shinto called Konkōkyō. For more on this remarkable Garden, including its relationship to Konkōkyō, see Ellen Phillips-Angeles and Jason M. Wirth.
8. Nietzsche, 307–8.
9. Bataille, 26.
10. Nietzsche, 3.
11. Graham Parkes, "Nietzsche's Care for Stone: The Dead, Dance, and Flying," in *Nietzsche's Therapeutic Teaching for Individuals and Culture*, edited by Horst Hutter and Eli Friedland (London and New York: Bloomsbury Academic, 2013), 181.
12. Parkes, "Nietzsche's Care for Stone," 189.
13. Graham Parkes, "Thinking Rocks, Living Stones: Reflections on Chinese Lithophilia," *Diogenes* 207. 52 (3) (2005): 76.
14. Parkes, "Thinking Rocks, Living Stones," 81.
15. Parkes, "Thinking Rocks, Living Stones," 84.
16. The *nembutsu* is the ritualistic invocation of the Amitabha Buddha, the Buddha of Infinite Light. In Japanese, one repeatedly chants "Namu Amida Butsu," hail to the Amitabha Buddha. It derives from the eighteenth-century vow in the so-called *Larger Sutra*, or the *Sutra on the Buddha of Immeasurable Life* as it is formally known.
17. Friedrich Nietzsche, *Kritische Studienausgabe*. Edited by Giorgio Colli and Mazzino Montinari (Munich and Berlin: Deutscher Taschenbuch Verlag and Walter de Gruyter. Volume 3, 1980), 343.
18. Suzuki, 17.
19. For more on Huiri's confrontation with Chan, see David W. Chappell, "From Dispute to Dual Cultivation: Pure Land Responses to Ch'an Critics," in *Traditions of Meditation in Chinese Buddhism*, edited by Peter N. Gregory (Honolulu: University of Hawaii Press, 1986), especially 169–74.

20 Thoreau, *Walden*, 96.
21 For more on the relationship of Nietzsche to East Asian Buddha Dharma, see Jason M. Wirth, *Nietzsche and Other Buddhas: Philosophy after Comparative Philosophy* (Bloomington: Indiana University Press, 2019).
22 Parkes, "Nietzsche's Care for Stone," 189.
23 Parkes's many influential works that bring Nietzsche in relationship to East Asian Buddhist thought include "Nature and the Human 'Redivinized': Mahāyāna Buddhist Themes in *Thus Spoke Zarathustra*," in *Nietzsche and the Divine*, edited by John Lippitt and Jim Urpeth (Manchester: Clinamen Press, 2000), 181–99; "Nietzsche and East Asian Thought: Influences, Impacts, and Resonances," in *The Cambridge Companion to Nietzsche*, edited by Bernd Magnus and Kathleen M. Higgens (Cambridge: Cambridge University Press, 1996), 356–83; "Nietzsche and Zen Master Hakuin on the Roles of Emotion and Passion," in *Nietzsche and the Gods*, edited by Weaver Santaniello (Albany: State University of New York Press, 2001), 115–34; "Nietzsche, Panpsychism and Pure Experience: An East-Asian Contemplative Perspective," in *Nietzsche and Phenomenology*, edited by Andrea Rehberg (Newcastle upon Tyne: Cambridge Scholars, 2011), 87–100; "Open Letter to Bret Davis: Letter on Egoism: Will to Power as Interpretation," *The Journal of Nietzsche Studies* 46 (1) (Spring 2015): 42–61; "Reply to Bret Davis: Zarathustra and Asian Thought: A Few Final Words," *The Journal of Nietzsche Studies* 46 (1) (Spring 2015): 82–8; and the classic edited volume, *Nietzsche and Asian Thought* (Chicago and London: University of Chicago Press, 1991).
24 Graham Parkes, *Composing the Soul: Reaches of Nietzsche's Psychology* (Chicago and London: The University of Chicago Press, 1994), 136.
25 See also Parkes, "Nietzsche's Care for Stone," 185: "Just as the alchemists strove to transform base metals into gold, the depth-psychologist in Nietzsche has to transform the base dross of experience into something invaluably radiant." Parkes, *Composing the Soul*, 166.
26 See Katsuki Sekida, translator with commentaries. *Two Zen Classics*: Mumonkan *and* Hekiganroku. Edited by A. V. Grimstone (New York and Tokyo: Weatherhill, 1977), 161–6.
27 This is the endless becoming of mountains and waters, being and time, that Dōgen called *kyōryaku*, which Abe Masao expansively translates as "passageless-passage." All of being is Buddha-nature, but Buddha nature is the temporal conditions of time. This is the truth of the impermanence of Buddha-nature: not that things are finite, but that they are ceaselessly emptied by time. As Abe explains, because "being and time are identical in terms of the manifestation [*genzen*] of the Buddha-nature" (Abe, 88), mountains walk, and the stone woman gives birth in the night.

References

Abe, Masao. 1992. *A Study of Dōgen: His Philosophy and Religion*. Edited by Steven Heine. Albany: State University of New York Press.
Bataille, Georges. 1991. *The Accursed Share*. Translated by Robert Hurley. New York: Zone Books.

Berthier, François. 2000. *Reading Zen in the Rocks: The Japanese Dry Landscape Garden*. Translated and with a Philosophical Essay by Graham Parkes. Chicago: University of Chicago Press.

Chappell, David W. 1986. "From Dispute to Dual Cultivation: Pure Land Responses to Ch'an Critics." In *Traditions of Meditation in Chinese Buddhism*. Edited by Peter N. Gregory, 163–97. Honolulu: University of Hawaii Press.

Dōgen. 2010. *Dōgen's Shōbōgenzō: Treasury of the True Dharma Eye*. Edited by Kazuaki Tanahashi. Boston and London: Shambhala.

Marx, Karl and Engels, Friedrich. 2002. *The Communist Manifesto*. Translated by Samuel Moore. New York: Penguin Classics.

Nietzsche, Friedrich. 1966. *Beyond Good and Evil*. Translated by Walter Kaufmann. New York: Random House.

Nietzsche, Friedrich. 1968. *Also Sprach Zarathustra, Nietzsche Werke: Kritische Gesamtausgabe*, Division six, Vol. 1, 307–8. Edited by Giorgio Colli and Mazzino Montinari. Berlin: Walter de Gruyter.

Nietzsche, Friedrich. 1980. *Kritische Studienausgabe*, Vol. 3. Edited by Giorgio Colli and Mazzino Montinari. Munich and Berlin: Deutscher Taschenbuch Verlag and Walter de Gruyter.

Parkes, Graham. 1994. *Composing the Soul: Reaches of Nietzsche's Psychology*. Chicago and London: The University of Chicago Press.

Parkes, Graham. 2005. "Thinking Rocks, Living Stones: Reflections on Chinese Lithophilia." *Diogenes* 207. 52 (3): 75–87.

Parkes, Graham. 2013. "Nietzsche's Care for Stone: The Dead, Dance, and Flying." In *Nietzsche's Therapeutic Teaching for Individuals and Culture*, edited by Horst Hutter and Eli Friedland, 175–90. London and New York: Bloomsbury Academic.

Phillip-Angeles, Ellen and Wirth, Jason M., eds. 2019. *Spirited Stone: Lessons from Kubota's Garden*. Seattle: Chin Music Press.

Sekida, Katsuki, translator with commentaries. 1977. *Two Zen Classics*: Mumonkan *and* Hekiganroku. Edited by Albert V. Grimstone. New York and Tokyo: Weatherhill.

Suzuki, Shunryū. 1971. *Zen Mind, Beginner's Mind*. New York and Tokyo: Weatherhill.

Tachibana, no Toshitsuna. 2001. *Sakuteiki: Visions of the Japanese Garden*. Translated by Jirō Takei and Marc P. Keane. Boston, Rutland, VT, and Tokyo: Tuttle Publishing.

Thoreau, Henry David. 2006. *Walden*. Edited by Jeffrey S. Cramer. New Haven and London: Yale University Press.

Unno, Taitetsu. 2002. *Shin Buddhism: Bits of Rubble Turn into Gold*. New York: Doubleday.

Wirth, Jason M. 2019. *Nietzsche and Other Buddhas: Philosophy after Comparative Philosophy*. Bloomington: Indiana University Press.

9

Philosophy as Petromania: Graham Parkes, Jane Bennett, and the (Not So) New Materialisms

Leah Kalmanson

In her Preface to *Vibrant Matter: A Political Ecology of Things*, Jane Bennett sets out to challenge the "habit of parsing the world into dull matter (it, things) and vibrant life (us, beings)."[1] A prominent voice among the so-called new materialists, Bennett says, "I will turn the figures of 'life' and 'matter' around and around, worrying them until they start to seem strange, in something like the way a common word when repeated can become a foreign, nonsense sound. In the space created by this estrangement, a *vital materiality* can start to take shape."[2] But, in the Chinese tradition at least, we need no such incantations to conjure the space for this vital materiality. Indeed, common English translations of *qi* (氣)—for example, "vital stuff," "psychophysical stuff," and "lively material"—take us close to Bennett's key concept and titular phrase. In this regard, JeeLoo Liu stresses that "[*qi*] is the *stuff* of animate and inanimate things alike."[3] Or as Daniel K. Gardner explains, "psychophysical stuff is the matter and energy of which the entire universe and all things in it, including functions and activities of the mind, are composed. It is the relative density and purity of each thing's psychophysical stuff that gives the thing its peculiar form and individual characteristics."[4] Roger Ames and David Hall describe *qi* as a "vital energizing field and its focal manifestations."[5]

We can compare this to Bennett's description of the "field" of vital materiality where "portions congeal into bodies … [,] an ontologically diverse assemblage of energies and bodies, of simple and complex bodies, of the physical and physiological."[6] In short, *qi* provides a framework that accommodates many of the moves that Bennett wishes to make—*qi* disrupts the divide between inert matter and "vibrant life" without being reducible to vitalism, animism, or panpsychism, as these theories are conventionally understood. Here I wish to complicate the lineage in which Bennett situates her own work and, by extension, the lineage of key figures in Bennett's book, such as Jacques Rancière and Bruno Latour, who have influenced the loosely defined fields of object-oriented ontology, speculative realism, and so-called new materialisms.

This alternative lineage includes Graham Parkes's work on Chinese "lithophilia" or "petromania," which takes seriously the practice of *fengshui* (風水)—literally, wind (*feng* 風) and water (*shui* 水)—as a kind of environmental diplomacy that can aid us in addressing the current ecological crisis. Such practices as *fengshui* range over various Chinese traditions, including both Daoism and the lineage of the Ru (儒), the latter term referring to the scholars or "literati" commonly known as "Confucians" in English. Here, we focus on Parkes's work in relation to certain advanced notions of *qi* that grew to prominence in the Song (960–1279) and Ming (1368–1644) dynasties under the influence of different fields of Ruist philosophy, including the *lixue* (理學) or "*li*-studies" of Zhu Xi (1130–1200) and what has come to be called the *qixue* (氣學) or *qi*-studies of an earlier figure Zhang Zai (1020–77).[7] Despite differences in emphasis, both fields attend to the principles or patterns (*li* 理) observable in the behavior of *qi*, and both are deeply invested in developing the appropriate contemplative and bodily practices required to align personal, social, and environmental patterns for optimal health and wellbeing. By engaging this Ruist philosophical heritage in the context of contemporary environmental issues, Parkes offers not only theoretical interventions but programs of practice for enacting better ecologies.

Bennett concludes her book with some reflections on the difficulties that she sees in implementing her "vital materialism" at the level of policy or practice. So, on the one hand, a comparative study of Parkes and Bennett on materiality can augment Bennett's work with this practical component. On the other hand, my ultimate focus is less comparative and more polemical. Clinton Godart has written: "When speaking about 'Asian philosophy,' the burden of proof is placed on the Asian traditions. Questions are posed such as 'was Confucianism philosophy,' not 'was Hegel a Confucianist' or 'did he complete the Way?' Thus Westernization has created a cultural imbalance of categories and representations."[8] In this essay, I ask: Is "new materialism" a type of *qi*-studies? This question does not ignore the cultural context in which *li*- and *qi*-studies developed, but it does recognize that Euro-American philosophies are not the only ones to presume that their major terms and categories have cross-cultural scope. By rejecting the idea that the burden of proof rests on the Asian or comparative philosopher to justify this application of *qixue*, we productively intervene in the academic practices that continue to define the "new" on Eurocentric models alone.[9]

Materialism as *Qi*-Realism

Parkes has devoted a number of works to the philosophical investigation of rocks and stones, from Zen rock gardens to the role of stone in Friedrich Nietzsche's writings.[10] Here we focus only on several essays that address physical matter in terms of *qi*. In "Thinking Rocks, Living Stone: Reflections on Chinese Lithophilia," Parkes directs attention to the long history of rock collecting and rock gardening in China: "The introduction to the 12th-century treatise by Du Wan, the *Cloud Forest Catalogue of Rocks*, which appears to be the world's first handbook of rock aesthetics, begins with the statements: 'The purest energy of the heaven-earth world coalesces into rock …. Its formations are wonderful and fantastic.'"[11] This "purest energy" is hence the most potent *qi*. Throughout the essay, Parkes refers to the general idea that all forms of physical matter are condensed *qi*, but rocks and stones are especially vibrant and fantastical concentrations. Immediately we see that the heft and rigidity of the stone do not indicate inertness but rather dynamism and power—the stone is not immobile but *unyielding*; its form is *decisive*; its living force is demonstrated precisely by its resolute materiality.

Parkes continues: "Since the human body is understood as a different configuration of the same energies, it is reasonable to assume that beneficial effects will flow from simply being in the presence of such rocks. The rock garden thereby becomes a site not only for aesthetic contemplation but also for self-cultivation and the enhancement of physical health."[12] This claim makes sense in the highly resonant world of *qi*, where the "vital stuff" of mental awareness and the "vital stuff" of the body and physical environment all mutually influence and respond to each other. Here, there are no fundamental differences between sentient and insentient, organic and inorganic, physical and biological, and so forth; there are only different configurations of *qi* at microcosmic and macrocosmic levels, which can, under optimal circumstances, be productively attuned to each other.

I borrow JeeLoo Liu's term "*qi*-realism" to account for this resonant activity of mind and matter. As Liu says, under *qi*-realism: "1. *Qi* is permanent and ubiquitous in the world of nature. There is nothing over and above the realm of *qi*. 2. *Qi* is real in virtue of its causal power. It constitutes everything and is responsible for all changes."[13] This *qi*-realism, as a framework for understanding the behavior of matter itself, helps guard against the accusation of anthropomorphism. As Parkes goes on to comment, "some readers may think that such unbridled enthusiasm for stone evidences a prevalence of primitive animism or, more charitably, anthropomorphic projection. Nothing could be farther from the truth."[14] Animism, as that term is used today, is still defined along the distinction between, in Bennett's words, "dull matter" and "vibrant life."[15] That is, the animist is thought to mistakenly imagine vibrant life where these is none and hence to incorrectly attribute an animating force to material objects.

Unlike animism, *qi*-realism does not attribute animate qualities to inanimate objects; rather, *qi*-realism helps us avoid the animate/inanimate distinction altogether. As Parkes explains, "By contrast with the western tendency to make a sharp distinction between the animate and inanimate, with rocks falling on the lifeless side of the divide, the ancient Chinese understand all natural phenomena, including humans, as configurations of an energy they call *qi*."[16] That is to say, *qi* is not the animating force somehow "inside" inert matter; rather, as the above discussion has already shown, animating forces (such as the mind) and physical matter (such as stones and rocks) are all different configurations of "vital stuff." To repeat Liu's definition, *qi* "constitutes everything and is responsible for all changes," and hence no additional "spirit" is needed to account for phenomena such as animation and sentience.

In this way, *qi*-realism is also distinct from panpsychism and vitalism, as these are conventionally understood. Bennett locates her precedents in "critical vitalists" such as Henri Bergson and Hans Driesch, whose notions of *élan vital* or entelechy "came very close to articulating a vital materialism."[17] But, as she says "they stopped short: they could not imagine a *materialism* adequate to the vitality they discerned in natural processes. (Instead, they dreamed of a not-quite-material life force.)"[18] As should be clear by now, *qi* is not a "not-quite-material life force" distinct from material objects; to the contrary, as dynamic concentrations of "vital stuff," material objects are themselves "lively" even when not combined with non-material phases of *qi*. As Roger Ames and David Hall explain, "*Qi* has to be distinguished from either 'animating vapors' or 'basic matter' because it cannot be resolved into any kind of spiritual-material dichotomy."[19]

Bennett sees similarities between her idea of vital materiality and certain "historical senses of the word *nature*" not as a "stable substrate of brute matter" but as an exuberant, generative fecundity, which she finds in thinkers such as Baruch Spinoza, Alfred North Whitehead, Nietzsche, and the American transcendentalists.[20] However, this undercurrent of *natura naturans* in Euro-American intellectual history proves elusive:

> Even if, as I believe, the vitality of matter is real, it will be hard to discern it, and, once discerned, hard to keep focused on …. What is more, my attention will regularly be drawn away from it by deep cultural attachments to the ideas that matter is inanimate and that real agency belongs only to humans or to God.[21]

Bennett is perhaps correct to say that vital materiality will be obscure to us if we remain immersed in Eurocentric cultural traditions alone. But, in Chinese traditions, *qi* informs the dominant paradigms—far from being elusive, it is impossible to miss.

A New Heritage for the New Materialists

As mentioned above, during the Song and Ming dynasties, theories of *qi* became increasingly complex. Questions were posed such as: Given the creative potency of *qi*, why does it configure itself into the world as we know it, as opposed to other possible arrangements? Do certain principles or patterns (*li* 理) govern the behavior of *qi*? If so, then what is the best approach to studying these *li*? Some philosophers seemed to suggest that *li* are governing principles that do exist on their own and can be studied as such. Others asserted that *li* have no independent existence but only describe the tendencies inherent in *qi* itself—this is the position of Liu's *qi*-realists. The famed philosopher Zhu Xi takes a position somewhere in between, speaking of the mutual dependence of the two terms: "In the universe there has never been any psychophysical stuff without principle nor any principle without psychophysical stuff [天下未有無理之氣, 亦未有無氣之理]."[22] Although Zhu Xi came to be known for promoting the method of *lixue* or *li*-studies, some scholars place Zhu's own position much closer to *qi*-realism, separating him from later developments in Chinese philosophical historiography that sought to neatly differentiate the various *qi*- and *li*-focused methods. For example, in *The Natural Philosophy of Chu Hsi*, Yung Suk Kim claims that, for Zhu, *li* as a general principle does not exist independently of *qi* at all; in this sense, the term functions not causally but descriptively: "*li* has little additional content beyond the object or phenomenon of which it is the *li*. In a sense, *li* is very much like a definition."[23] Only later, says Kim, did *li*-studies become associated with the idea that *li* can be treated as an independent principle.[24]

As noted above, Bennett roots her work in the "critical vitalism" of figures such as Driesch and Bergson. Much like the *li*-focused and *qi*-focused philosophers, these vitalists debated whether "vital force could have any existence apart from the bodies in which it operated."[25] Although these theorists come close to accounting for the vibrancy of matter in the way that Bennett envisions, ultimately, "for them becomings include a moment of transcendence in the form of *élan vital* or entelechy."[26] We should note, however, that *lixue* and *qixue* relocate the terms of this debate: rather than investigating the relation between matter and non-material forces, these fields investigate the different configurations that arise out of vital stuff, whether these are material, non-material, or somewhere in between. This places us very close to where Bennett wishes to be, that is, in "a universe of this lively materiality that is always in various states of congealment and diffusion, materialities that are active and creative without needing to be experienced or conceived as partaking in divinity or purposiveness."[27] In the field of *qi*, non-material *qi* is not simply a force that invigorates material *qi*, but rather all the various configurations that are vigorous—and as such, they can interact in any number of creative and novel ways. In this sense, the relation between *li* and *qi* is not analogous to the relation between *élan vital* and matter, and, especially for the *qi*-realists, *li* is not "a moment of transcendence." As Liu explains:

> What separates Neo-Confucian *qi*-realists from the school of Zhu Xi is their insistence on the status of *li* as the inherent, not transcendent, principle of *qi*'s movements. *Li* is not a formal cause of *qi* as Zhu Xi depicts the relationship; there is also no top-down determination from the realm of *li* to the realm of *qi*. The fluctuation of *qi* itself has inevitability, or we might say, an internal logic. This internal logic is the principle (*li*) inherent in *qi*, and it is both the regulative principle of how things *are* in nature and the normative principle to which humans *ought to* conform.[28]

I wish to suggest that Bennett, in picking her way through the thicket of materialism and vitalism in Euro-American thought, has unknowingly veered close to strands of discourses whose roots are in China's Song dynasty and whose branches extend throughout East Asia (the Four-Seven debates and Horak debates of Joseon-period Korea, for example, and the various responses to *lixue* philosophy in Tokugawa-period Japan). She seeks not only a philosophical approach to overcoming the conventional dualisms we associate with the Western tradition, but also a political approach to the way things *ought* to be on the field of vital materiality.

Fengshui as Politics of Wind and Water

Qi-realism can provide Bennett this normative lens on her political agenda. "Why advocate the vitality of matter?" she asks; "Because my hunch is that the image of dead or thoroughly instrumentalized matter feeds human hubris and our earth-destroying fantasies of conquest and consumption."[29] She aims to radically re-envision the political ecosystem to include "nonhuman agencies," suggesting that we "devise new procedures, technologies, and regimes of perception that enable us to consult nonhumans more closely, or to listen and respond more carefully to their outbreaks, objections, testimonies, and propositions."[30] Here, she credits Bruno Latour and his theory of "actants" as laying the groundwork for her vision of "*distributive* agency" in the field of vital materiality.[31] But, again, I wish to complicate her lineage by locating a precursor in Parkes's studies of *fengshui* (風水) in the context of ecological crisis.

The practice of *fengshui* informed almost every building project in China up to the nineteenth century and remains in common use today in China and the larger East Asian world. Parkes quotes at length from the book *Everyday Life in China* by the missionary Edwin Joshua Dukes (1847–1930), in which Dukes expresses his disdain for the practice. As Parkes describes, "in discussing the way *fengshui* involves the whole community (including its dead) in the planning of any new building project, the author waxes so ironical that his attempt to condemn unintentionally commends (at least to the ecologically sensitive reader of today)."[32] The passage from Dukes runs as follows:

> It will suggest itself at once to the reader that if we ignorant European outsiders were to live where we choose in China, to build as we like, to make roads and railways, to erect telegraph posts, to quarry stone wherever we saw any to our fancy, to delve recklessly into the bowels of the earth for coal, we should, in the opinion of the Chinese, be like 'a maniac scattering dust' and 'a fury slinging flame.' … No vengeance would be too dire to execute upon the rash mortal who could disregard the interest of his fellow-creatures in such a manner.[33]

Reflecting on the current situation, Parkes comments that "one can only regret that economic pressures in a rapidly modernizing China have pretty much destroyed this traditional inclination to take into account the neighbors—living and dead, animate and inanimate—when deciding where and how deeply to encroach upon the environment."[34]

Returning to Bennett's project, we might wonder whether we need "new procedures, technologies, and regimes of perception" to attune ourselves to nonhuman agencies, or whether we simply need greater awareness of longstanding and readily available systems of infrastructure planning such as *fengshui*. Practitioners of *fengshui* possess techniques for managing the often-competing interests of humans, animals, plants, minerals, environmental features, and the non-material configurations of *qi* known as ghosts (*gui* 鬼) and spirits (*shen* 神). *Fengshui* is a political negotiation—a kind of ecological diplomacy—requiring the utmost sensitivity, precision, and care.

Such diplomatic techniques also inform the construction and maintenance of the rock gardens so popular in many parts of East Asia. When done properly, *fengshui* produces a space where all material agencies flourish in health. Parkes expresses the sentiment in this way: "Once the dichotomy between the animate and inanimate is seen as somewhat arbitrary, along with the borders between the animal and vegetal and mineral realms, the well-arranged garden can be experienced as a field in which the energies constituting the human body are harmonized with the *qi* of all other inhabitants of the place."[35] We need to keep in mind, however, that *fengshui* is not a matter of restoring lost harmony, which is to say, the field of *qi* is not necessarily human-centered and does not naturally configure itself in ways conducive to human flourishing. On the contrary, *qi* can just as easily harm or destroy us. To avoid negative outcomes, the spaces of cities, towns, homes, gardens, and so forth all require careful diplomacy in the planning stages and careful, daily maintenance once established. In this way, the politics of *fengshui* is amenable to Bennett's pragmatic approach to conflict:

> For while every public may very well be an ecosystem, not every ecosystem is democratic. And I cannot envision any polity so egalitarian that important human needs, such as health or survival, would not take priority …. The political goal of vital materialism is not the perfect equality of actants, but a polity with more channels of communication between members.[36]

Again, she credits this idea to what Latour calls a "vascularized" collective, and indeed Latour's work is widely influential (and deserves more attention in the field of Asian philosophy). One of my goals in this short essay is a more vascularized picture of our collective academic lineages, where lines of influence can not only challenge the usual modern-postmodern narrative but can be traced outside the orbit of that narrative altogether.

Vitalism as Practice

If we take *fengshui* as one relevant practice that emerges from this vascularized lineage, then we may next ask how a person becomes a skilled practitioner. Within her philosophical and political projects, Bennett poses fundamental questions about the quality of human life among all the material agencies that contribute to the world we live in. She spends time reviewing Driesch's use of *élan vital* to explain how organisms come to have the form that they do: "Driesch describes this directing power inside the organism as a kind of gatekeeping function: entelechy decides which of the many formative possibilities inside the emergent organism become actual. In (what will come to be known as) the stem cells of the sea urchin, for example, there is 'an enormous number of possibilities ….'"[37] As Bennett quotes him, Driesch observes the potential of these stem cells and asks if "something else *can* be formed than actually is formed, why then does there happen in each case just what happens and nothing else?"[38] Today, we know that not *élan vital* but the stem cell's genetic code is responsible for directing its development. In this sense, we have a scientific answer to the question of why the stem cell behaves the way it does, but we do not have an existential solution. Out of the primordial field of formless *qi*, why does our cosmos emerge? Why this cosmos, and not some other? If we can understand the *li* of the *qi* of reality as we know it, then we will better understand what it means to live, as human beings, in these conditions.

Crucial to all the strands of Song and Ming philosophical discourse is the claim that self-cultivation—to be exact, the cultivation of the *qi* of the human heart-mind (*xin* 心)—is necessary to the study of *li*. Principle (*li*) will not be discerned if the heart-mind is in a state of disarray. Analogous to the well-tended garden or the well-planned city, the *qi* of the heart-mind must be developed and maintained. It is at this place of understanding that Parkes's "petromania" has the most to contribute to Bennett's project and, by extension, to other new (or, somewhat new) materialisms. Parkes concludes his essay "Winds, Waters, and Earth Energies: *Fengshui* and Sense of Place" not with a series of claims about *fengshui* or a series of arguments to prove its validity, but with an overview of practices traditionally thought to develop the sensibilities of the well-trained *fengshui* practitioner. In other words, he exhorts us not to "believe in" *qi* but to cultivate its potency for ourselves. The preparatory *fengshui* practices he enumerates include studying the *qi*-manipulation techniques of the best landscape painters, learning small-scale gardening methods (in preparation for more complex environmental diplomacy on a larger scale), and practicing physical exercises to increase the potency of the *qi* of our own bodies and minds.[39]

In his thinking, Parkes is aligned with East Asian traditions broadly where philosophy has always been most crucially about practices. It is no wonder that Zhu Xi rooted his Song-era philosophical and pedagogical reforms upon a chapter of the *Book of Rites* (*Liji* 禮記) known as the "Great Learning" ("Daxue" 大學), which informs us that proper alignment throughout the family, the state, and the whole world begins with mental orderliness (心正) and self-cultivation (身脩).[40] According to Zhu, students cannot hope to understand the meaning of a classical text, let alone the commentarial tradition, without engaging in certain body-mind practices to settle and attune the *qi* of the heart-mind: "Now, if you want to read books, you must first settle the mind to make it like still water or a clear mirror [今且要讀書，須先定其心，使之 如止水，如明鏡]."[41] Certain techniques, such as deep breathing, quiet-sitting (*jingzuo* 靜坐), and textual recitation, will have a positive effect on the *qi* of the heart-mind and hence aid in scholarship. Moreover, in the lineage of the Ru, scholarly activity itself is a type of *qi*-manipulation technique, which molds the heart-mind and increases its potency. The larger goal is not simply the comprehension of principle (*li*) via scholarship, but the enactment of philosophy in daily life such that principle pervades the conditions under which we live. The regulated heart-mind, the properly tended garden, the well-oriented city, the peaceful society, and the vigorously healthy landscape all mutually reinforce each other through a positive feedback loop, as it were, in and through the activity of *qi*.

At the conclusion of *Vibrant Matter*, Bennett raises the question of how to enact the ecological reform she sees as the outgrowth of philosophical re-orientation around her new materialism. She even poses the question in the same vein as the *qixue* and *lixue* philosophers: "Are there more everyday tactics for cultivating an ability to discern the vitality of matter?"[42] Although she worries about the risks of "superstition," she suggests that we might hesitantly indulge in anthropomorphism, that is, to allow ourselves to believe in the sentience of things as a way to overcome the old habits of modernity and its dualisms. In response to her suggestion, Graham Parkes and the Song-Ming philosophers would all point out in their own ways that *habits* cannot be countered with *beliefs*. Rather, reforming our habits will require the adoption of better practices, and the Chinese tradition is nothing if not an extraordinarily rich repository of practices for cultivating the persons who can cultivate better worlds.

Bennett ends her book with a declaration of belief, or what she calls "a kind of Nicene Creed for would-be vital materialists":

I believe in one matter-energy, the maker of things seen and unseen. I believe that this pluriverse is traversed by heterogeneities that are continually doing things. I believe it is wrong to deny vitality to nonhuman bodies, forces, and forms, and that a careful course of anthropomorphization can help reveal that vitality, even though it resists full translation and exceeds my comprehensive grasp. I believe that encounters with lively matter can chasten my fantasies of human mastery, highlight the common materiality of all that is, expose a wider distribution of agency, and reshape the self and its interests.[43]

Readers of East Asian philosophies in general will know that philosophical discourse is sometimes practiced by swapping poems—a kind of non-Socratic dialogue in verse. In the spirit of such exchange, I offer Bennett's creed the following reply from the Warring States text *Guanzi*:

If you mold your *qi* to become like spirit [搏氣如神],
Your grasp of all things will be complete [萬物備存].
Can you mold it [能搏乎]? Can you focus it [能一乎]?
Without resorting to divination, can you foresee fortune and misfortune [能無卜筮而知吉凶乎]?
Can you stop [能止乎]? Can you hold back [能已乎]?
Rather than seeking it in others, can you find it in yourself [能勿求諸人而得之己乎]?
Think about it! Think about it [思之思之]!
Again, think about it [又重思之]!
If you think about it but do not reach it [思之而不通],
Ghosts and spirits will help you reach it [鬼神將通之].
This is not because they possess brute strength [非鬼神之力也].
It is because the purest *qi* goes the farthest [精氣之極也].[44]

Notes

1 Jane Bennett, *Vibrant Matter: A Political Ecology of Things* (Durham: Duke University Press, 2010), vii.
2 Bennett, Vibrant Matter: *A Political Ecology of Things*, vii.
3 Stephen C. Angle and Justin Tiwald, *Neo-Confucianism: A Philosophical Introduction* (Cambridge: Polity Press, 2017); Daniel K. Gardner, translation and commentary, *Learning to Be a Sage* by Chu Hsi (Berkeley: University of California Press, 1990); Philip J. Ivanhoe, *Three Streams: Confucian Reflections on Learning and the Moral Heart-Mind in China, Korea, and Japan* (Oxford: Oxford University Press, 2016); JeeLoo Liu, "In Defense of Qi-Naturalism," in *Chinese Metaphysics and Its Problems*, edited by Chenyang Li and Franklin Perkins (Cambridge: Cambridge University Press, 2013), 33.
4 Gardner, *Learning to Be a Sage*, 90.
5 Roger Ames and David Hall, trans., *Dao De Jing: A Philosophical Translation* (New York: Ballantine Books, 2003), 61.

6   Bennett, *Vibrant Matter*, 117.
7   I thank Sarah Mattice for helping me navigate the appropriate alternatives to the term "neo-Confucianism."
8   Gerard Clinton Godart, "'Philosophy' or 'Religion'? The Confrontation with Foreign Categories in Late Nineteenth Century Japan," *Journal of the History of Ideas* 69 (1) (January 2008): 71–91.
9   Parts of my discussion of Bennett's work appearing here were revised and included in my essay "Speculation as Transformation in Chinese Philosophy: On Speculative Realism, 'New' Materialism, and the Study of *Li* (理) and *Qi* (氣)" (Leah Kalmanson, "Speculation as Transformation in Chinese Philosophy: On Speculative Realism, 'New' Materialism, and the Study of *Li* (理) and *Qi* (氣)," *Journal of World Philosophies* 3 (1) (2018): 17–30).
10 See, for example, "The Role of Rock in the Japanese Dry Landscape Garden" and "Nietzsche's Care for Stone: The Dead, Dance, and Flying."
11 Graham Parkes, "Thinking Rocks, Living Stone: Reflections on Chinese Lithophilia," *Diogenes* 52 (3) (2005): 78.
12 Parkes, "Thinking Rocks, Living Stone: *Reflections on Chinese Lithophilia,*" 81.
13 JeeLoo Liu, "The Is-Ought Correlation in Neo-Confucian Qi-Realism: How Normative Facts Exist in Natural States of Qi," *Contemporary Chinese Thought* 42 (1) (Fall 2011): 60–77.
14 Parkes, "Thinking Rocks, Living Stone," 78.
15 Bennett, *Vibrant Matter*, vii.
16 Parkes, "Thinking Rocks, Living Stone," 78.
17 Bennett, *Vibrant Matter*, 63.
18 Bennett, Vibrant Matter: *A Political Ecology of Things*, 63.
19 Ames and Hall, trans., *Dao De Jing*, 61.
20 Bennett, *Vibrant Matter*, 119.
21 Bennett, Vibrant Matter: *A Political Ecology of Things*, 119.
22 See passage numbered 6 at http://ctext.org/zhuzi-yulei/1/zh#n586150. I consulted Gardner, trans. 1990, 90. Zhu Xi, *Zhuxi yulei*, in *Chinese Text Project*, edited by Donald Sturgeon. 2011. http://ctext.org/zhuzi-yulei/zh.
23 Yung Suk Kim, *The Natural Philosophy of Chu Hsi 1130–1200* (Philadelphia: American Philosophical Society, 2000), 26.
24 Kim, *The Natural Philosophy of Chu Hsi 1130–1200*, 27
25 Bennett, *Vibrant Matter*, 66.
26 Bennett, Vibrant Matter: *A Political Ecology of Things*, 93.
27 Bennett, Vibrant Matter: *A Political Ecology of Things*, 93.
28 Liu, "The Is-Ought Correlation in Neo-Confucian Qi-Realism," 65.
29 Bennett, Vibrant Matter: *A Political Ecology of Things*, ix.
30 Bennett, Vibrant Matter: *A Political Ecology of Things*, 108.
31 Bennett, Vibrant Matter: *A Political Ecology of Things*, viii–ix.
32 Parkes, "Thinking Rocks, Living Stone," 85.
33 Qtd. in Parkes, "Thinking Rocks, Living Stone," 85.
34 Parkes, Graham. 2005. "Thinking Rocks, Living Stone: Reflections on Chinese Lithophilia," 85.
35 Parkes, Graham. 2005. "Thinking Rocks, Living Stone: Reflections on Chinese Lithophilia," 84.
36 Bennett, *Vibrant Matter*, 104.
37 Bennett, *Vibrant Matter*, 72.

38 Qtd. in Bennett, *Vibrant Matter*, 72.
39 Graham Parkes, "Winds, Waters, and Earth Energies: *Fengshui* and Sense of Place," in *Nature across Cultures: Views of Nature and the Environment in Non-Western Cultures*, edited by Helaine Selin (Dordrecht: Springer, 2003), 202–6.
40 "Daxue" 大學, in *Chinese Text Project*, edited by Donald Sturgeon. 2011. https://ctext.org/liji/da-xue/zh.
41 See passage numbered 12 at https://ctext.org/zhuzi-yulei/11/zh. I consulted Gardner, trans. 1990, 145. Zhu Xi, *Zhuxi yulei*.
42 Bennett, *Vibrant Matter*, 119.
43 Bennett, Vibrant Matter: *A Political Ecology of Things*, 122.
44 *Guanzi* 2011, "Nei Ye," passage 6, https://ctext.org/guanzi/nei-ye#n48609. I would like to thank Jennifer Liu for her help in translating this passage in the context of my contribution to her 2021 guest-edited special issue of the *Journal of the Pacific Association for the Continental Tradition* (see Leah Kalmanson, "A Visit to the Local God: Reclaiming the Diversity of the Divine," *Journal of the Pacific Association for the Continental Tradition* 4 (2021): 22–44). Other versions of this translation appear in W. Allyn Rickett, *Guanzi: Political, Economic, and Philosophical Essays from Early China*, vol. 2 (Princeton: Princeton University Press, 1998), 50–1; Harold D. Roth, *Original Tao: Inward Training and the Foundations of Taoist Mysticism* (New York: Columbia, 1999), 82; and Robert Eno's version available copyright-free at http://www.indiana.edu/~p374/Inner_Enterprise.pdf.

References

Ames, Roger and Hall, David, trans. 2003. *Dao De Jing: A Philosophical Translation*. New York: Ballantine Books.

Angle, Stephen C. and Tiwald, Justin. 2017. *Neo-Confucianism: A Philosophical Introduction*. Cambridge: Polity Press.

Bennett, Jane. 2010. *Vibrant Matter: A Political Ecology of Things*. Durham: Duke University Press.

Chu Hsi [Zhu Xi]. 1990. *Learning to Be a Sage*. Translated and with a commentary by Daniel K. Gardner. Berkeley: University of California Press.

"Daxue" 大學. 2011. In *Chinese Text Project*, edited by Donald Sturgeon. https://ctext.org/liji/da-xue/zh.

Gardner, Daniel K., translation and commentary. 1990. *Learning to Be a Sage*. Edited by Chu Hsi. Berkeley: University of California Press.

Godart, Gerard Clinton. 2008. "'Philosophy' or 'Religion'? The Confrontation with Foreign Categories in Late Nineteenth Century Japan." *Journal of the History of Ideas* 69 (1) (January): 71–91.

Guanzi. 2011. In *Chinese Text Project*, edited by Donald Sturgeon. http://ctext.org/guanzi/nei-ye.

Ivanhoe, Philip J. 2016. *Three Streams: Confucian Reflections on Learning and the Moral Heart-Mind in China, Korea, and Japan*. Oxford: Oxford University Press.

Kalmanson, Leah. 2018. "Speculation as Transformation in Chinese Philosophy: On Speculative Realism, 'New' Materialism, and the Study of *Li* (理) and *Qi* (氣)." *Journal of World Philosophies* 3 (1): 17–30.

Kalmanson, Leah. 2021. "A Visit to the Local God: Reclaiming the Diversity of the Divine." *Journal of the Pacific Association for the Continental Tradition* 4: 22–44.
Kim, Yung Suk. 2000. *The Natural Philosophy of Chu Hsi 1130–1200*. Philadelphia: American Philosophical Society.
Liu, JeeLoo. 2011. "The Is-Ought Correlation in Neo-Confucian *Qi*-Realism: How Normative Facts Exist in Natural States of *Qi*." *Contemporary Chinese Thought* 42 (1) (Fall): 60–77.
Liu, JeeLoo. 2013. "In Defense of Qi-Naturalism." In *Chinese Metaphysics and Its Problems*, edited by Chenyang Li and Franklin Perkins, 33–53. Cambridge: Cambridge University Press.
Parkes, Graham. 2000. "The Role of Rock in the Japanese Dry Landscape Garden." In *Reading Zen in the Rocks: The Japanese Dry Landscape Garden*, edited by François Berthier, translated by Graham Parkes, 85–155. Chicago: University of Chicago Press.
Parkes, Graham. 2003. "Winds, Waters, and Earth Energies: *Fengshui* and Sense of Place." In *Nature across Cultures: Views of Nature and the Environment in Non-Western Cultures*, edited by Helaine Selin, 185–205. Dordrecht: Springer.
Parkes, Graham. 2005. "Thinking Rocks, Living Stone: Reflections on Chinese Lithophilia." *Diogenes* 52 (3): 75–87.
Parkes, Graham. 2013. "Nietzsche's Care for Stone: The Dead, Dance, and Flying." In *Nietzsche's Therapeutic Teaching for Individuals and Culture*, edited by Horst Hutter and Eli Friedland, 175–90. London: Bloomsbury.
Rickett, W. Allyn. 1998. *Guanzi: Political, Economic, and Philosophical Essays from Early China*, Vol. 2. Princeton: Princeton University Press.
Roth, Harold D. 1999. *Original Tao: Inward Training and the Foundations of Taoist Mysticism*. New York: Columbia.
Zhu Xi. 2011. "Zhuxi yulei." In *Chinese Text Project*, edited by Donald Sturgeon. http://ctext.org/zhuzi-yulei/zh.

10

A Mind Possessed: The Rhythmic Body and Entraining the Social Mind

Bradley Douglas Park

Here is the prefiguring of the thinker as dancer, who sees through the illusion of firm ground but avoids plunging into the abyss by cultivating a lightness of foot that is responsive to the rhythms of natural phenomena.[1]

The body occupies a special place in the ways of thinking and being that is Graham Parkes. A core commitment to the body, and by extension to the earth, guides the philosophical movement of his thinking and clearly informs his various engagements with Nietzsche, Japanese Buddhism, Daoism, and ecology. In one way or another, his thought circles back through the body. This return marks an effort to critically retrieve corporeality from the oblivion of thought and the wispy metaphysics of spirit, which have dominated so much of the Western philosophical tradition. As a teacher, Parkes's insistence on the constitutive significance of the body has profoundly shaped my own thought path, such that the body has become the central thematic of my own thinking.

In his speech "On the Despisers of the Body," Nietzsche's Zarathustra offers a teaching that can serve as a touchstone for Parkes's philosophical orientation and a shared point of contact with the likes of Nietzsche, Dōgen, and Zhuangzi:

> Body am I through and through, and nothing besides; and soul is merely a word for something about the body. … But the greater thing—in which you do not want to believe—is your body and its great reason: it does not say I, but does I. … Behind your thoughts and feelings, my brother, stands a mighty commander, an unknown wise man—his name is Self. In your body he dwells, he is your body. There is more reason in your body than in your finest wisdom.[2]

The body roots us to the earth, and not as an incidental function of mass; rather, for Nietzsche, the body is a configuration of drives expressing an ancient legacy of life upon the earth and a prehistoric dance with gravity. This history of living inscribed within the body's drives, this bio-logos as it were, roots us ineluctably into the earth and weaves us into a community of other bodies—mineral, vegetal, animal, and human— within which we find ourselves and which is found within us. The body embeds us in a situation from within, and thereby radically in touch with the reality around us. Thus, for Nietzsche, the standpoint of a healthy body yields a clear and direct relation to itself and its place: "Listen rather, my brothers, to the voice of the healthy body: a more honest and purer voice than this. More honestly and purely does the healthy body talk, being complete and four-square and it talks of the sense of the earth."[3] For Nietzsche, it is only the sick body, the decadent body, the nihilistic body that needs to distance its "self" from the immediate reality of *this* flesh and the corresponding world in which it finds itself. This distancing is enacted through the simultaneous positing of a *metaphysical self* (for example, an eternal soul) along with an *Other-world* serving as the soul's true and proper home. This distance is then thoroughly purified via the absolute metaphysical concepts of substance, identity, and permanence. As a consequence, the lived reality of the breathing, sweating, and bleeding human being is philosophically abstracted and reduced to the point of view of the representational mind—the *cogito*.

The motivation for enacting this distance, the desire for this kind of transcendence, ultimately expresses an existential rejection of the reality of time, change, and death. If one is going to insist on the self's permanence through time and the reality of a future state, then it follows that the spirit must be metaphysically annexed from its everchanging body. Nietzsche's counter-insistence, that one is "body through and through, and nothing besides," needs to be read as containing within it an affirmation of time, change, contingency, and death.

The body is essentially finite and temporal. But it is not simply passive vis-à-vis the flow of time; it is not a mere patient within the flux. The body generates its own time. It gives rise to timing, to the temporality of response, from out of the depths of its physical-mechanical structure and the endogenous dynamics of its metabolic and nervous systems.[4] And while our bodies are living (dissipative) systems persisting far from equilibrium through the operational closure enacted by biology, the intrinsic timings of the body constitute new horizons of sensitivity—new edges of influence and perturbation. It is the body's rhythms, its innate temporal biases, that define its aboriginal openness to world, which is to say, its propensity toward entangling itself with other temporalities.[5] The body is a field of resonance in an open and pulsing dance with its environment. It continually coordinates its timing in relation to other dynamic processes across a wide array of time scales, and from highly local to global features of its situation. Our body solicits entrainment and then calibrates itself through this history. Indeed, it is these basic temporal entrainments that make mind possible at all. Our being is rhythmic, and thus our body-mind dances.

In his book, *Composing the Soul*, Parkes engages a passage from Plato's *Ion* that nicely contrasts the reflective Apollonian mind to the immersion of the Dionysian dancers and the lyric poets, who are not in their "right minds" by virtue of becoming possessed by rhythm.

> Just as those carried away by Corybantic frenzy are not in their right minds when they dance, so also the lyric poets are not in their right minds when they make these fine songs of theirs. But when they launch into melody and rhythm, they are frantic and possessed, like Bacchic dancers who draw honey and milk from rivers when they are possessed but cannot when they are in their right minds.[6]

Despite the intimate clarity of action and purpose granted by such rhythmic possession, "right minds" are equated with the distanced lucidity that attends reflective self-awareness. Through the entrainment of rhythm, the dancer and the poet sink beneath the lucidity of an aloof representational consciousness and into the darker origins of an embodied self, a somatic consciousness, that is intimately and pre-reflectively immersed in its environment. When in deep solidarity with its situation, the body draws and drafts from its milieu, receiving nourishment, energy, inspiration, "honey and milk" from the river-like flow of its Bacchic possession.

Nietzsche's insistence on the "great Reason" of the body, then, represents a Heraclitean and Dionysian challenge to the Platonic conception of the "right mind" and the philosophical priority given to self-consciousness:

> The features Nietzsche mentions as characteristics of Dionysian music, whether the dithyrambic chorus or the overtures of Wagner, are likewise melody and rhythm. However, while Socrates' saying that the lyric poet is "inspired and out of his mind" and "without intelligence." (*Ion* 534b–d) is a criticism intended to demonstrate the limitations of the poet's powers, the Nietzsche of *The Birth of Tragedy* follows Schopenhauer and Wagner in extolling the unconscious inspiration and deprecating reflection and conscious intellection.[7]

The reflective brightness of this Apollonian perspective provides for a sharply individuated experience, a phenomenology of separation, containment, of inhibition, and of staying within boundaries: self-control as self-negation. But for Nietzsche, Apollonian-Platonic "right mind" represents a reactive and nihilistic mind, a mind decadently turned against itself, a mind self-consciously neutralized by the heavy reticence of its own reflection. In contrast, the dark buoyancy of the Dionysian body flows spontaneously along the contours of well-attuned instincts and timely movements. The freedom of its dance leaps forth from the feeling of its own power and clarity, and not from a blinking paralysis before boundaries and moments of decision. A body-mind thoroughly at home in its situation, Nietzsche's "healthy body," follows the sway of world in an intimacy that (re)minds us *in* our animality.

> But what helps one cross light-footedly, stepping instinctively and without calculation, is perhaps less the assistance of a god than imagination aided by a kind of somatic memory: the recollection of a natural inborn ability inherited from the ancestors—and, ultimately, from our animal ancestors—what later Nietzsche will call our "preschooling" in animality. In contrast to the disembodied *anamnēsis* of Plato, this kind of recollection is a memory of *the body*, the recall of abilities embodied there.[8]

Importantly, the poet and the dancer know the way forward in their frenzy and possession; they relinquish the self-negation of the executive ego and release themselves into an ecstatic coupling with their situation through time, which is to say, a Dionysian lust for intimacy, resonance, and the blurring of boundaries.

To look deeply into Nietzsche's insistence on the intrinsic wisdom of the body, and its thoroughgoing temporality, is to see a radically rhythmic structure disposed to dynamically couple and coordinate with other rhythmic processes. Further, the body's rhythmic structure constitutes the primal mode of openness underlying the emergence of mind, precisely because it is this disposition toward dynamic coupling, particularly toward interbodily entrainment, that enables our most aboriginal forms of social imitation and coordination. These basic forms of social coordination are then essential for bootstrapping a mind capable of higher-order representational thought. Thus, contra Plato, it is not that the Dionysian mind of the rhythmic body is anathema to our more Apollonian modes of intelligence; rather, the rhythmic body constitutes the very ground of its possibility as the ontological and developmental foundation for "right mind."

Part One: Externalism and Dynamic Coupling

Central to the philosophical view presented herein is (1) a refusal to frame such foundational coordinative processes in *informational* terms, and (2) a refusal to rely on explanatory strategies committed solely to *centralized, internal* processes of neural regulation. The concept of information (and surrogate notions, such as content, representation, idea, computation) and contemporary commitments to brain-centrism remain continuous with core modernist assumptions, and ultimately represent "rebranded" expressions of an underlying substance ontology. To the extent that such commitments remain unchallenged, theory-construction in the sciences remains uncritically in the sway of a modernist ontology and a philosophically untenable view of the subject.[9] More pointedly with respect to the problem of information, the central objection is that information itself is a higher-order concept and, as such, is inconsistent with a strictly naturalistic explanation of mind. Information is not a basic and natural physical feature of our world, a feature with straightforward causal efficacy that can be part of a coherent naturalistic explanation from the ground up.[10]

Stated more positively, the view herein is motivated by the kind of anti-Cartesian externalism articulated via notions, such as Heidegger's "being-in-the-world" or Merleau-Ponty's "flesh," and should be read as an effort to develop these lines of thinking in relation to current empirical data about our embodiment and sociality. The hope is that this interdisciplinary approach can offer a finer-grained analysis of such externalism that not only provides more clarity and specificity, but one that can also be compelling to a broader sweep of disciplinary perspectives.

A constitutive element in bodily coordination is timing. And while the mainstream view presumes that timing is centrally determined, in truth, the temporality of the body is not reducible to a master internal clocking mechanism inside the skull. Timing, rather, is distributed across the body, dancing upon its very surface, and pervading its deepest structures. Oscillatory processes of dynamic coupling, such as entrainment and resonance, offer alternative ways to think about the primordial openness of the rhythmic body that do not rely on informational exchange.[11] *Dynamic coupling* refers to causal influence, often complex and subtle, between systems, wherein the state of one system affects the time-evolution of others. In such coupling, there is discernible causal influence revealed through time, that is, through systematic correlations in the dynamic changes across coupled systems, even when the "lines" of this causal influence are often obscured behind a tangle of nonlinear and recursive causal loops. *Entrainment* specifically describes the process "by which two oscillating systems, which have different periods when they function independently, assume the same period, or integer-ratio related periods, when they interact."[12] *Resonance* functions across several domains—mechanical-acoustical resonance, electrical resonance, optical resonance, orbital resonance, and molecular resonance—and is ubiquitous wherever vibrations or waves are involved. Mechanical-acoustical resonance represents the most paradigmatic instance, wherein a periodic (oscillating) force is transmitted to another system at a frequency that approaches the resonant (natural or *eigen*) frequency of that system, thus stimulating the system to increase the amplitude of oscillation at this natural frequency, that is, to sympathetically resonate. More technically, the correct periodic timing (oscillatory frequency) of the transfer of energy from the stimulating system can dramatically reinforce and amplify the oscillation of the sympathetic system when it coincides with its own periodic conversion between potential and kinetic energy. This is to say that a small nudge can have great effect, if the nudge is timely.

Indeed, if the very notion of informational-representational content is inconsistent with naturalism, then any naturalistic account of mind cannot presuppose the concepts of representation or information. The resonant picture of interbodily entrainment, in contrast, does not depend on the notion of "content" and therefore offers a more viable avenue for constructing a rigorously naturalistic account of the emergence of mind.

Part Two: The Problem of Meaning (or Getting Intentionality off the Ground)

Meaning is not grounded metaphysically via transcendent essences qua Plato, nor is it to be found simply lying among the facts of the world. Sociopragmatist responses to the question of meaning, such as those proffered by the classical Confucians, early Heidegger, and the mature Wittgenstein, converge around the idea that meaning is produced in and through the social transmission of practices.[13] That is, they yield an ontology wherein "what something is" is defined by the way that these practices convey the *importance* of things, or even better, *how they matter* within these practices. This "mattering" governs our pre-philosophical understanding of the possibilities implicitly defining the sense of our situation, or more specifically, an understanding of how the things in our situation can interface with our bodies and be relevant to our tasks.

It is at this point that the Confucian-Heideggerian-Wittgensteinean story is furthered by the work of Maurice Merleau-Ponty, whose thinking about motor-intentionality and the body schema deepens the explication of our being-in-the-world. Or to put the point differently, Merleau-Ponty clarifies the fact that Heidegger's pre-philosophical understanding does not subsist as an abstract mental phenomenon nor in some nebulous general will of *das Man*; rather, it resides in the sensorimotor habits of our socialized-socializing bodies.

Describing such understanding as pre-philosophical (or pre-ontological or pre-thematized or pre-objective) is to say that the inculcation of our deepest and most constitutive practices is never a top-down "mental" process whereby we can autonomously survey habituation as it unfolds. Rather, it is this understanding that orients us as such, and thus gives us a personal view and a situation. Indeed, it is only through the bodily inculcation of these practices that an autonomous mind can arise at all. Primitive forms of motor intentionality reach out from a body coming to grips with its world long before more abstract modes of intentionality become projected ahead of a self-transparent cogito. These primitive forms ground the more abstract intentional capacities scaffolded in and through representational structures. The bodily incorporation of basic practices and their attending normative dispositions represent a cultural process of transmitting mind, that is, of training a body to have a mind.

If this basic line of reasoning is correct—that social practices provide a shared normative structure sufficient to give rise to original intentionality (shared preunderstanding; meaning) and that the bodily incorporation of these normative structures is the process of coming to have a mind (or, and it amounts to the same thing, to having a world)—what remains of the question of meaning beyond Merleau-Ponty's account? We arrive at a different question: How is it the case that a *new* body can be influenced by a *practiced* body, such that mind can be transmitted? The short answer is to say that it is via an innate propensity of new bodies to *imitate* the ways of practiced bodies. But then our question becomes—what makes imitation possible?

This question is nontrivial since we do not have recourse to any shared meaning in the construction of our account of imitation, since it is precisely the origin of shared meaning that the appeal to imitation seeks to explain. Thus, pointing toward shared perceptual orientation, for example, is just to smuggle shared meaning into perception and to further beg the question. In psychological circles, this issue of shared meaning in perception is discussed in terms of the development of "joint attention." Joint attention does not mean that you and I merely happen to be attending to the same object; rather, it denotes the fact that this process of attending is, in part, laterally co-structured by a shared awareness that *we* are attending to the same object. In other words, seeing *with* another is to share a meaningful world. It is to *see from* the same world, and not merely to *look at* the same objects. Indeed, it is telling that the psychological literature around joint attention clearly recognizes the two intersecting philosophical problems at the core of shared perception: the problem of other minds and the problem of meaning.[14] Hence, for many developmental psychologists, the capacity for joint attention has come to serve as an obvious criterion for what it means to have achieved a fully mature, competent mind. But if this is the case, then joint attention cannot be used to account for imitation.

In a previous engagement with the Daoist notion of *ganying* ("sympathetic resonance"), I argued that Daoists view the body-mind as a *resonant field*.[15] This perspective offers a productive alternative to container metaphors of mind with their corresponding dependency on notions of isolated content or information. To frame the mind as a resonant field is to insist that our body is a sensitive, resonant, and open structure through time. Building on that framing of mind, the central thought being argued for here is that the processes giving rise to the most elementary forms of imitation across bodies ought to be conceived in terms of resonant relationships, wherein dynamically coupled bodies systematically covary across time in ways that produce shared significance. Our bodies have evolved to dynamically couple with their environment, particularly with other human bodies. It is via a first-order history of bodily covariance, that is, imitation and social coordination with caregivers, that the shared norms underlying attention, intentionality, meaning, and mind are introduced, practiced, and inculcated. It is via our competent participation, initiative, and improvisation within these socially constituted norms of perception and behavior that renders us into a "who," and thereby gains us entrance into a second-order world of meaning, sense, and information wherein X can come to mean Y for *someone*.

Part Three: The Rhythmic Body

To begin appreciating how the social mind emerges from processes of rhythmic entrainment, it is first necessary to think the body in an adequate manner. This sketch of the rhythmic body starts from the specific composition and mechanical dynamics of the musculoskeletal body, which are fundamental in shaping the body's timing. In the latter half, I turn to contemporary psychological and developmental literature to help flesh out this sketch and further explicate the emergence of the social mind through interbodily entrainment.

At the center of this view of the rhythmic body is an appreciation of our skeletal structure as a nested series of fixed length pendulums. To take this pendula structure seriously is to recognize that each pendulum comes with its own timing structure determined by its length. This intrinsic periodic length is also referred to as a pendulum's *natural frequency* or *eigenmode*.[16]

Through my own training as a musician to generate a strong rhythm out of my body, I have come to appreciate that the rocking movement of the body column constitutes the foundational pendulum or oscillator of the skeleton, while the primary limbs of the body—themselves pendulums suspended from the spinal column—are further composed of smaller and smaller pendulums (or modes).[17] For example, forearm movements anchored at the elbow mark a shorter pendulum or faster oscillator nested within the larger rhythmic frame of the rocking torso, while the micro-movements and angular adjustments at the wrist and fingertips constitute very short, quick pendulums allowing for even finer grained timing.

Moreover, these fine motor structures are coupled to the larger macro-movements of the torso in resonantly reinforcing phase relationships. In other words, the body's vertical column serves as the fundamental frequency while the peripheral limbs, outward to the fingers, function as faster mode oscillators standing in positive harmonic relations to the fundamental frequency. These positive harmonic relations can then preserve and focus the distribution of mechanical energy through the mode shape of the entire system. As a result, there emerges a vertical stability and integration of the nested rhythmic structure that might be called the body's *eigenmode*. This mode shape of rhythmic coherence is wholly different from the mental task of trying to tether fine movements of the hand to inner representations of time through acts of mental counting, for example. This grounding of timing in the body's concrete movements and our proprioceptive grip on the relative phase relationships of these nested pendulums underwrites the common description of timing as *feel*. Feel denotes a holistic kinesthetic sense of our dynamic comportment constituting the phenomenological background of performance, and it grounds the first-person expressiveness of musical timing, phrasing, emphasis, and intonation. Feel describes the temporal pulse of our torso-limb synchrony, as well as the corresponding possibilities for regulating the flow of mechanical energy through the body from core to periphery—the flow of force animating a musician's distinctive sense of *touch* at the point of contact with the instrument.

In *Ways of the Hand: The Organization of Improvised Conduct*, Paul Sudnow provides a phenomenological chronicling of his learning to improvise on piano. A key moment of insight in that process is when Sudnow learns to imitate the distinctive phrasing of the jazz pianist, Jimmy Rowles. His insight speaks directly to the significance of bodily feel and to the primacy of the torso in organizing the prosody of the body:

> For months, night after night I'd watch him move from chord to chord with a broadly swaying participation of his shoulders and entire body. I'd sympathetically feel him delineate waves of movement, some broadly encircling, other subdividing the broadly undulating strokes with finer rotational ones, so that as his arm reached out to get from one chord to another it was as if some spot on his back circumscribed a very small figure at the same time, as if at slow tempos this was the way to bolster a steadiness to the beat .... I found that for getting a song to feel like his, his observable body idiom served as a roughly useful guide. In the very act of swaying gently and with elongated movements, the lilting, stretching, almost oozing quality of his interpretations could be at least vaguely evoked.[18]

In this passage, Sudnow chronicles how the slower, fundamental frequency pattern originating in the spinal column constitutes the "broad" frame "delineating" and "encircling" the "finer rotational" movements of the hand. In other words, there exists a definite dynamic-mechanical coordination between the undulating "spot on his back," which lines the whole rhythmic activity, and the reach of his hand. And while he does not explicitly characterize the point in terms of the nested pendulum structure of the body, Sudnow comes very close to this view at times: "In an anchored heel, you could see only the up-and-down movements of the foot, but in the accompanying head rotation and shoulder swaying, you could see a circularly undulating flow of motion, a pushing and releasing, a thrust and relaxation."[19] Sudnow notes that the snappy movements of the foot (that is, the heel pendulum) at the periphery emanate from the core movements of the "head and shoulder swaying" (that is, the spinal pendulum), which excites the total "undulating flow of motion." The timing of the hand issues from the torso precisely because the displacement of the body column in gravity energizes the entire rhythmic pulse of the activity by codetermining finer movements downstream. A master musician like Rowles has learned to exploit the affordances of gravity by allowing the mechanical energy generated from shifting the body's center of mass to shape the spatial, temporal, and force contours of peripheral movements. The shaping of these contours is concretely realized through the specific biomechanics of the body, wherein the particularities of bone length, muscle and tendon elasticity, joint angles, and so forth are all relevant to the ways in which punctuated displacements of core mass ripple through the body. Exploiting gravity, however, is not only about the efficient use of energy; rather, the systematic correlations between core shifts and peripheral contours can, as Sudnow realizes, be integral to the rhythmic signature underlying the unique "voice" of a musician. It was precisely in the context of trying to emulate Rowles's distinctive phrasing—the "lilting, stretching and almost oozing quality of his interpretations"—that Sudnow came to notice the significance of his back for the expressiveness of his hands.

Sudnow's insights about the rhythmic relationship between body core and periphery is anticipated in Abby Whiteside's classic work, *On Piano Playing: Indispensables of Piano Playing–Mastering the Chopin Etudes and Other Essays*. The pedagogical foundation of Whiteside's method is the cultivation of "basic rhythm" through the whole body:

> But the one fact must be repeated incessantly, because so many mistakes have been made on this score, is that the center controls the periphery; it can never be the other way around. The body governs the fingers in playing piano, and no amount of coaching in finger dexterity will ever lead to the easy beauty in playing that must be our objective. The fingers in themselves have no power of coordination. The *body* must be taught, and the fingers will find their way under the guidance of this central control.[20]

Whiteside holds that the "center controls the periphery" and that the "body governs the fingers" just as Sudnow noted how the undulating center circumscribes the phrasing of the hands. But this center is neither an informational nor an executive center. Rather, Whiteside points to the displacement of body mass as holistically regulating movement, power, and expression: "Rhythm stems from the point of resistance to the application of power. It creates its magic by a follow-through activity which involves a balancing of weight of the entire body. The point of resistance when we are on our feet is the floor; when we are seated it is the chair seat."[21] Here, Whiteside is simply tracing the flow of energy from foundation to fingertips, but elsewhere she explicitly locates the fount of this energy in the pendular mechanics of the seated body column:

> The torso, which contacts resistance at the chair seat with the ischial bones, and is the fulcrum which makes the power of the top arm effective, never sits back stodgily and lets the arm do the work, as it were. Rather the torso is so vitally balanced that it participates in all the actions of the arm, and creates an outlet for the emotional response to the music.[22]

Finally, it is important to stress that this core movement of the torso is, according to both Sudnow and Whiteside, the root of *emotional expressiveness*; it is not just a brute metronymic regularity keeping time for an eloquent periphery, but a vital pulse implicating the whole person in the prosody of affect.[23]

Part Four: Oscillation vs. Information

Research into beat induction (that is, rhythm perception) and rhythmic coordination further support the idea that the perception of rhythm, and the corresponding capacity to entrain our timing with respect to rhythmic affordances, is not explicable solely in terms of a centralized process of neural regulation.[24] Rather, it is a process that *constitutively* involves externalization, precisely because it is dependent upon the musculoskeletal structure of our bodies and the specific dynamics and history of our movement. Rhythmic perception is a sensorimotor process grounded in the specificity of our body. Indeed, not only is rhythmic coordination a sensorimotor process, but it is also directly tied into vestibular perception and balance on the vertical plane. The pendular-oscillatory structure of our embodiment, which has been developing through the musical analyses of Sudnow and Whiteside, has thus far been centered on the generation, distribution, and control of our *own* bodily timing. But this pendular-oscillatory structure is also implicated in the perception of extra-bodily rhythms, as well as in our capacity for entraining our timing with this exteriority—a rather unique ability that may be limited to a handful of species that engage in vocal mimicry.[25]

In their discussion of beat induction, Neil Todd and Christopher Lee contrast a centralized neural-clock model and their sensorimotor approach that pointedly underscores the difference in theoretical presuppositions:

> Such models [neurodynamic] ultimately create a mind-set which looks inwards within the brain for autonomous cells or anatomical structures which have the property of a clock ... In contrast the sensory-motor [sic] approach creates a mind-set which looks outwards to the interaction of the body with the environment and to the distributed representation of the body and its environment within the brain.[26]

More specifically, "beat induction and temporal tracking" are taken to be "form[s] of sensory-guided action" that are "mediated by the motion of an internal representation of the musculoskeletal system."[27] One does not need to accept Todd and Lee's focus on "internal representation" to appreciate the force of this contrast. What is important about their view is that bodily mediation of rhythm perception and coordination pertains to the actual dimensions of our body, and thereby the natural mode shape attending the pendular length of our limbs. This mediation is evidenced by "a link between preferred dance tempo and the dimensions of the body, consistent with a natural frequency or eigenmode explanation of preferred tempo."[28] Indeed, the tempo range of locomotion movements, which fall between 300 and 900 ms, and heartbeat rates of 60–150 bpm or 400–1,000 ms, corresponds with the pulse rates we are capable of discerning (ones that are not too slow or fast to constitute a rhythmic pattern).[29] And we spontaneously tend toward step durations of 500–550 ms when walking, which corresponds to generally favored tempo rates of 110–120 bpm.[30] In other words, our bodies are rhythmically biased toward certain tempos and embed us within a preferred tempo range. We rely on the *eigen*-timing of our own bodily movements to help us "find the beat" so-to-speak.

A dynamical systems perspective offers a helpful vantage for further articulating the rhythmic coupling of the sensorimotor body with the environment. Even more strongly than the sensorimotor approach of Todd and Lee, which recoups centralization by locating the distributed bodily representations all in the brain, the dynamical perspective categorically rejects centralization and the concept of information. Janeen Loehr and Caroline Palmer (2009) frame a parallel contrast between "information processing approaches," which take the "timing of sequence elements in production [as] independent of the movements used to produce them,"[31] and "dynamical systems approaches," which view "timing [as] an emergent property of those movements."[32] The dynamical perspective distributes timing across intra-bodily systems, environmental affordances, and even across time itself by taking the history of a particular action as constitutive of future timing possibilities. For example, the biomechanical interdependencies of our bodies, such as those between our fingers, and the actual movement trajectories used to produce an action affects the timing of subsequent actions.[33] The issue here is not about the ontological status of information per se, which was addressed earlier, but about the methodological disposition of informational approaches toward inappropriately abstracting, isolating, and centralizing what are actually embedded, embodied, and distributed processes. The particularities of our bodies and the history of our activity matter. Endogenous adaptive oscillators at the level of our central nervous system play a role in priming the tempo of our movements, but our active engagement with processes in our environment enacts a recalibration of periodicity and phase via dynamics of entrainment: "Aspects of rhythm perception rely upon neural oscillations that resonate with rhythmic stimuli."[34] While Large and Snyder are thinking here solely in terms of neural resonance, the biomechanical resonances belonging to the body's pendular structure on the horizontal and vertical planes need to be considered.

Beyond the bodily mediation of timing by virtue of our musculoskeletal structure, it also appears that our immediate reflexive poise within gravity (vertical balance is also a pendular structure) turns out to be integral in perceptually discerning rhythmic pulse: "Rhythm perception is a form of vestibular perception."[35] In other words, our ability to perceive and synchronize our bodies vis-à-vis an environmental rhythm is actively mediated by our kinesthetic-proprioceptive-vestibular grip on our own bodies and movement. Phillips-Trainor et al. showed that vestibular stimulation alone affects the perceptual discrimination of pulse. They conclude that "The early development of the vestibular system, and infant delight at vestibular stimulation when bounced to a play song or rocked to a lullaby, suggest that we are observing a strong, early vestibular-auditory interaction that is critical for the development of human musical behavior."[36]

Part Five: Rhythmic Entrainment, Social Coordination, and the Emergence of Mind

We now turn to the centrality of musicality and rhythmic entraining for underwriting our social dynamics in infancy and for supporting sociality and language acquisition.

Coming apart and getting together in time is inherent to animate form. Processes go up and down, back and forth. They have a rhythm to them. Oscillation or rhythm is the cornerstone of life, of everything that isn't dead, of all systems that are not at equilibrium. Although rhythmical behaviors can be quite complicated, we have the deep impression that the principles underlying them possess a beautiful simplicity. Rhythms are known to confer positive functional advantages for the organism. Among these are spatial and temporal organization, prediction of events, energetic efficiency, and precision of control.[37]

This final section stresses the affective salience of the Other and the forms of dynamical coupling emerging as a result; these spontaneous and rudimentary entrainments come to engender jointly enacted transactional patterns or "rituals" that, eventually, can be initiated by either participant. Such ritualized patterns of interaction are what structurally scaffold the development of attention beyond the dyad, while a deference toward what is important to important Others motivates these reaches of attention. Indeed, the phenomena of gaze-cuing, gaze-following, and social referencing, which represent the earliest adventures of extra-dyadic attention, testify to the primacy of the Other in the development of attention. The infant's growing discriminations about significance are intensely mediated by the interest and affective response of Others to aspects of their experience, while a growing repertoire of shared ritual provides a progressively individuating platform for the full determinacy of joint attention.

The psychological research of Colwyn Trevarthen and Daniel Stern is grounded in micro-analytic observations of face-to-face bodily interactions between mother and child in a semi-naturalistic context. From this methodological perspective, Trevarthen and Stern emphasize the correspondences in *time*, *intensity*, and *form* that can be observed inside these dyadic interactions. These correspondences structure the processes of mutual regulation that play out in the mimetic call-and-response behavior between caregiver and infant.[38] The dynamics of this relation are best characterized in terms of the synchronies of affect, movement, and vocalization, which in turn are leveraged to realize more complex forms of mutual entrainment.[39]

Trevarthen is particularly interested in what he calls *synrhythmic regulation*, or the ongoing mutual adjustments in wellbeing between mother and child. Synrhythmic regulation represents a natural extension from the *amphoteronomic regulation* of the fetus' vital functions during pregnancy to the post-uterine regulation of the infant's needs through caregiving practices. Against this background of direct biological co-regulation, synrhythmic co-regulation concerns the *mind* of the infant via the rhythmic coordination of bodily behavior and affect: "A mutual control of the dynamics of *psychological* interest, of eager consciousness and purposes, is achieved by coordination of the rhythms of visual attention, vocalizations and gestures."[40] Synrhythmic regulation extends the more primitive metabolic-affective regulatory functions and practices into the sensorimotor transactions played out within the dyad, such that emotional undercurrents infuse these relational exchanges wherein the "engagement between mother adult and infant depends on a fine emotional appraisal of the timing of response of the partner, with both infant and mother expecting a sympathetic and creative contribution from the other in intersynchrony with what they offer."[41] These exchanges, moreover, cannot be conceived as discrete, piece-meal actions, because they only gather their significance in the context of the imitation-repetition and call-and-response structure, which determines whether a given behavior represents a meaningful social move via its place within a jointly constituted history.[42]

At the center of his analysis, Stern stresses "vitality contour" or "vitality affects" to denote the ongoing metabolic-affective dynamics functioning in the background of our living: "These are the temporally contoured feelings that accompany all experience. As an experience unfolds—say, as you watch someone smile at you—there are micro-shifts in the quality and intensity of the act, and of the feeling evoked in you. These shifts trace a timeline."[43] As Stern notes, an affective contour attends experience as such, because homeodynamic regulation necessarily underlies all our sensorimotor life. However, the gaze of the Other is especially salient from the outset, neurologically and emotionally; this special salience originates in several evolved neurobiological sensitivities to Others, such as basic eye-detection and face-recognition mechanisms. Moreover, the caregiver's gaze upon the infant is often accompanied by caressing touches, rocking, and vocal tones aimed at soothing the infant; such care prompts changes in the endocrine system, such as the release of oxytocin, which further promotes affiliative feelings. In short, our rich socio-biological inheritance renders our bodies constitutively sensitive to the proximity of the Other and bootstraps the phenomenological revelation of Others as uniquely salient amongst the stuff of the world.

Vasudevi Reddy, who shares Trevarthen's basic view and methodology, cites Merleau-Ponty's claim that "I discover vision, not as 'thinking about seeing,' to use Descartes's expression, but as a gaze at grips with a visible world, and that is why for me there can be another's gaze."[44] She turns to Merleau-Ponty here in support of her central thesis that "Mutual gaze—or receiving gaze to oneself—is an emotionally salient event."[45] Sartre's well-known analysis of shame in *Being and Nothingness*, which he presents as an ontological proof against solipsism, is intended to capture this same notion: "It is shame or pride which reveals to me the Other's look and myself at the end of the look. It is the shame or pride which makes me *live*, not *know* the situation of being looked at."[46] Whereas Sartre is drawing on the mature social emotions of shame and pride in his account of intersubjectivity,[47] Reddy's research analyzes the ways that attention and emotion are fundamentally intertwined within an infant's emerging capacity for social engagement. But both thinkers are interested in the primacy, ontologically and methodologically, of *living through* these emotional encounters ahead of *knowing them* qua theory of mind. This is precisely why Reddy accents the emotional repercussions of the Other's gaze as the primary process by which the Other's attentional perspective *as such* first becomes conspicuous for the infant:

> The perception of the object directedness of someone looking at you is dependent on the experiencing of emotional reactions to that looking. In other words, the object directedness of the looking is experienceable. If it is experienced (and it may not always be), then the directedness of the look is available through emotional and proprioceptive changes.[48]

In other words, the metabolic-emotional significance of the Other's looking-upon one's self makes the Other's looking-toward significant prior to any theory of mind. It is this primal significance that motivates interest in following the gaze of the other: "The implication of this argument is that it is only if gaze is first felt to the self that another's gaze toward other things in the world—to one's own body or acts or to objects distant in space or to events distant in time—can be understood as attending in the way that we commonly understand the term."[49] Against more behaviorist-flavored accounts, Reddy contends that infants play coy, pretend, tease, and joke, all in the absence of a theory of mind. Indeed, if it is only via these kinds of second-person affective dynamics that elements within a mental ontology, such as attention, can become explicitly thematic, then such second-person affective engagements necessarily underwrite the emergence of joint-attention and a third-person theory of mind:

> To begin with, an infant joins with another person's subjective state through sharing or otherwise co-ordinating feelings in contexts of one-to-one primary intersubjectivity. Another person's feelings, as expressed through bodily manifestations such as facial expressions, bodily gestures, and emotional toned vocalizations, have the power to affect an infant, as well as an older child, as well as an adult. … Therefore identification involves potential *movement* in psychological orientation … [and] jointness comes with being moved just enough to sense the psychological orientation of the other in oneself, but as the other's. This happens through intersubjective engagement that is emotional in source and emotional in quality.[50]

Within the context of these affective dynamics, imitation is foundational. Imitation is basic to second-person transactions between infant and caretaker, and thus is a critical point of discussion in developmental theory. Meltzoff's and Moore's[51] famous work on neonate facial imitation is so important because it indicates that an infant arrives into the world constitutively open to the body of *human* Others, especially faces, in and through the activation of her own bodily possibilities. Moreover, it is not just that we are born sensitive to faces, but we are particularly sensitive to the *contingency* of faces in interaction.[52] This point is important, because it is not that the face of the Other is merely a perceptually interesting but otherwise inert object. The face of the Other is compelling because it is participatory, dramatic, and correlates dynamically through time as it affectively registers the infant's salience and indexes the infant's own behavior through its tightly coupled responsivity. In other words, the infant's interest in the Other, and in imitation of the Other, plays out within the context of *responsive* contingency, which again highlights the salience of the temporal dynamics in imitation and agency.

As Shaun Gallagher and Daniel Hutto note, this phenomenon of neonate facial imitation reveals that our fundamental biological inheritance includes a primal perceptual *contrast* between the self and an Other, as well as a primal sensorimotor responsivity to Others, or an "intermodal tie between [the] proprioceptive sense of one's body and the face that one sees."[53] The simultaneous distinction and overlap between bodies are important to acknowledge when thinking about the role of imitation in social development, because it resists the tendency to view imitation as a purely passive process. Indeed, this is precisely why Merleau-Ponty describes mimesis as a process via which alterity codetermines the primal shape of one's subjectivity, and thus as an "invasion" and an "assuming":

> Mimesis is the ensnaring of me by the other, the invasion of me by the other; it is that attitude whereby I assume the gestures, the conducts, the favorite words, the ways of doing things of those whom I confront … [it] is the power of assuming conducts or facial expressions as my own … I live in the facial expressions of the other, as I feel him living in mine.[54]

This point is closely related to Trevarthen's contention that while we certainly need to recognize the importance of imitation in development, it is also necessary to include the infant's own insistent and intrinsic motivations for doing so: "Animal agency is never just imitative, never just instructed, and never just protected. It does not just react to stimuli."[55] To be perfectly clear, however, this recognition of the infant's primal autonomy does not entail a projection of an internal mental perspective on the infant, such as conscious intentions, since that is precisely what we seek to explain! Rather, it is to appreciate that imitation is sub-served by the endogenous physical and affective dynamics of the infant's body that solicit spontaneous entrainment with the Other. For example, the endogenous rhythms patterning the spontaneous movements of the infant can be seen as priming the infant for becoming imitatively entrained to the movement of an adult: "The timing of the saccadic eye rotations and of the reach and grasp movements of a newborn approximates that of an adult looking or reaching to grasp an object, and the basic pulse is the same for the two forms of activity."[56] This biological correspondence in motoric pulse invites entrainment, while the background affective state of the infant provides the motivational context such that imitative entrainment is not purely mechanical process driven by external stimuli. By focusing on affective context and the endogenous organization of movement, we retain a notion of autonomy that blocks the reduction of imitation to mere behaviorism, while also avoiding the attribution of articulate mental states to the infant. A primal autonomy first belongs to the organization of the body as an autopoietic system before it belongs to the organization of the mind as a space of reasons.

Indeed, for Trevarthen, subjective agency grows out of the appropriation of the dynamical possibilities afforded by our specific corporeality: "We need to understand the special innate sentience of human agency, which requires a unique proprioception of the many 'degrees of freedom' and 'polyrhythms' of the human body."[57] To conceive of the body polyrhythmically is to cast it as a nested, codetermining field of oscillating processes that are open to environmental coupling across a host of different descriptive levels and timescales. Trevarthen links this polyrhythmic conception of agency directly to James Gibson's sensorimotor view of perception, which is similarly anchored in the dynamics of contingency and covariance; Trevarthen, however, extends this basic picture into intersubjectivity.[58] Elsewhere, Malloch and Trevarthen cite with approval what they characterize as Paul Byers's "neat description of sympathy," which is framed in terms of rhythmic entrainments taking place *between* bodies rather than *inside* minds: "The information carried by interpersonal rhythms does not move directly from one person to another. Thus, information cannot easily be conceptualized as messages since the information is always simultaneously shared and always about the state of the relationship."[59] Byers's use of information here is unfortunate because his characterization of sympathy is clearly not consistent with the notion of informational content, which is precisely why such spontaneous, mutual coordination "cannot easily be conceptualized as messages." Indeed, such rhythmic sympathy is not information-like at all, but a resonant process of dynamic coupling. It is easy, however, to see the deep affinity between Byers's view here and Trevarthen's conception of intersubjective sympathy:

> All social cooperation depends on this mysterious intersubjective sympathy. … Impulses to move excite, not exactly "mirror" effects in the motor centers of other brains, but negotiable "sympathetic" impulses in the emotional systems of other individuals. By this means the couple, family, group, or society gains the properties of an "organism" of higher adaptive order, the cohesion of which depends on the matching and mutually supportive emotions of its members. It depends on "sympathy."[60]

Daniel Stern's view is similarly rooted in the dynamics of vitality and the bonding interactions that attend our practices of caregiving, wherein infant and caregiver mutually regulate the infant's interests and affective state. For Trevarthen and Stern, the intertwined dynamics of movement and affect are fundamental, while the emergence of correspondences and determinate forms from out of these dynamic interactions constitutes a generative elaboration on these sub-personal dimensions of intersubjectivity. Stern is categorical about the importance of appreciating the *temporality* of the body:

> Behaviors, thoughts, feelings, actions have a musical quality. Each behavioral, or affective, or even cognitive phrase—that is, the shortest meaningful chunking—has a contour in time. Behaviors are not discrete, on-off events. They unfold and describe temporal profiles as they do so. "Time-shapes" include forms such as fading, acceleration, explosion, effortfulness, hesitancy, tentativeness, boldness. For the most part psychology has ignored temporal dynamics.[61]

This emphasis on temporal dynamics invites an analysis of social coordination in terms of processes of entrainment, wherein psychological activity, bodily behavior, and vital biological functions can become sympathetically coupled and co-regulating across various phase relationships. It is thus not surprising that Trevarthen and Stern stress *rhythm*, *synchrony*, and *musicality* in order to articulate the proto-elements of significance that emerge through intersubjective correspondences in *timing, intensity,* and *form*.[62] Indeed, in his "Introduction" to the 2002 edition of *The First Relationship*, Stern reflects on the fact that "after so much observation at the micro-local level of mother-infant interaction, metaphors from music and dance not only crept into my writing but also became a way for me to think about what I saw."[63] He goes on to note his full realization:

I have come to realize that the metaphors for temporal dynamics I originally used in this book are more than metaphors, and that the vitality affects described in 1985 have an applicability far beyond mother-infant intersubjectivity. Vitality affects exist in all subjective experience, at all ages, and in all domains and modalities. My work years ago with choreographers and dancers is echoed now in the observation that our subjective experience has more in common with music than with a digital code.[64]

Along with Stern, Trevarthen appeals to the musicality of affective coordination as the primal mode of communing with Others, rather than to the communication (transmission) of mental content: "Most importantly, we communicate with [others] not just in space, but *in time* by intimate coordination with matching rhythms of movement. We move with other human beings in the 'musicality' of volatile feelings."[65] Indeed, Trevarthen's notion of "primary intersubjectivity" is largely grounded on the sympathetic mirroring of basic affective and kinesthetic transitions, such that the "matching of rhythm or pulse of movement offers a powerful correspondence by which mind states may be coupled";[66] these phasic couplings are lived out through vitality contours shaping our motivations toward further engagement and experimentation with new correspondences. Indeed, Trevarthen argues for a continuity between these root entrainment structures involved in infant sociality, such as imitation and turn-taking, and mature, higher-order domains of meaning production, such as drama, poetry, dance, and music:

> The intimate involvement in movement is engaging patterns of motor regulation and changing affective states in the two persons according to intrinsic neural processes that are present in both adult and infant, and recognizable in interpersonal communication among adults, especially in the artistic forms of drama, poetry, dance, and music. The concept of the essential "musicality" of basic forms of creative and cooperative human engagement in "vitality dynamics" has assisted development of a theory of primary motives for language learning, and a means of relating rational processes of practical and realistic cognition in individual minds and their communication by symbolic means with the emotionally creative, embodied, and highly social rituals of art.[67]

The musicality of primal intersubjectivity was first made thematic in Mary Catherine Bateson descriptions of "proto-conversations" between caregiver and infant.[68] Similar to Trevarthen, Stern, and Reddy, Bateson's account emphasizes the turn-taking, coordination of timing, and affective involvement in these proto-conversations:

> The study of timing and sequencing showed that certainly the mother and probably the infant, in addition to conforming in general to a regular pattern, were acting to sustain it or to restore it when it faltered, waiting for the expected vocalization from the other and then after a pause resuming vocalization, as if to elicit a response by a sort of delighted, ritualized courtesy and more or less sustained attention and mutual gaze. Many of the vocalizations were of types not

described in the acoustic literature on infancy, since they were very brief and faint, and yet were crucial parts of the jointly sustained performances.[69]

Stephen Malloch has subsequently applied sophisticated techniques of acoustical measurement, including spectography, pitch and timbre analysis to more precisely microanalyze the *music* of mother-child dynamics across multiple dimensions. The results have culminated in a theory of "communicative musicality" anchored in the constructs of *pulse*, *quality*, and *narrative*.

"Pulse" refers to the rhythmic structure of discrete vocal and gestural interactions that enable temporally coordinated behavior, while "quality" refers to the "modulated contours of expression through time" (that is, responsive changes in volume, pitch, and timbre), or stated more generally, the "attributes of direction and intensity of the moving body."[70] Malloch and Trevarthen explicitly link the phenomenon of modulating quality to Stern's notion of "vitality contour," thus highlighting the affective roots of this musical exchange. "Narrative," then, designates the larger chunks of shared meaning enacted through the temporally extended, systematic coordination of pulse and quality. Trevarthen provides a succinct summary of this line of research, the "[a]pplication of musical acoustic measurement to the interplay of vocal sounds between a young infant and her mother has been used to reveal the dynamic, and dramatic, form of a narrative created by their cooperation in mutual awareness. Their 'musicality' appears to reveal the source of human meaning in companionship."[71] Trevarthen and Malloch take this primal bodily sympathy as the origin of human culture and meaning: "Human beings have a specially adapted capacity for sympathy of brain activity that drives cultural learning."[72] While the narrow focus on brain activity here is problematic, the basic notion that primal bodily sympathy drives the cultural bootstrapping of mind is consonant with the perspective developed herein, assuming that this sympathy is understood as a distributed bodily sensitivity toward spontaneously covarying with other human bodies in time.

Conclusion

Questions about the origin of meaning and the emergence of mind coincide. To have a mind in the full-blooded sense is to participate in shared meaning (which is to have been shaped by shared meaning). The origin of shared meaning, however, cannot be explicated through an ontology that relies on information or representational content, because such content necessarily presumes the perspective of a mind. Thus, any such account is doomed to remain question-begging.

Socio-pragmatist accounts of the origin of meaning, such as those proffered by the Confucians, Wittgenstein, and Heidegger, hold that meaning arises out of practices. These practices are normative structures transmitted via processes of imitation, which initiate and underwrite the process of achieving competence with respect to our cultural practices. I find these accounts convincing. The problem of shared meaning then becomes the problem of accounting for these processes of imitation and normative training without falling back into mentalistic categories, such as information, representation, and computation. Instead, a valid naturalistic account of these imitative processes needs to be explicated solely via basic physical processes.

Interbodily entrainment offers the most fruitful explanatory route because it is grounded in direct causal coupling. By more clearly appreciating the pendula-oscillatory structure of our musculoskeletal body, as well as the oscillatory patterns of neural and metabolic processes, it is easy to conceptualize the body as a dynamically pulsing, polyrhythmic structure inviting entrainment. On this basis, it has been argued that our bodies are temporal fields disposed to couple with other rhythmic systems—especially the bodies of other persons—via spontaneous events of entrainment and resonance. Through entrainment we train new bodies to have a world, and thereby a mind. These processes are intrinsically coordinating and constitute the fundamental ground for various forms of socio-emotional attunement and imitation across a diversity of phase relationships—from synchrony to turn-taking. These primal attunements, imitations, and rituals open the path toward the development of language, mature intersubjectivity, and thereby the higher-order capacities of the representational mind.

Nietzsche's metaphysically deflationary claim presented above, namely, that he is "body through and through, and nothing besides," has guided the philosophical story presented herein—a story that has sought to rigorously avoid the lure toward modern-day spiritualism, that is, informational content. Rather than tarrying with such intellectualist notions, I have insisted that our bodies come primed to dance, and that we become minded by virtue of learning to dance with other bodies in the world. This story also reveals why it was a mistake for Plato to simply contrast the "possessed" minds of the Dionysian poets with the "right" minds of the Apollonian philosophers, since the rhythmically possessed body-mind constitutes the ontological and developmental ground for the emergence of the representational mind. Indeed, if my view is correct, then the rhythmic body is fundamental to the possibility of being-in-the-world, and thus for acquiring the second-order capacities that would seem to define Plato's conception of "right mind." Hence, the Apollonian phenomenology of boundaries and separation masks the fact that "right mind" is never as self-possessed nor as independent as it likes to think of itself. It, too, is always and already possessed by rhythm.

Notes

1   Graham Parkes, *Composing the Soul: Reaches of Nietzsche's Psychology* (Chicago: University of Chicago Press, 1994), 144. Reprinted by arrangement with University of Chicago Press.
2   Friedrich Nietzsche, *Thus Spoke Zarathustra: A Book for Everyone and Nobody*. Translated by Graham Parkes (Oxford: Oxford University Press, 2008), 30.
3   Nietzsche, *Thus Spoke Zarathustra*, 29.
4   Throughout, "metabolic" and "homeodynamic" will be used as umbrella terms for our visceral processes (thereby including the circulatory, digestive, endocrine, exocrinal, renal, hematopoietic, immunological, reproductive, and respiratory systems). Much could be said about the endogenous rhythms shaping these various

systems, but for the sake of this essay, it is explanatorily cleaner to parse our biology into three basic rhythmic orders: the musculoskeletal, the nervous, and the metabolic.

5   The use of "bias" here should be read in the rehabilitated Gadamerian sense of prejudice. The use of bias rather than prejudice is intended to differentiate descriptive levels. Here I am focused on bodily biases anchored purely in oscillatory patterns, and not prejudices with propositional content per se, such as beliefs or understandings.
6   Parkes, *Composing the Soul*, 75.
7   Parkes, *Composing the Soul*, 75.
8   Parkes, *Composing the Soul*, 144–5.
9   This is not a claim that philosophy of mind is unilaterally grounded in empirical theories. It is to say, however, that for the majority of cognitive scientists and philosophers of mind, the brain alone is taken to be a sufficient condition for mind while brain function becomes articulated through the metaphor of information processing.
10 John Searle, *The Mystery of Consciousness* (New York: The New York Review of Books, 1997); Daniel Hutto and Erik Myin, *Radicalizing Enactivism* (Cambridge: MIT Press, 2013).
11 Dynamical theorists have shown that simple configurations of oscillators can effectively model linear and nonlinear, dynamical processes at the bodily and neural level, which lends further credence to the overall explanatory strategy adopted herein; namely, that basic oscillatory processes defining our embodiment can lead to deeply complex forms of dynamic coupling.
12 Janeen Loehr, Edward Large, and Caroline Palmer, "Temporal Coordination and Adaptation to Rate Change in Musical Performance," *Journal of Experimental Psychology: Human Perception and Performance* 37 (4) (2011): 1293.
13 John Haugeland, *Having Thought: Essays on the Metaphysics of Mind* (Cambridge: Harvard University Press, 1998).
14 Developmental psychology, however, offers further resources for structuring our approach to the transmission of mind. There is a generally accepted trajectory of development from mutual gaze to performances of gaze-following and, ultimately, to jointly attending. *Mutual gazing* describes processes of sensory-affective immersion within the caregiver-infant dyad, while *gaze-following* (or *gaze-cuing*) speaks to the way infants begin to follow the head and eye movements of the caregiver beyond the dyad and out toward the wider environment; children become active participants in gaze-following as they learn to direct the gaze of others through pointing, and so on. Wherein early forms of gaze-following marks a merely behavioral form of orienting, that is, a purely exterior tracking of the direction of the other's looking, true joint attention denotes a mental orienting or recognition of the Other's point of view per se.
15 Bradley Park, "The Resonant Mind: Daoism and Situated-Embodied Cognition," *Newsletter: Asian and Asian-American Philosophers and Philosophies* 13 (1) (2013).
16 An *eigenmode* refers to the natural vibrational pattern of an oscillating system wherein its various parts all move sinusoidally together at the same frequency and in fixed phase relationships, such that their respective amplitudes covary (increasing or decreasing) systematically. This motion described by an eigenmode or any other natural mode is resonance. A *mode shape* refers to the resultant amplitudes and phase relationships of the various components of a system.

17 This point was first brought to my attention by my friend and banjo guru, Dwight Diller of Pocahontas County, West Virginia.
18 David Sudnow, *Ways of the Hand: A Rewritten Account* (Cambridge: MIT Press, 2001), 74–5.
19 Sudnow, *Ways of the Hand*, 74.
20 Abby Whiteside, *On Piano Playing: Indispensables of Piano Playing and Mastering the Chopin Etudes and Other Essays* (Portland: Amadeus Press, 1997), 4.
21 Whiteside, *On Piano Playing*, 8.
22 Whiteside, *On Piano Playing*, 8–9.
23 These "phenomenological" analyses by Sudnow and Whiteside are further supported by Nicolai Bernstein's neurophysiological account of motor control and dexterity. Bernstein presents a hierarchically structured model of motor coordination where coarser motor structures and the rhythms of the largest muscles groups and primary joints support the finer movements and rhythms realizing dexterity further out at the periphery, for example, at the hands and fingertips. Nicolai Bernstein, *Dexterity and Its Development*. Edited by Mark Latash and Michael Turvey (New York: Routledge, 1996).
24 In this context, entrainment refers to "spatiotemporal coordination resulting from rhythmic responsiveness to a perceived rhythmic signal." Jessica Phillips-Silver, C. Athena Aktipis, and Gregory Bryant, "The Ecology of Entrainment: Foundations of Coordinated Rhythmic Movement," *Music Perception: An Interdisciplinary Journal* 28 (1) (2010): 3–14.
25 Most animals engage in rhythmic behavior, but only a small subsection of (vocally mimicking) species can rhythmically entrain to a diversity of externally determined tempos. Adena Schachner, Timothy Brady, Irene Pepperberg, and Marc Hauser, "Spontaneous Motor Entrainment to Music in Multiple in Multiple Vocal Mimicking Species," *Current Biology* 19 (10) (2009): 831–6.
26 Neil Todd and Christopher Lee, "The Sensory-Motor Theory of Rhythm and Beat Induction 20 Years On: A New Synthesis and Future Perspectives." *Frontiers in Human Neuroscience* (2015): 7. https://doi.org/10.3389/fnhum.2015.00444
27 Todd and Lee, "The Sensory-Motor Theory of Rhythm and Beat Induction 20 Years On," 4.
28 Todd and Lee, "The Sensory-Motor Theory of Rhythm and Beat Induction 20 Years On," 8.
29 Laurel Trainor, Xiaoqing Gao, Jing-jian Lei, Karen Lehtovaara, and Laurence Harris, "The Primal Role of the Vestibular System in Determining Musical Rhythm," *Cortex* 45 (1) (2009): 41.
30 Bigitta Burger, Marc Thompson, Geoff Luck, Suvi Saarikallio, and Petri Toiviainen, "Hunting for the Beat in the Body: On Period and Phase-locking in Music-induced Movement," *Frontiers in Human Nueroscience* 8 (903) (2014): 2. https://doi.org/10.3389/fnhum.2014.00903.
31 R. A. Schmidt, "A Schema Theory of Discrete Motor Skill Learning," *Psychological Review* 82 (1975): 225–60; "Motor Schema Theory after 27 Years: Reflections and Implications for a New Theory," *Research Quarterly for Exercise and Sport* 74 (2003): 366–75; A. M. Wing, "Voluntary Timing and Brain Function: An Information Processing Approach," *Brain and Cognition* 48 (2002): 7–30.
32 J. A. S. Kelso, *Dynamic Patterns: The Self-organization of Brain and Behavior* (Cambridge, MA: MIT Press, 1995); "Self-organizing Dynamical Systems," in *International Encyclopaedia of Social and Behavioral Sciences,* edited by N. J. Smelser

and P. B. Baltes (Amsterdam: Pergamon, 2001), 13844–50; Janeen Loehr and Palmer, Caroline, "Sequential and Biomechanical Factors Constrain Timing and Motion in Tapping," *Journal of Motor Behavior* 41 (2) (2009): 128.

33 Loehr and Palmer, "Sequential and Biomechanical Factors Constrain Timing and Motion in Tapping," 128.

34 While I do not find their neurocentrism compelling, I am sympathetic with Large and Snyder's dynamic-resonant approach, rather than an informational approach to rhythmic perception in music: "the pulse and meter arise as the result of neural oscillations resonating to rhythmic stimulation." Edward Large and Joel Snyder, "Pulse and Meter as Neural Resonance," *Annals of the New York Academy of Sciences* 1169 (1) (2009): 48, 46.

35 Todd and Lee, "The Sensory-Motor Theory of Rhythm and Beat Induction 20 Years On," 1.

36 Indeed, the full significance of being oriented in gravity has been grossly under-thematized in philosophy and the sciences: "The vestibular system is a very primitive system that emerges early in both phylogeny and ontogeny and that determines the organization and development of the other senses. ... Phylogenetically the vestibular system was the first sensory system to develop in evolution (Walls 1962) and ontogenetically it is the first system to develop in the womb (Romand 1992), suggesting that a sense of orientation in the gravitational field is more fundamental to perception than is vision and hearing." Trainor et al., "The Primal Role of the Vestibular System in Determining Musical Rhythm," 41; Jessica Phillips-Silver and Laurel Trainor, "Feeling the Beat: Movement Influences Infant Rhythm Perception," *Science* 308 (7727) (2005): 1430.

37 Scott Kelso, "The Complementary Nature of Coordination Dynamics: Self Organization and Agency," *Nonlinear Phenomena in Complex Systems* 5 (4) (2002): 366.

38 Rhythmic convergence in social transactions and likeability correlate with subconscious mimicry and rhythmic synchronizations.

39 The development of these resonances, which is to say the construction of new biological and behavioral attunements, enriches Husserl's notion of "pairing" (*paarung*) beyond the limited notion of kinesthetic mirroring.

40 Colwyn Trevarthen, "The Generation of Human Meaning: How Shared Experience Grows in Infancy," in *Joint Attention: New Developments in Psychology, Philosophy of Mind, and Social Neuroscience*, edited by Axel Seeman (Cambridge: MIT Press, 2011), 95.

41 Trevarthen, "The Generation of Human Meaning," 88.

42 Infants track personal histories.

43 Daniel Stern, *The First Relationship: Infant and Mother* (Cambridge: Harvard University Press, 2009), 13.

44 Maurice Merleau-Ponty, *Phenomenology of Perception*. Translated by Colin Smith (London: Routledge and Kegan Paul, 1962), 409, as cited in Vasudevi Reddy, "A Gaze at Grips with Me," in *Joint Attention: New Developments in Psychology, Philosophy of Mind, and Social Neuroscience*, edited by Axel Seeman (Cambridge: MIT Press, 2011), 138.

45 Reddy, "A Gaze at Grips with Me," 142.

46 Jean-Paul Sartre, *Being and Nothingness*. Translated by Hazel Barnes (New York: Philosophical Library, 1956), 261.

47 "All our endeavors and ambitions in society are regulated by them [pride and shame] (Scheff 1988)." Trevarthen, "The Generation of Human Meaning," 91.
48 Reddy, "A Gaze at Grips with Me," 142.
49 Reddy, "A Gaze at Grips with Me," 144.
50 R. Peter Hobson, "What Puts the Jointness into Joint Attention?" in *Joint Attention: Communication and Other Minds. Issues in Philosophy and Psychology*, edited by Naomi Eilan, Christoph Hoerl, Teresa McCormack and Johannes Roessler (Oxford University Press, 2005), 185.
51 A. N. Meltzoff and M. K. Moore, "Imitation of Facial and Manual Gestures by Human Neonates," *Science* 198 (4312) (1977 Oct 7): 74–8. doi: 10.1126/science.897687. PMID: 897687; "Imitation, Memory, and the Representation of Persons," *Infant Behavior & Development* 17 (1) (1994): 83–99. https://doi.org/10.1016/0163-6383(94)90024-8
52 A range of authors have come to this conclusion. See especially Ann Bigelow and Phillipe Rochat, "Two-Month-Old Infants' Sensitivity to Social Contingency in Mother-infant and Stranger-infant Interaction," *Infancy* 9 [3] [2006]: 313–25; Kjell Stormark and Hanne Braarud, "Infants' Sensitivity to Social Contingency: A 'Double Video' Study of Face-to-Face Communication to between 2-and-4 Month Olds and their Mothers," *Infant Behavior and Development* 27 (2) (2004): 195–203; Jacqueline Nadel, Isabelle Carchon, Claude Kervella, Daniel Marcelli, and Denis Réserbat-Plantey, "Expectancies for Social Contingency in 2-Month-olds," *Developmental Science* 2 (2) (1999): 164–73; Phillipe Rochat, Ulric Neisser, and Viorica Marian, "Are Young Infants Sensitive to Interpersonal Contingency?" *Infant Development and Behavior* 21 (2) (1998): 355–66; L. Murray and C. B. Trevarthen, "Emotional Regulation of Interactions between 2 Month Olds and Their Mothers," in *Social Perception in Infants*, edited by T. M. Field, and N. A. Fox (New Jersey: Ablex, 1985).
53 Shaun Gallagher and Daniel Hutto connect this primary intersubjectivity to what they call "direct perceptual understanding" qua intercorporeality. They argue that this "direct perceptual understanding" of Others has its origins in (1) neonate imitation; (2) emotional coordination; (3) and a transactional understanding of purposiveness on the basis of bodily coordination. (Sean Gallagher and Daniel Hutto, "Understanding Others through Primary Interaction and Narrative Practice," in *The Shared Mind: Perspectives on Intersubjectivity*, edited by Jordan Zlatev, Timothy Racine, Chris Sinha, and Esa Itkonen [Philadelphia: John Benjamins Publishing, 2008], 17–38].)
54 Maurice Merleau-Ponty, *The Primacy of Perception and Other Essays on Phenomenological Psychology, the Philosophy of Art, History, and Politics*. Edited by James Edie. Translated by William Cobb (Evanston: Northwestern University Press, 1964), 118, 145, 146, as cited in Peter Hobson, and Hobson, Jessica. "Joint Attention or Joint Engagement? Insights from Autism," *Joint Attention: New Developments in Psychology, Philosophy of Mind, and Social Neuroscience* (2011): 115–36.
55 Colwyn Trevarthen, "Stepping Away from the Mirror: Adventures of Companionship." Reflections on the Nature and Emotional Needs of Infant Intersubjectivity," in *Attachment and Bonding: A New Synthesis*, edited by C. Sue Carter, Lieselotte Ahnert, K. E. Grossmann, Sarah B. Hardy, Michael Lamb, Stephen Porges, and Norbert Sachser (Cambridge: MIT Press, 2005), 58.
56 Trevarthen, "The Generation of Human Meaning," 80.
57 Trevarthen, "The Generation of Human Meaning," 79.

58 James Gibson developed an "ecological" view of perception wherein primary perceivables are environmental "affordances" rather than sense data or physical properties. To perceive an affordance is to perceive a possibility for action, such as, to perceive something as "stand-on-able" or "hide-behind-able," which is to perceive a pragmatic relation between our body and the environment. This entails that perception is a sensorimotor activity that involves exploring and coming to grips with the systematic covariances between movement and sensory contingencies. (James Jerome Gibson, *The Senses Considered as Perceptual Systems* [Houghton Mifflin, 1966]; James Gibson, *The Ecological Approach to Visual Perception* (London: Houghton Mifflin, 1979).)

59 Byers 1976, 160, as cited in Stephen Malloch and Colwyn Trevarthen, *Communicative Musicality: Exploring the Basis of Human Companionship* (Oxford: Oxford University Press, 2009), 2.

60 Colwyn Trevarthen and Kenneth J. Aitken, "Infant Intersubjectivity: Research, Theory, and Clinical Applications," *The Journal of Child Psychology and Psychiatry and Allied Disciplines* 42 (1) (2001): 3–48; Jean Decety and Thierry Chaminade, "Neural Correlates of Feeling Sympathy," *Neuropsychologia* 41 (2) (2003): 127–38; Trevarthen, "Stepping Away from the Mirror," 59.

61 Stern, *The First Relationship*, 13.

62 For more on the "musicality" of movement and nonverbal vocal communication in infant development, see Malloch and Trevarthen, *Communicative Musicality*, 2; Mechthild Papoušek and Hanuš Papoušek, "Musical Elements in the Infant's Vocalization: Their Significance for Communication, Cognition, and Creativity," *Advances in Infancy Research* (1981).

63 Stern, *The First Relationship*, 13.

64 Stern, *The First Relationship*, 13–14.

65 Trevarthen, "The Generation of Human Meaning," 74.

66 Trevarthen, "Stepping Away from the Mirror," 64.

67 J. Panksepp and C. Trevarthen, "The Neuroscience of Emotion in Music," in *Communicative Musicality: Exploring the Basis of Human Companionship*, edited by S. Malloch and C. Trevarthen (Oxford University Press, 2009), 105–46; Ellen Dissanayake, "Root, Leaf, Blossom, or Bole: Concerning the Origin and Adaptive Function of Music," in *Communicative Musicality: Exploring the Basis of Human Companionship,* edited by S. Malloch and C. Trevarthen (2009), 17–30; Ellen Dissanayake, "Bodies Swayed to Music: The Temporal Arts as Integral to Ceremonial Ritual," in *Communicative Musicality: Exploring the Basis of Human Companionship*, edited by S. Malloch and C. Trevarthen (2009), 533–44; Daniel N. Stern, *Forms of Vitality: Exploring Dynamic Experience in Psychology, the Arts, Psychotherapy, and Development* (USA: Oxford University Press, 2010); Papousek and Papousek, "Musical Elements in the Infant's Vocalization"; Malloch and Trevarthen, *Communicative Musicality*; Colwyn Trevarthen, "The Musical Art of Infant Conversation: Narrating in the Time of Sympathetic Experience, without Rational Interpretation, Before Words," *Musicae Scientiae* 12.1_suppl (2008): 15–46; Per Aage Brandt, "Music and How we became Human—A View from Cognitive Semiotics: Exploring Imaginative Hypotheses," in *Communicative Musicality: Exploring the Basis of Human Companionship,* edited by S. Malloch and C. Trevarthen (Oxford University Press, 2009), 31–44; Bjorn Merker, "Ritual Foundations of Human Uniqueness," in *Communicative Musicality: Exploring the Basis of Human*

Companionship, edited by S. Malloch and C. Trevarthen (Oxford University Press, 2009), 105–46; Trevarthen, "The Generation of Human Meaning," 87.
68   Similar to Trevarthen, Stern, and Reddy, Bateson's account emphasizes the turn-taking, coordination of timing, and affective involvement in these proto-conversations: "The study of timing and sequencing showed that certainly the mother and probably the infant, in addition to conforming in general to a regular pattern, were acting to sustain it or to restore it when it faltered, waiting for the expected vocalization from the other and then after a pause resuming vocalization, as if to elicit a response by a sort of delighted, ritualized courtesy and more or less sustained attention and mutual gaze. Many of the vocalizations were of types not described in the acoustic literature on infancy, since they were very brief and faint, and yet were crucial parts of the jointly sustained performances" (Mary Bateson, "The Epigenesis of Conversational Interaction: A Personal Account of Research Development," in *Before Speech: The Beginning of Interpersonal Communication*, edited by Margaret Bullowa (Cambridge: Cambridge University Press, 1979), 65; Bateson 1975.
69   Bateson, "The Epigenesis of Conversational Interaction," 65.
70   Malloch and Trevarthen, *Communicative Musicality*, 4.
71   Trevarthen, "The Generation of Human Meaning," 76.
72   Trevarthen, "Stepping Away from the Mirror," 55.

References

Bateson, Mary. 1979. "The Epigenesis of Conversational Interaction: A Personal Account of Research Development." In *Before Speech: The Beginning of Interpersonal Communication*, edited by Margaret Bullowa, 63–78. Cambridge: Cambridge University Press.

Bernstein, Nicolai. 1996. *Dexterity and Its Development*, edited by Mark Latash and Michael Turvey. New York: Routledge.

Bigelow, Ann and Rochat, Phillipe. 2006. "Two-month-old infants' Sensitivity to Social Contingency in Mother-infant and Stranger-infant Interaction." *Infancy* 9 (3): 313–25.

Burger, Bigitta, Thompson, Marc, Luck, Geoff, Saarikallio, Suvi, and Toiviainen, Petri. 2014. "Hunting for the Beat in the Body: On Period and Phase-locking in Music-induced Movement." *Frontiers in Human Nueroscience* 8 (903): 1–16. https://doi.org/10.3389/fnhum.2014.00903.

Gallagher, Sean and Hutto, Daniel. 2008. "Understanding Others through Primary Interaction and Narrative Practice." In *The Shared Mind: Perspectives on Intersubjectivity*, edited by Jordan Zlatev, Timothy Racine, Chris Sinha, and Esa Itkonen, 17–38. Philadelphia: John Benjamins Publishing.

Gibson, James. 1979. *The Ecological Approach to Visual Perception*. London: Houghton Mifflin.

Haugeland, John. 1998. *Having Thought: Essays on the Metaphysics of Mind*. Cambridge, MA: Harvard University Press.

Hutto, Daniel and Myin, Erik. 2013. *Radicalizing Enactivism*. Cambridge, MA: MIT Press.

Kelso, Scott. 2002. "The Complementary Nature of Coordination Dynamics: Self Organization and Agency." *Nonlinear Phenomena in Complex Systems* 5 (4): 364–71.

Large, Edward and Snyder, Joel. 2009. "Pulse and Meter as Neural Resonance." *Annals of the New York Academy of Sciences* 1169 (1): 46–57.

Loehr, Janeen and Palmer, Caroline. 2009. "Sequential and Biomechanical Factors Constrain Timing and Motion in Tapping." *Journal of Motor Behavior* 41 (2): 128–36.
Loehr, Janeen, Large, Edward, and Palmer, Caroline. 2011. "Temporal Coordination and Adaptation to Rate Change in Musical Performance." *Journal of Experimental Psychology: Human Perception and Performance* 37 (4): 1292–309.
Malloch, Stephen and Trevarthen, Colwyn. 2009. *Communicative Musicality: Exploring the Basis of Human Companionship*. Oxford: Oxford University Press.
Merleau-Ponty, Maurice. 1962. *Phenomenology of Perception*. Translated by Colin Smith. London: Routledge and Kegan Paul.
Merleau-Ponty, Maurice. 1964. *The Primacy of Perception and Other Essays on Phenomenological Psychology, the Philosophy of Art, History, and Politics*. Edited by James Edie. Translated by William Cobb. Evanston: Northwestern University Press.
Murray, Lynne and Trevarthen, Colwyn. 1985. "Emotional Regulation of Interactions between Two-Month-Olds and Their Mothers." In *Social Perception in Infants*, edited by Tiffany Field and Nathan Fox, 177–97. New York: Ablex Publishing.
Nadel, Jacqueline, Carchon, Isabelle, Kervella, Claude, Marcelli, Daniel, and Réserbat-Plantey, Denis. 1999. "Expectancies for Social Contingency in 2-Month-olds." *Developmental Science* 2 (2): 164–73.
Nietzsche, Friedrich. 2008. *Thus Spoke Zarathustra: A Book for Everyone and Nobody*. Translated by Graham Parkes. Oxford: Oxford University Press.
Papoušek, Mechthild and Papoušek, Hanuš. 1981. "Musical Elements in the Infant's Vocalization: Their Significance for Communication, Cognition, and Creativity." *Advances in Infancy Research* 1: 163–224.
Park, Bradley. 2013. "The Resonant Mind: Daoism and Situated-Embodied Cognition." Newsletter: *Asian and Asian-American Philosophers and Philosophies* 13 (1): 9–14.
Parkes, Graham. 1994. *Composing the Soul: Reaches of Nietzsche's Psychology*. Chicago: University of Chicago Press.
Phillips-Silver, Jessica and Trainor, Laurel. 2005. "Feeling the Beat: Movement Influences Infant Rhythm Perception." *Science* 308 (7727): 1430.
Phillips-Silver, Jessica, Aktipis, C. Athena, and Bryant, Gregory. 2010. "The Ecology of Entrainment: Foundations of Coordinated Rhythmic Movement." *Music Perception: An Interdisciplinary Journal* 28 (1): 3–14.
Reddy, Vasudevi. 2005. "Before the 'Third Element': Understanding Attention to Self." In *Joint Attention: Communication and Other Minds: Issues in Philosophy and Psychology*, edited by Naomi Eilan, Christoph Hoerl, Theresa McCormack, and Johannes Roessler, 85–109. Oxford: Oxford University Press.
Reddy, Vasudevi. 2011. "A Gaze at Grips with Me." In *Joint Attention: New Developments in Psychology, Philosophy of Mind, and Social Neuroscience*, edited by Axel Seeman, 137–57. Cambridge, MA: MIT Press.
Rochat, Phillipe, Neisser, Ulric, and Marian, Viorica. 1998. "Are Young Infants Sensitive to Interpersonal Contingency?" *Infant Development and Behavior* 21 (2): 355–66.
Sartre, Jean-Paul. 1956. *Being and Nothingness*. Translated by Hazel Barnes. New York: Philosophical Library.
Schachner, Adena, Brady, Timothy, Pepperberg, Irene, and Hauser, Marc. 2009. "Spontaneous Motor Entrainment to Music in Multiple in Multiple Vocal Mimicking Species." *Current Biology* 19 (10): 831–6.
Searle, John. 1997. *The Mystery of Consciousness*. New York: The New York Review of Books.

Stern, Daniel. 2009. *The First Relationship: Infant and Mother*. Cambridge, MA: Harvard University Press.
Stormark, Kjell and Braarud, Hanne. 2004. "Infants' Sensitivity to Social Contingency: A 'Double Video' Study of Face-to-Face Communication to between 2-and-4 Month Olds and their Mothers." *Infant Behavior and Development* 27 (2): 195–203.
Sudnow, David. 2001. *Ways of the Hand: A Rewritten Account*. Cambridge, MA: MIT Press.
Todd, Neil and Lee, Christopher. 2015. "The Sensory-Motor Theory of Rhythm and Beat Induction 20 Years On: A New Synthesis and Future Perspectives." *Frontiers in Human Neuroscience* 9: 1–25. https://doi.org/10.3389/fnhum.2015.00444.
Trainor, Laurel, Gao, Xiaoqing, Lei, Jing-jian, Lehtovaara, Karen, and Harris, Laurence. 2009. "The Primal Role of the Vestibular System in Determining Musical Rhythm." *Cortex* 45 (1): 35–43.
Trevarthen, Colwyn. 2005. "Stepping Away from the Mirror: Adventures of Companionship. Reflections on the Nature and Emotional Needs of Infant Intersubjectivity." In *Attachment and Bonding: A New Synthesis*, edited by Carol Sue Carter, Lieselotte Ahnert, Klaus E. Grossmann, Sarah B. Hardy, Michael Lamb, Stephen Porges, and Norbert Sachser, 55–84. Cambridge, MA: MIT Press.
Trevarthen, Colwyn. 2011. "The Generation of Human Meaning: How Shared Experience Grows in Infancy." In *Joint Attention: New Developments in Psychology, Philosophy of Mind, and Social Neuroscience*, edited by Axel Seeman, 73–113. Cambridge, MA: MIT Press.
Whiteside, Abby. 1997. *On Piano Playing: Indispensables of Piano Playing and Mastering the Chopin Etudes and Other Essays*. Portland: Amadeus Press.

Part Four

Saving the Last Dance

11

Staying True to the Earth in Zarathustra, Zhuangzi, and Zen

Timothy J. Freeman

> *Here Zarathustra fell silent for a while and looked with love upon his disciples. Then he continued to talk thus: —and his voice was transformed.*
> *"Stay true to the earth for me, my brothers, with the power of your virtue! May your bestowing love and your understanding serve the meaning of the sense of the earth! Thus I bid and beseech you."*
> Thus Spoke Zarathustra, *"On the Bestowing Virtue"*[1]

Staying True to the Earth

So often these days it feels like the scene in Wim Wenders's Odyssean epic film *Until the End of the World* when, in the middle of a kiss, the engine suddenly cuts out in their small single engine plane, leaving the protagonists Claire and Sam adrift over the Australian outback. "It's the end of the world," Claire concludes, understanding that the engine failure was likely the result of an electromagnetic pulse from the explosion of an out-of-control nuclear satellite.[2] Of course, it turned out not to be the end of the world. The apocalyptic setting of the famously long film just added a sense of urgency to Wenders's primary concerns in exploring the blinding power of images, the importance of dreams, and the search for love and the meaning of existence. At the end of the film, set sometime in the beginning of the twenty-first century, Claire is an astronaut, orbiting the earth as an ecological observer. What the film could not have anticipated is what an ecological observer orbiting the earth would see today—fires burning forests across the globe, parched drought-stricken land masses, devastating floods, massive storms, and the melting of sea ice in the Arctic Ocean. Of all the signs of climate change scientists are most alarmed by what has been taking place in the Arctic. The dramatic increase in temperature, the loss of sea ice, and the release of vast quantities of methane, all suggest we are dangerously close, perhaps already past the tipping point of climate change—and thus like Wenders's protagonists, powerless and adrift, hurtling over a desolate landscape toward the end of the world.

One could say that Nietzsche saw this coming. Not that he anticipated the problem of climate change, but his late writings are marked by an ever-increasing urgency, warning of an unparalleled crisis facing humanity that is the result of the underlying values of Western culture. The longing to free the soul from the prison of the body and earthly existence expressed in Socrates's last words and in the subsequent development of Christianity devalued this life on earth. For Nietzsche, this longing also led to a profound misunderstanding in which human beings did not understand themselves, the natural world, or their relationship to the rest of nature. With the human soul understood as separate from the body, and all other living things reduced to soulless machines, human beings became the only beings that mattered, with all the rest of nature merely serving human interests. With this longing for eternal life in another world, the earth becomes not our home but a wasteland, something to be used up and left behind.

Graham Parkes has long emphasized the importance of Nietzsche's project of a revaluation of all values, summed up in Zarathustra's exhortation to stay "true to the earth," for environmental philosophy.[3] Environmental philosophers, however, have sometimes challenged the relevance of Nietzsche's thought for environmental philosophy. Some contend that even though Nietzsche may have sought a perspective that is loyal to the earth, his critique of truth and his perspectivism inevitably lead to an untenable relativism which undermines any basis for an ecologically sound philosophy.[4] There is also the widespread view, which Parkes calls attention to, that "Nietzsche is such a strong advocate of will to power as domination and exploitation that one cannot sensibly count him as a contributor to environmental philosophy."[5] Parkes attempts to meet these objections with a "green" reading of Nietzsche. To begin with, Parkes emphasizes "Nietzsche's definitive pronouncement" criticizing anthropocentrism in the late writings: "The human being is by no means the crown of creation: every creature is, alongside the human, at a similar level of perfection."[6] Parkes also points to a passage from *The Genealogy of Morals* which he finds especially "ecologically prescient": "Our whole attitude toward nature today is *hubris*, our raping of nature by means of machines and the unthinking resourcefulness of technicians and engineers."[7]

Since his seminal essay in suggesting the resonances between Nietzsche's thought and Daoism, "The Wandering Dance: *Chuang Tzu* and *Zarathustra*," Parkes has emphasized the importance of "a transformation of our ideas of self and world— and thereby of ourselves" in Nietzsche's thought, Daoism, and Zen.[8] In subsequent essays Parkes has also drawn attention to a few passages in Nietzsche's writings that suggest an experience, similar to that found in Daoism and Zen, of "seeing things as they are."[9] Parkes has also challenged the reading of will to power that would be inconsistent with environmental philosophy, contending that "Nietzsche's philosophy of nature, his understanding of the natural world and human existence as interdependent processes and dynamic configurations of will to power, can contribute to grounding a realistic, global ecology that in its loyalty to the earth may be capable of saving it."[10]

Parkes's work is important and exemplary in showing what comparative philosophy can offer—in drawing attention to the possible resonances between Nietzsche's thought, Daoism, and Zen, Parkes challenges us to rethink what we know about these disparate philosophies. His work has been even more important in emphasizing the relevance of such a reflection in this time of ecological crisis when the very future of life on earth is at stake. In "The Wandering Dance," Parkes explains that his reflection "is intended as a prolegomena to a wider and deeper study."[11] In a long and distinguished career Parkes has certainly widened and considerably deepened this study. In this essay, I hope to contribute to this further study by reflecting on some of the crucial issues raised in Parkes's work. In the first section, I take up Parkes's suggestion of the resonance between Nietzsche's thought and Daoism, focusing on Parkes's suggestion of an experience of "seeing things as they are" in both *Zarathustra* and the *Zhuangzi*. The second section takes up the comparison between Nietzsche's thought and Buddhism, focusing on the problem posed by the notion of will to power, as well as the notion of "seeing things as they are" in Zen. The closing section takes up a reflection on Parkes's emphasis on the importance of a "psychical transformation" in Nietzsche's thought, Daoism, and Zen. This involves a reflection on the idea of eternal recurrence, the key idea in Zarathustra's call to stay true to the earth, and its possible resonances with Daoism and Zen.

Zarathustra and Zhuangzi

One of the more obvious resonances between Nietzsche's thought and Daoism is a common critique of anthropocentrism. For the Daoist philosophers, the Confucian focus on human beings was too narrow; they emphasize trying to take a wider view to see human beings in the perspective of the vast (*da* 大), the vastness of "the heavens and the earth" (*tiandi* 天地).[12] In contrast to the view expressed in *Genesis* that the Earth and all of its creatures were created for human beings, Parkes points out that the Daoist philosophers emphasize that human beings are "irrevocably subject to the powers of Heaven and Earth" and thus must approach the task of governing by "following the ways of nature."[13] In the *Daodejing*, most similes for *dao*, as Parkes observes, are drawn from nature; human beings are encouraged to be more like water, thawing ice, or an uncarved block of wood. The Daoist view, Parkes concludes, "is not only that human beings will flourish if they emulate natural processes, but also that this happens primarily because the best ruler is the most consummate emulator—of water especially."[14] Parkes draws an affinity between this Daoist view and Nietzsche's project of re-naturalizing human beings, and thus overcoming the dualism that separates human beings and nature, as well as the anthropocentrism which conceives nature as existing to serve human interests. Parkes calls attention to a similar use of imagery drawn from the natural world, both in the Daoist texts and in *Zarathustra*. In "The Wandering Dance," Parkes emphasizes that *Zarathustra* and *Zhuangzi* are "first and foremost works of *imagery.*"[15] "Beyond being works of the philosophical imagination," Parkes continues, "both texts share the same kinds of images. The primary source of imagery is the natural world: the elements—sky, earth, fire, and water; the sun, moon, and stars; the climate, weather, and seasons; and the realms of plant and animal."[16] Thus, just as the Daoist texts recommend emulating nature in a decidedly non-anthropocentric view, Parkes contends that Zarathustra's teaching of the *Overhuman* is "profoundly relevant for ecological thinking" since it "signifies a way of being that is attained by 'overcoming' the human, which, as the rest of Zarathustra shows, requires that one go beyond the merely human perspective and transcend the anthropocentric view."[17]

The most crucial question raised here concerns just what Parkes means by suggesting a transcendence "beyond the merely human perspective." In a recent essay, Parkes suggests that his comparison between Nietzsche and Zhuangzi "might highlight aspects of their thought that have generally gone unnoticed—especially on the question of whether and how perspectives beyond the human might be attainable."[18] Of course, one of the most distinctive features of Nietzsche's thought is his perspectivism. In the preface to *Beyond Good and Evil*, Nietzsche suggests that Plato's fundamental error, the error that made the history of Western thought the "history of an error," was the mistake of "denying perspective, the basic condition of all life."[19] Parkes turns to an important passage from the *Genealogy* in which Nietzsche emphatically emphasizes this basic condition of all life, highlighting the part where Nietzsche goes on to suggest that the nearest we can get to any objectivity is to multiply our perspectives:

> There is only a perspective seeing, only a perspectival "knowing"; the more affects we are able to put into words about a thing, the more eyes, various eyes we are able to use for the same thing, the more complete will be our "concept" of the thing, our "objectivity."[20]

Parkes then wonders, "multiplying perspectives all around is enlightening—but can't we thereby go further to some kind of perspectiveless experience?"[21] One of the main themes of Parkes's work in recent years has been the contention that both in Nietzsche's writings and in the *Zhuangzi* one can find suggestions of just such an experience, one that would allow, as he puts it, "knowing things as they are in themselves."[22] In support of this interpretation, Parkes highlights a few passages in the *Zhuangzi* describing an experience "in the broad light of Heaven," comparing this with the experience described in the section titled "Before the Sunrise" in *Thus Spoke Zarathustra*.

Before examining Parkes's reading of these passages, it is worth noting that in the early "Wandering Dance" essay we do not find the suggestion that there is ever any pulling away from perspectivism, either in Nietzsche's writings or in the *Zhuangzi*. There we find Parkes drawing the connection between Nietzsche, "who emphasizes experience is always necessarily perspectival," and Zhuangzi, who "does not believe that we could ever attain a kind of 'perspectiveless seeing.'"[23] It turns out the problem arises, not because we see things from perspective points of view, but only "when we become *fixated* in a particular perspective."[24] Parkes notes that both thinkers address this problem through the dream. In *The Joyous Science*, Nietzsche develops the notion of the philosopher as lucid dreamer: "I have suddenly awakened in the middle of this dream, but only to the consciousness of dreaming, and that I *must* continue to dream lest I perish, just as the sleepwalker must continue to dream lest he slip and fall."[25] Zhuangzi also suggests the philosopher as lucid dreamer when he mocks Confucius and other philosophers who think they are awake, closing his riposte with the famous butterfly dream in which one can no longer distinguish between dreaming and waking life.[26] In "The Wandering Dance," Parkes embraces the perspectivism in both thinkers and explains that Zhuangzi's butterfly dream makes the point, "relevant also to Nietzsche's perspectivism, that when one is in a certain perspective it is impossible to see it *as a perspective*. Only when we are placed in a different perspective can we appreciate the limitations of our former standpoint."[27] The problem is not that we are dreamers, but rather, as Parkes explains, "the refusal to admit that we are dreamers, to become aware of the extent to which the 'real world' is projected by human needs and desires, and to celebrate this creative activity by both seeing through and playing with it at the same time."[28]

This play with different perspectives is what the wandering dance is all about. Parkes draws attention to the notion of "wandering" (*yóu* 遊) in the title of the first chapter of the *Zhuangzi*, translated as "free and easy wandering," "going rambling without a destination," or "wandering far and unfettered," and also points out a connotation with "dance" in the cognate term (*yóu* 游) meaning "to dance, float, swim about in water."[29] The stories in the chapter, Parkes explains, "conduct the reader through a variety of perspectives ranging from the vegetative through the animal to the human, all point up the limitations of adopting a fixed standpoint."[30] In another essay a little later, Parkes explains that the point of Zhuangzi's perspectivism is to get us to see that "all value judgements are relative insofar as they are made from a particular perspective, and that particular perspectives are by their nature narrow and limited in comparison with the openness of heaven or the way."[31]

In the "Wandering Dance" Parkes emphasizes that Zarathustra is also a wanderer and a dancer. Throughout the narrative, Zarathustra proceeds to wander, Parkes continues, "from place to place, trying out the perspectives of mountain top and valley, underworld and ocean."[32] Parkes points out that the "tightrope walker" is literally a "tightrope dancer" (*Seiltänzer*); and this, he suggests, is one of the keys to the whole text: "This corresponds to the dance as a central image in *Zarathustra* and an indispensable capability of the overman. The overman must be dancer because through realizing the relativity of all perspectives, he knows that there is no longer any firm ground on which to take a stand."[33] At this point Parkes seems to fully embrace a perspectivism in both Zhuangzi and Nietzsche in which it would not make sense to speak of a perspectiveless experience that would enable "knowing things as they are in themselves."

In subsequent writings, Parkes seems to want to pull both Nietzsche and Zhuangzi back from perspectivism, at least slightly, in emphasizing a "transperspective experience." He begins to suggest this as he turns his attention to defending Nietzsche as an ecological thinker. In his characterization of the development of Nietzsche's thought, Parkes sees a tension developing in the middle period of his writings where there is, on the one hand, a growing awareness of how our conceptions of nature are "conditioned by various kinds of fantasy projections," and yet also a recognition of the need to withdraw these projections: "The tension between a view that understands fantasy projection as an ineluctable (if occasionally see-throughable) aspect of the human condition and one that allows for a seeing of the world of nature as it is in itself, apart from human projections on to it persists to the time of *Zarathustra*."[34] Parkes thinks Nietzsche is suggesting a "withdrawal of at least some kinds of projection," when he asks, "When may we begin to *naturalize* ourselves by means of the pure, newly discovered, newly redeemed nature?"[35] We have misunderstood the relationship between human beings and nature because we have misunderstood both human beings and nature. The task of re-naturalizing the human being requires a new understanding of nature, and involves a twofold process, as Parkes explains, "to strip away the fantastic metaphysical interpretations of human origins that have obscured human nature, and to confront human beings with nature itself, similarly stripped of human projections."[36]

The key passage in *Zarathustra* Parkes turns to as also suggesting this experience of nature stripped of human projections is Zarathustra's blessing in "Before the Sunrise": "But this is my blessing: to stand over each and everything as its own Heaven, as its round roof, its azure bell and eternal security."[37] Parkes finds that Zarathustra's blessing, in liberating all things from their bondage under purpose, "frees them from any universal teleology, whether stemming from divine providence or the projection of a scientific view of progress, in order to let them be—or rather, come and go—in what Nietzsche calls the 'innocence of becoming.'"[38] This "Before the Sunrise" passage is of crucial importance, as Parkes explains elsewhere: "since it seems to go beyond Nietzsche's customary perspectivism and allows for an experience of the world that is not merely 'from our little corner' but from a horizon that transcends anthropocentric views."[39] In another text Parkes finds this blessing to resonate with both Daoism and Zen in allowing things to be just as they are:

> Just as the Daoist sage and the Zen master are able to experience events in the "self-so-ing" of their spontaneous unfolding, so Zarathustra's blessing lets each particular thing generate its own horizons, arising and perishing just as it does. In terms of environmental ethics, to experience in this way allows one to appreciate the intrinsic value of the natural world absolutely.[40]

In his recent book on climate change, Parkes suggests that in this passage "Zarathustra sounds very much like a Zen master or Daoist sage" since this blessing frees things "from being bound up in our instrumental view of them as things made or adapted for human purposes."[41]

Parkes finds a resonance with Zarathustra's blessing in the "Autumn Floods" dialogue in the Outer Chapters of the *Zhuangzi* where the sage is described as able to "penetrate the pattern of the myriad things" by "fathoming the beauty of heaven and earth" and thus have "a full view of heaven and earth."[42] Parkes also points to a passage in the Inner Chapters where Zhuangzi suggests the importance of knowing the difference between the human and Heaven: "'To know what is Heaven's doing and what is man's is the utmost in knowledge. Whoever knows what Heaven does lives the life generated by Heaven. Whoever knows what a man does uses what his wits know about to nurture what they do not know about."[43] Parkes draws out the comparison with Zarathustra's blessing:

> Just as the Daoist sage (a precursor of the Zen master) is able to broaden his perspective to the point where he is able to "illumine all things in the light of

heaven," and by acting in a way harmonious with heaven and earth can "help the ten-thousand things be themselves", so Zarathustra's blessing lets each particular thing generate its own horizons and be (or, rather, *become*: arise and perish) just as it is.[44]

Sometimes Parkes seems to acknowledge that there is no transcending perspectivism in Nietzsche's task of broadening perspectives: "This is not a transcending toward some God's eye perspective or view from nowhere, but rather a broadening of the human world view to include an appreciation of the perspectives of the natural phenomena with which we share the world."[45] Yet in the very same text, Parkes goes on to emphasize that even though Nietzsche "is certainly concerned with our interpretations of and projections on to the natural world, but this does not mean that we can never know nature 'as it is in itself.'"[46] In this essay, and in a more recent one, Parkes thinks Nietzsche elaborates on the idea of knowing things as they are in themselves, rather than as human awareness construes them, when he writes, in the notebooks from 1881: "The task: to see things as they are!"[47] Parkes seems to suggest here that Nietzsche's task of seeing things "as they are" involves transcending perspectivism.

Parkes contends that Nietzsche's task of "seeing things as they are" draws a comparison with Zhuangzi's recommendation of the "fasting of the heart-mind (*xin* 心)."[48] As Parkes explains, this is a "matter of emptying the mind of what we human beings bring to our engagement with the world in the way of prejudices and preconceptions, inclinations and aversions, all of which get in the way of our experiencing what is actually going on … and lets one experience through the openness of *qi*, 'the presence of beings.'"[49] Drawing together these passages from Nietzsche and Zhuangzi, Parkes contends both thinkers suggest an experience going beyond merely seeing from multiple perspectives, to a "perspectiveless experience" in which one is able to know "things as they are in themselves, rather than as human awareness construes them."[50]

As Parkes has made quite clear, Nietzsche surely does emphasize overcoming the narrow anthropocentric view that has shaped so much of the human comportment toward the natural world; and since the notion of the *Overhuman* involves overcoming the human in some sense, it is obvious that Nietzsche emphasizes overcoming "merely" human anthropocentric perspectives. But does Parkes really mean to suggest something of a return to the notion of *nature as origin*, the view that is the target of the poststructuralist critique of the traditional notion of "nature"?[51] At one point Parkes explains that he is responding to the problem posed by the poststructuralist deconstruction of "nature," the view, as he puts it, that "nature is always socially constructed, so we can never reach anything like 'pure' nature in itself, apart from human factors that condition all experience of it."[52] The problem, of course, is that the poststructuralist critique of the notion of *nature as origin* owes so much to Nietzsche.

In the preface to *The Joyous Science*, the text where Parkes finds Nietzsche suggesting the task of confronting human beings "with nature itself, similarly stripped of human projections," Nietzsche makes a bit of a risqué joke calling into question the very notion of a "naked truth," emphasizing that we "should cherish the *modesty* with which nature has concealed herself behind enigmas and iridescent uncertainties."[53] One would be hard pressed to find a better, more succinct statement of the poststructuralist critique of the conception of nature as origin. One might also recall the famous fragment from Heraclitus, "Nature loves to hide," which Nietzsche is surely playing on here."[54] One should cherish the modesty of nature, concealing herself behind enigmas and iridescent uncertainties; and, by implication, one should be more modest with respect to nature, giving up the "youthful madness" as Nietzsche puts it, to see nature stripped of her veils. What may be the most radical aspect of Nietzsche's thought—and the one aspect most often missed—is the modesty of his thought. Is not the very notion of seeing nature, as it is in itself, exactly what Nietzsche is here finding indecent?

Nietzsche continues this play with the "woman-truth" in the preface to *Beyond Good and Evil* where he again makes fun of philosophers, this time portraying them as lovesick suitors, clumsy in their pursuit of the woman-truth, and left standing around all "dispirited and discouraged" because they never understood the woman-truth, never understood that "she has not allowed herself to be won."[55] This is where Nietzsche goes on to suggest that the problem with these lovesick philosophers is that they were seduced by Socrates and thus fell into Plato's error of "denying *perspective*, the basic condition of all life." In contrast to this, Nietzsche's "philosophers of the future," returning now to the end of the preface to *The Joyous Science*, will be those who understand they are artists.

In order to emphasize a transperspectival experience allowing for "knowing things as they are," Parkes ends up deemphasizing the creative activity of the philosopher he had earlier celebrated in "The Wandering Dance." Parkes wonders, "what are we to make of Nietzsche's occasional praise of creative experience and repudiation of 'mirror'-like perception?"[56] Nietzsche's praise of creative experience, however, hardly seems occasional, as the conception of the philosopher as artist seems so crucially important in Nietzsche's thought from *The Birth of Tragedy* to the last writings. Take, for example, this passage from *Beyond Good and Evil* in which Nietzsche uses an analogy drawn from painting to suggest the philosopher as artist: "Is it not sufficient to assume degrees of apparentness and, as it were, lighter and darker shadows and shades of appearance—different 'values,' to use the language of painters? Why couldn't the world *that concerns us*—be a fiction?"[57] The modesty of Nietzsche's thought emphasizes that the world that concerns us is a fiction, a product of an active interpretation. There may be narratives, stories we tell ourselves about the point of it all and the nature of nature, but there is no "ultimate and real" story or "metanarrative."[58]

Here we find ourselves at the crux of the issue—in response to the point where Nietzsche suggests that we should see that the world *that concerns us* is a fiction, Parkes suggests this is so only most of the time. He points to other passages in which Nietzsche seems to suggest an extraordinary experience in which one is able to withdraw these projections or fictions and see things as they really are. In the latest essay Parkes puts the issue this way:

> Suppose we were able through reflection or practice to withdraw these humanizing projections: Aren't we still stuck in the human perspective, experiencing the world from the locus of this particular human body and perceptual apparatus? How can we come to know the evanescent natural beings that Nietzsche calls for us to honor and affirm? Know them as they really are, and not just as how they appear to us as human beings.[59]

Which is it for Nietzsche—is the world that concerns us a fiction only most of the time, or all of the time?[60]

The emphasis on art from the earliest to the last writings is indicated by the prominence of the figure of Dionysus in Nietzsche's thought. In what might be regarded as his last words, the closing line of his autobiography *Ecce Homo*, Nietzsche writes, "Have I been understood?—*Dionysus against the crucified one.*"[61] One might get some sense of what he means by this opposition from what he says about *The Birth of Tragedy* in the preface that he attached to the second edition. There he makes clear that his first book is opposed to the Christian teaching which is "hostile to art" because of its "vengeful antipathy to life itself: for all of life is based on semblance, art, deception, points of view, and the necessity of perspectives and error."[62] At the end of his career, in *The Antichrist*, Nietzsche condemns the Christian interpretation of the meaning of the "life of Christ" for its arrogance in assuming that its narrative is the "truth" and not just an interpretation. In that narrative, the meaning of the life of Christ is symbolized by the image of the crucified one—the death on the cross was the promise of eternal life in heaven for the believer.[63] The "crucified one" in Nietzsche's last words is perhaps an image both for the Christian interpretation expressing the longing for another world, and for this hostility to art, this inability to recognize its own interpretation as an interpretation. Against this denial of art, Nietzsche's last words point to Dionysus, a figure always connected with art, indeed, with the highest aim of art in Nietzsche's thought. Nietzsche's last words would then suggest that if one wants to understand him, one must understand this opposition between "Dionysus" and "the Crucified"— the opposition between the philosopher as artist, modest with respect to the woman-truth, in contrast to the philosopher who longs to see nature stripped of her veils.

The Birth of Tragedy is often regarded as merely illustrating Nietzsche's youthful Romanticism when he suggests that the Dionysian experience reveals the truth of reality behind the veils. It may offer a preview of his mature thought, however, in the suggestion that what the Dionysian experience reveals is not the truth of reality as it is in itself—*nature as origin*—but rather, the abysmal truth that there is no truth of reality as it is in itself. In the crucial passage, Nietzsche explains that in the Dionysian experience "*Excess* [*Das Übermass*] revealed itself as truth."[64] All of our truths, Nietzsche suggests, are the result of the Apollonian drive to carve the figure out of the stone—the drive to make sense of the chaos of existence; Dionysian insight, however, reveals truth as *excess*—despite all our attempts to make sense of existence, it always exceeds all those attempts as it is always capable of being interpreted otherwise. Here is nature, not as origin, as solid ground, but as abyss. The preview of Nietzsche's mature thought in *The Birth of Tragedy* lies in confronting the abyss that is revealed in the Dionysian experience.[65]

Later, Nietzsche's confrontation with this abysmal truth is developed most powerfully in the "death of God," a metaphor for the collapse of the traditional notion of truth as the ground that has served as a foundation of Western thought since Plato. As this notion of truth is symbolized by the sun in Plato, the "death of God" is like unchaining the earth from its sun, opening up an abyss in which we are falling, without direction, "as through an infinite nothingness."[66] Later in the text Nietzsche describes the "death of God" to be like an "eclipse of the sun," that leads inevitably to the collapse of "our entire

European morality."⁶⁷ If nothing is true, all is permitted. This is, of course, what leads some environmental philosophers to dismiss Nietzsche as an ecological thinker, and it is perhaps also why Parkes attempts to pull Nietzsche back from his perspectivism, back from the emphasis on art and creative experience, back from confronting the abyss. In the "Wandering Dance" essay, however, Parkes draws attention to Zarathustra's confrontation with the abyss: "Every apparently firm ground (*Grund*) is, for Nietzsche, an abyss (*Abgrund*): 'Where does man not stand at the edge of abysses?' Is to see not itself—to see abysses?"⁶⁸ Parkes emphasizes that for Nietzsche and for Zhuangzi, "the appropriate response to the realization of the relativity of all standpoints is to develop lightness of foot and learn to dance over the abyss."⁶⁹

Seeing is seeing abysses, Nietzsche emphasizes, because seeing always involves perspective points of view, and the world is always interpretable otherwise. Nietzsche suggests this in another well-known passage from *The Joyous Science*: "The world has once more become 'limitless' [*unendlich*] to us, in so far as we cannot deny the possibility that it *contains limitless interpretations*."⁷⁰ In the aphorism just prior to the madman's announcement of the "death of God," Nietzsche suggests the sea as an image for this "infinity" or "limitlessness" of perspectivism: "We have left dry land and put out to sea! ... there will be hours when you realize that it is infinite, and that there is nothing more terrible than infinity [*Unendlichkeit*]."⁷¹ Instead of turning to an experience of things as they are in themselves, Nietzsche's response to the crisis opened up by the "death of God" is to suggest the courage needed to dance over the abyss, the courage also of an intrepid seafarer venturing out into the open sea:

> In fact, we philosophers and "free spirits" experience the news that "the old God is dead" as if illuminated by a new dawn; our hearts are overflowing with gratitude, astonishment, presentiment, expectation—at last the horizon seems free again, even if it is not be bright; at last our ships can set sail again, ready to face any danger; every venture of the knowledge-seeker is permitted again; the sea, *our* sea, lies open again before us; perhaps there has never been such an "open sea."⁷²

Nietzsche's response to the crisis of nihilism is then this courage of the seafarer, the courage to continue to venture out into the open sea and attempt to make sense of existence, all the while knowing that all around us there is only the open sea and no solid ground, since the world is always capable of being interpreted otherwise. We must continue seeking knowledge, knowing full well that the world that concerns us is a fiction, that we are artists, that we are dreaming, and must continue to dream lest we perish. The notion of the philosopher as lucid dreamer—"I must continue to dream lest I perish"⁷³—is echoed in another, much discussed line from the late notebooks: "We possess *art* lest we *perish of the truth*."⁷⁴ In those notes Nietzsche emphasizes art as the "countermoverment to nihilism,"⁷⁵ and in this we hear an echo of the thesis of *The Birth of Tragedy* that art is the "saving sorceress" necessary to go on living after Dionysian insight into the abysmal, tragic character of existence.⁷⁶ Is the key to the overcoming of nihilism in Nietzsche's pulling back from perspectivism and the emphasis on art, or is it rather in recognizing the importance of the philosopher as artist, recognizing that the world that concerns us is always necessarily a fiction?

Does Zhuangzi ever really pull back from perspectivism and emphasize an experience of seeing things as they are? One might wonder, first, whether the very notion of the "mutuality and collaterality" of "heaven and earth" and human beings precludes the very possibility of seeing "heaven and earth" as it is in-itself? The passage from the Outer Chapters[77] where Parkes wants to emphasize the notion of having a "full view of Heaven" seems to really only emphasize overcoming the anthropocentrism that reduces "heaven and earth" to a mere resource for human use. In the passage from the Inner Chapters where Parkes wants to call attention to Zhuangzi's emphasis on knowing the difference between Heaven and the human being, Zhuangzi goes on to admit there is a problem here: "So how could I know whether what I call the Heavenly is not really the Human? How could I know whether what I call the Human is not really the Heavenly."[78] Is Zhuangzi suggesting that we can really distinguish the Heavenly (nature) from the human, or is he emphasizing the modesty we should have in all our efforts to understand the vastness of "heaven and earth"? This notion of the vastness (*da* 大) of "heaven and earth" draws a comparison with Nietzsche's imagery of the limitlessness (*Unendlichkeit*) of the sea and the modesty of the philosopher as lucid dreamer. When Zhuangzi ridicules Confucius and others for thinking they are awake when they are still dreaming, he admits "when I say you're dreaming, I am dreaming too."[79] In "The Wandering Dance," Parkes draws our attention to this passage, explaining that this "should shake our confidence ... that we know the true nature of the 'I' who supposedly 'does this and that.'"[80] Should not Zhuangzi's dreaming also shake our confidence that we can ever wake from the dream and get a "full view of Heaven," knowing the true nature of things as they are in themselves?

Another issue that Parkes brings up in "The Wandering Dance" concerns the opposites or polarities of *yin* and *yang*, which is such an interesting feature of Chinese thought. Parkes brings this up in discussing the connection between the acceptance of change and the radical perspectivism in both Zhuangzi and Nietzsche's thought: "A philosophy that acknowledges the relativity of opposites tends to be a *perspectivism* (how things appear depends on your point of view, your place on the continuum) as well as a philosophy of flux."[81] Parkes suggests that yinyang thought is one of the important "illustrious precursors" of the "dynamic perspectivisms" of Zhuangzi and Nietzsche: "Such a philosophy of flux leads naturally to a perspectivism: the opposites of *yin* and *yang* are intimately linked, each depending on the other in order to be what it is and having the germ of the other immanent in; what is going on depends on what has been going on and where the process is heading."[82] One of the most striking features of Daoism is the emphasis on the *yin*, on the feminine, empty, dark, and yielding, over the masculine, full, bright, and aggressive *yang*. The yin emphasis is introduced in the famous opening line of the text in which it is acknowledged at the outset that the *dao* cannot be put into words.[83] Zhuangzi's recommendation of the emptying or fasting of the heart-mind, which Parkes draws attention to, also suggests this movement toward yin. One explanation for the yin emphasis in Daoism is drawn from Chinese medicine.[84] If one's condition is out of balance due to an excess of yin, a yang remedy is needed, while a yin remedy is needed in response to an excess of yang. The yin emphasis in Daoism might then be understood as a response to a time out of joint due to an excess of the yang, as the Warring States Period surely must have been. In "The

Wandering Dance," Parkes draws attention to a passage in which Zhuangzi calls out the "failing to understand the pattern of heaven and earth, and the myriad things as they are. It is as though you were to take heaven and your authority and do without earth, take Yin as your authority and do without the Yang, that this is impractical is plain enough."[85] Is this all that Parkes is getting at in suggesting an experience of seeing things as they are in the *Zhuangzi*? At least at this point, Parkes does not suggest this experience of seeing the myriad things as they are involves pulling back from Zhuangzi's radical perspectivism. Is the modesty of Zhuangzi's perspectivism, and the experience of seeing the myriad things as they are, the yin and yang of Zhuangzi's thought?

As for Nietzsche, in the passage from the notebooks where Parkes emphasizes the task of seeing things as they are, Nietzsche explains that the means to do this is "to be able to see with a hundred eyes, from many persons!"[86] Here Nietzsche seems to suggest that seeing "things as they are" involves recognizing that we only see things as they are from perspective points of view. Rather than contrasting with the perspectivism in which Nietzsche emphasizes that there is "only a perspective seeing," this passage is consistent, emphasizing that the means to seeing things as they are is to see from multiple perspectives. As Parkes had explained in "The Wandering Dance," if one becomes fixated in one perspective, one can fail to recognize it as a perspective. One might be deluded into thinking that one sees reality as it is in itself apart from its appearance. The more we are able to see from different perspectives, the more we will be able to recognize that we only see from perspective points of view.[87]

Zarathustra and Zen

It is well known that Nietzsche had a pessimistic and incomplete understanding of Buddhism. In *The Antichrist* Nietzsche expresses the hope that his condemnation of Christianity has not involved an injustice toward Buddhism. He says that Buddhism is "a hundred times more realistic than Christianity" in that the concept of "god" had already become irrelevant, and in its psychological approach to the problem of suffering as opposed to the "struggle against sin."[88] It is also much healthier than Christianity in showing no signs of *ressentiment*. Of the Buddha, Nietzsche writes that "he does not ask his followers to fight those who think otherwise: there is nothing to which his doctrine is more opposed than the feeling of revenge, antipathy, *ressentiment*."[89] And yet, because Nietzsche understood *nirvāṇa*, as Schopenhauer thought, to be the final goal of extinction, he concluded that Buddhism was like Christianity in being nihilistic, hostile to life, a religion of *décadence*, and thus not loyal to the Earth. In *Beyond Good and Evil*, Nietzsche describes the thought of eternal recurrence as a joyful affirmation of the world, contrasting this with the "most world-denying of all possible ways of thinking," which he sees in the philosophy of Schopenhauer and the Buddha.[90]

Despite Nietzsche's negative view of Buddhism, Parkes has drawn affinities between Nietzsche's thought and the Buddha's central teachings of interdependence (*pratītyasamutpāda*), impermanence (*anitya*), and "no-self" (*anātman*), and especially with Mahāyāna Buddhism, with which Nietzsche was unfortunately not aware. When *nirvāṇa* is understood, not as a liberation from this world, but rather, as another way of being here, there is, as Parkes puts it, a "consequential reverence for this world," and this is where "the interesting resonances with Nietzsche's thinking begin."[91] Bret Davis has challenged Parkes's attempt to find a resonance between Nietzsche's thought and Mahāyāna Buddhism, and Zen especially, finding Nietzsche's central idea of will to power to be incompatible with the "standpoint of *śūnyatā*" in Zen. As Davis puts it, "in Nietzsche's affirmation of the egoism of will to power, then, we run up against a formidable limit to the search for 'ironic affinities' with Buddhism."[92] Davis argues that it is the Buddhist path, particularly the way of Zen, which offers "a great affirmation of living *otherwise than willing.*"[93] Davis explains that the standpoint of *śūnyatā* "demands first of all a radical negation of the will."[94] The standpoint of will to power, Davis contends, thus falls short of the standpoint of non-ego on the field of *śūnyatā*, which "requires breaking through all such transmutations of self-centered willing."[95] The crux of Davis's reading that Nietzsche falls short of Zen is his understanding of will to power as the willful craving that the Buddha had identified as the cause of suffering: "To the extent that the will to power could be understood as a form of *taṇhā*, a critique of the will to power would lie at the very heart of Buddhism."[96]

Parkes contends that Davis has misunderstood Nietzsche "as advocating the 'egoism of will to power'" and that this misunderstanding has led him to "consistently overlook or ignore key aspects of his [Nietzsche's] thinking that are consonant with Buddhist ideas."[97] As Parkes explains, "a major theme of Nietzsche's psychology, from *The Birth of Tragedy* to *Twilight of the Idols*, is the rejection of the ego as a convenient but ultimately unnecessary fiction."[98] "Throughout his career," Parkes points out, "Nietzsche regards the I as something that stands in the way of one's becoming what one is."[99] The crude reading of will to power as a desire for power can be rejected because the "will" in "will to power" is not a self-conscious ego. Although he was concerned about the negative consequences, the *décadence*, that can result from the "disintegration of the ego," Nietzsche "never talks about the task of constructing an ego."[100]

There is no point in even considering whether there is an overcoming *of* will to power in Nietzsche's thought, Parkes explains, because "the will to power is the whole world, and '*there is nothing outside the whole!*'"[101] Parkes here calls attention to the famous passage from the notebooks where Nietzsche describes the world as a dynamic play of forces and then concludes "*This world is the will to power—and nothing besides!* And you yourselves are also this will to power—and nothing besides!"[102] This conception of the entire world as "will to power and nothing besides" is not "an instance of anthropocentrism," Parkes explains, "since Nietzsche has just desubstantialized the 'soul' into a configuration of forces ('a social structure of the drives and affects') ... and demonstrated 'will' to be a complex function of forces issuing from a social structure of multiple 'souls' within the body."[103] In undermining the concept of a substantial self, Nietzsche echoes the no-self doctrine in Buddhism. As Parkes explains, "[a]ll this corresponds to the idea of 'no-self' (*anātman*) that is central to Buddhism and which, on the basis of a radically relational ontology, applies equally to the *I* and to things."[104] Nietzsche's various passages on the will to power suggest that the universe as a whole and all living things within it from the smallest organisms to the most complex human beings are this play of forces.[105]

Perhaps the most challenging passage in thinking through the resonances between Nietzsche's thought and Zen, and the relevance of Nietzsche's thought for environmental philosophy, is the passage from *Beyond Good and Evil* where he emphasizes that "life simply *is* will to power."[106] Davis cautions against "any postmodern or comparative attempt to skip lightly over such passages."[107] One could, of course, simply reject or resist what Nietzsche says here. It is worth noting what Nietzsche writes to a friend in the summer of 1888 that "it is not at all necessary or even desirable to side with me; on the contrary, a dose of curiosity, as if confronted with some unfamiliar plant, and an ironic resistance would be an incomparably *more intelligent* position to adopt."[108] Just prior to this troubling passage about will to power, Nietzsche writes that "truth is hard."[109] One might find what he says next too hard, too dangerous a plant to handle; nevertheless, one might easily provide an analysis explaining the whole climate catastrophe as the result of this hard truth: "Life is *essentially* appropriation, injury, overpowering of what is alien and weaker; suppression, hardness, imposition of one's own forms, incorporation and at least, at its mildest, exploitation [*Ausbeutung*]."[110] He continues to say that this "exploitation" is not a character of primitive societies that humanity has evolved out of; nor is this true only of corrupt societies, aberrations from the refined norm of modern advanced civilization. This "exploitation," Nietzsche explains, "belongs to the *essence* of what lives, as a basic organic function; it is a consequence of the will to power, which is after all the will of life." All of life, he explains, strives "to grow, spread, seize, become predominant" precisely because "life simply *is* will to power."[111] One might like to resist this thought and argue that Nietzsche was wrong in this supposition that all of life is this will to power; but when one considers the totality of the human impact upon the earth—the near exponential population growth, continual depletion of resources, the appropriation and overpowering of alien, that is, nonhuman and weaker species for food and other resources, the constantly increasing need for energy, the ever-increasing release of greenhouse gases into the atmosphere—it is hard to avoid the conclusion that Nietzsche may have been right in this hard truth about life. One might say that the suggestion that we are now living in the Anthropocene is a confirmation of this hard truth. It doesn't really resolve the problem posed by this passage if the will to power is not the desire of a self-conscious ego. If the underlying drive of all life is this force of exploitation as described in this passage, how can humanity avoid the ecological catastrophe that is impending due to the human exploitation of the earth?

It would be easier if will to power were merely the desire or craving the Buddha identified as the cause of suffering and could be extinguished. It seems clear that for Nietzsche there is no life without will to power. Staying true to the earth is then not about extinguishing will to power, but rather its transformation. When Zarathustra implores us to stay true to the earth, he adds "with the power of your virtue."[112] This echoes the famous passage where the will to power is first introduced, when Zarathustra explained that the virtues of a people—the tablets of good and evil—are "the voice of its will to power."[113] If our values are expressions of will to power, as Parkes explains, "it all comes down to a question of will to power, conflicts between competing interpretations and world-views."[114] More recently, Parkes explains that Nietzsche's conception of will to power entails that everything is "a configuration of interpreting will to power" and thus "is at every moment construing all other things and is the product of their manifold interactions."[115] If then "nothing can twist free from the world 'as the will to power and nothing besides' and still be," Parkes draws the conclusion that what is needed is "a transformation of the interpreting will to power."[116] If everything is a configuration of interpreting will to power, however, in what sense does it make sense to speak of "seeing things as they are?"

In what sense does it make sense to speak of "seeing things as they are" in Buddhism? The expression "seeing things as they are" can be found in just about all Buddhist traditions. In *The Pali Canon*, "seeing things as they are" means, to put it simply, to see things unfold in their interdependence, according to the three marks of impermanence (*anitya*), no-self (*anātman*), and suffering (*duḥkha*). In the *Dhammapada*, a crucial passage suggests that "seeing things as they are" involves recognizing how suffering follows as a result of the mental constructs which shape the way the world shows up for us.[117] The Buddha goes on to suggest that *nirvāṇa* follows from seeing this, recognizing we can change the mental constructs that shape the way the world shows up for us. "Seeing things as they are" is then not about seeing things as they are in themselves apart from the mental constructs that shape the world that concerns us, but rather seeing this process by which the world that concerns us is created. Here, perhaps *nirvāṇa* is already another way of being here. In the Buddha's *Fire Sermon*, *nirvāṇa* might also be understood as another way of being here if one understands that the point of the Buddha's teaching is not extinguishing the fire, but rather changing the fuel with which we burn.[118]

In the development of Mahāyāna Buddhism, Nāgārjuna and the Mādhyamaka school similarly emphasize overcoming the conceptual fabrications (*prapañca*) that lead to suffering, while the Yogācāra school focuses on a profound transformation in the deepest depths of consciousness. The distinctive Yogācāra doctrine of *vijñapti-mātra* (perception or cognition-only), often understood as a sort of Buddhist Idealism, might rather be compared with Nietzsche's notion that the world that concerns us is a fiction.[119] The crucial question concerning Yogācāra concerns just what the point of the practice of yoga (*yogācāra*) might be. Yogācāra is known for its depth psychology, its analysis of eight levels in the ocean of consciousness. In addition to the five sense consciousnesses and the mind-consciousness recognized in the Abhidharma analysis, Yogācāra recognized two subliminal levels of consciousness, the afflicted suffering-consciousness that is always going on below the surface, and then the root or store-house consciousness (*ālāya-vijñāna*) at the bottom of the ocean. Stored in this root-consciousness are impressions from previous experiences, from other lifetimes, which form the seeds scenting the whole ocean of consciousness. The aim of the practice of yoga is to bring about a revolution in the deepest depths of consciousness, in the root consciousness, so that the afflictions, arising in consciousness like ocean waves, are brought to an end. Some contend that this revolution leads to a cessation of the process of *vijñapti-mātra*, enabling one to see reality as it is, in its suchness (*tathatā*), apart from all interpretation.[120] Is the practice of yoga about the cessation of this process, however, or its transformation?[121]

The Yogācāra analysis about what takes place in the depths of the ocean of consciousness draws a comparison with Nietzsche's depth psychology. In *Composing the Soul*, Parkes explores Nietzsche's psychology and points out that "the ocean is a major premise in *Zarathustra*" and "the sea is a fine analogue for the complex relation of the individual soul to the play of will to power that makes up the world."[122] This suggests that will to power is not a form of craving (*taṇhā*), as Davis contends, but is instead analogous to this ocean of consciousness in the Yogācāra analysis. Zarathustra's teaching of the Overhuman is about a transformation in the depths of the soul. Perhaps will to power is analogous to the fire with which everything is burning in the *Fire Sermon*, and the overcoming of humanity is not about extinguishing the fire, but changing the fuel with which we burn.

Yogācāra had a profound influence upon Zen, and the question of what is meant by "seeing things as they are" is a fundamental question in Dōgen's Zen, in both the "Genjōkōan" and in the "Sansuikyō" ("The Mountain and Waters Sūtra") reflections in the *Shōbōgenzō*. A Zen term for "seeing things as they are" is *kenshō* (見性) combining *ken* (seeing) and *shō* (nature). It is often translated as "seeing one's (true) nature," that is, the Buddha-nature within the heart-mind. One can thus appreciate the importance of this fundamental question in the famous lines from the "Genjōkōan": "To study the Buddha Way is to study the self./To study the self is to forget the self./ To forget the self is to be enlightened by the myriad things of the world."[123] This fundamental question concerning *kenshō* is also crucial in "The Mountains and Waters Sūtra," which is characterized by the translator Shokaku Okumura as a commentary on the "Genjōkōan." In one of the crucial passages, Dōgen emphasizes the Yogācāra notion of *vijñapti-mātra* at play in all our seeing:

> In general, then, the way of seeing mountains and waters differs according to the type of beings [that sees them]. In seeing water, there are beings who see it as a jeweled necklace. This does not mean, however, that they see a jeweled necklace as water. How, then, do we see what they consider water? Their jeweled necklace is what we see as water. Some see water as miraculous flowers, though it does not follow that they use flowers as water. Hungry ghosts see water as raging flames or as pus and blood. Dragons and fish see it as a palace or a tower, or as the seven treasures or the *mani* gem. [Others] see it as woods and walls, or as the Dharma nature of immaculate liberation, or as the true human body, or as the physical form and mental nature. Humans see these as water. And these [different ways of seeing] are the conditions under which [water] is killed or given life.[124]

What Dōgen says here draws a comparison with Nietzsche's view that the world that concerns us is a fiction. But neither in Nietzsche's thought, nor Dōgen's, does this entail that we should rest content with our fictions, our limited perspectives. For Dōgen, the different ways of seeing are not all the same, not equally valid, as they are the conditions under which all things—the "water" in Dōgen's reflection—are killed or given life. Surely, we have to become aware of the consequences of our perspectives and thus, perhaps, become capable of changing our perspectives; but does Dōgen ever suggest a "perspectiveless experience" in which one is able to see "things are they are in themselves"? Dōgen raises this very question: "Although we say there is water of various types, it would seem there is no original water."[125] A little later in the text Dōgen goes on to explain that when "those who study Buddhism seek to learn about water, they should not stick to [the water of] humans; they should go on to study the water of the way of the buddhas."[126] What is "the water of the way of the buddhas?" Okamura explains that the key to understanding this point is when Dōgen explains: "The Buddha has said, 'All things are ultimately liberated. They have no abode.'"[127] This, Okamura explains, is Dōgen's expression for the wisdom of the *Heart Sūtra*.[128] Here is where the "Mountains and Waters Sūtra" helps to explain the "Genjōkōan." To study the Buddha Way is to study the self. One must begin by becoming aware of the self and all the ways one has come to see things as a result of karmic consciousness. But then one must forget this self, not stick to the water of humans, but study the water of the way of the buddhas—to understand that all things have no abode, are empty of inherent existence, existing instead in interdependence with all things. Is this what Parkes means in drawing our attention to the Zen sense of seeing things as they are, seeing how everything arises in interdependence, how "everything in existence is related to everything else?"[129] However, as it turns out, it is for this reason—that everything is empty of separate existence—that Dōgen was suspicious of the term *kenshō*.[130]

Bret Davis is perhaps helpful in explaining what he refers to as the *karmic editing* process by which the world that shows up for us is created. While one might think that one is just seeing things as they are, meditation is "a practice of emptying the mind of this conceit that our own edited version of reality is the only unbiased and therefore valid one."[131] Davis goes on to explain: "We cannot prevent our mind from creating our world, but we can wake up to the fact that this is what is happening." As Parkes puts it in "The Wandering Dance," what "we wake up to is the realization that we are always bound by some perspective: this awakening is itself a perspective—but one that acknowledges and embraces the multiplicity of all possible perspectives."[132] Once we become aware of this process, we might be able to edit those fictions which shape the world that concerns us. As Davis puts it, "our experience of the world is always limited and perspectival, but it can be more or less egoistic or empathetic, more or less closed- or open-minded, more or less rigidly assertive or flexibly responsive."[133]

If we are responsible for the way in which the world that concerns us shows up for us, then what is most crucial in staying true to the earth is that transformation of ourselves that Parkes called attention to in "The Wandering Dance." This is why Parkes has emphasized the importance of imagery in Nietzsche, Daoism, and Zen, as a "philosophy presented in images," he explains, "works on the reader's psyche by inviting the kind of participation in their play that effects a psychical transformation more radical than just a change of mind."[134] The key to the psychical transformation in *Thus Spoke Zarathustra* is the strange thought of eternal recurrence, which is closely related to the notion of *amor fati*. The Kyoto school philosopher Keiji Nishitani, one of the first to draw our attention to resonances between Nietzsche's thought and Zen, emphasized the importance of eternal recurrence and *amor fati*. In *Religion and Nothingness*, Nishitani explains that Nietzsche's thought of eternal recurrence is "one of the currents in Western thought to come closest to the Buddhist standpoint of *śūnyatā*."[135] In an earlier essay in *The Self-Overcoming of Nihilism*, Nishitani explains that it was "in such ideas as *amor fati* and the Dionysian as the overcoming of nihilism that Nietzsche came the closest to Buddhism, and especially to Mahāyāna."[136] In the introduction to that volume, Parkes explains that Nishitani's Zen standpoint "brings into relief a nexus of issues surrounding the core of Nietzsche's thought: The idea of eternal recurrence in its connections with the notion of *amor fati*, love of fate."[137] How, then, is the thought of eternal recurrence, along with the related notion of *amor fati*, the key to staying true to the earth?

The *Kōan* of Eternal Recurrence

In the autobiography, Nietzsche famously relates the story of when the thought of eternal recurrence came to him:

> Now I shall relate the story of *Zarathustra*. The basic conception of the work—the *thought of eternal recurrence*, this highest attainable formula of affirmation—belongs to the August of 1881: it was dashed off on a sheet of paper with the caption "6000 feet above man and time." On that day I was walking through the woods by Lake Silvaplana, not far from Surlei I stopped next to a massive block of stone that towered up in the shape of a pyramid. Then this thought came to me—.[138]

In his most recent text, Parkes explains the importance of this place: "If you experience the actual natural settings *where Zarathustra* was composed this very much enhances your next reading of the text."[139] On this point I would concur. It is certainly one of the most beautiful places on this earth. If one were to walk those paths along the lake and experience the "*azure blue* solitude in which this work lives," it is easy to understand Zarathustra's call to stay true to the earth.[140] In the most recent work, Parkes contends that the thought of eternal recurrence is about seeing things as they are and accepting what is given in nature. Here, Parkes takes up Lawrence Lampert's suggestion of the connection between the thought of eternal recurrence and Zarathustra's blessing:

> Zarathustra's blessing on things is a sheltering vault of blue sky, a letting be, an allowing, a sparing. Because the heavens do not speak … man is free to speak the blessing on things that they be just as they are. His blessing does not do violence to things but allows them to become themselves, luminous and intense in their evanescence …. Eternal return is the teaching that lets beings be.[141]

Parkes emphasizes that Nietzsche's "philosophy of will to power demands a self-transformation on the part of human beings in modern times," but this transformation involves becoming "more accepting of what is *given*" and capable of "letting nature hold sway."[142] In accepting what is given, Parkes suggests, as mentioned earlier, that Zarathustra's blessing as well as the teachings of the Daoist sage and Zen master open an experience which "allows one to appreciate the intrinsic value of the natural world absolutely."[143] Nietzsche, however, seems to explicitly reject the very notion of "intrinsic value": "Whatever has *value* in the present world has no intrinsic or natural value [*das hat ihn nicht an sich*]—there is no such thing—but rather the value which has been given [*gegeben*] and bestowed [*geschenkt*] upon it, and it was *we* who gave and bestowed! We alone have created the world *which is of any concern to man!*"[144] This passage anticipates not only the passage from *Beyond Good and Evil* where Nietzsche suggests that the "world *that concern us*" is a fiction, but also the play with giving and bestowing that is such a central theme in *Zarathustra*, most crucially in the exhortation to stay true to the earth.

This theme of the gift, of giving and bestowing, shines forth in the text through the image of the golden sun. The sun always gives or bestows its light; and gold, Zarathustra explains, has the highest value only as an image or "allegory of the highest virtue," which he goes on to explain is "the bestowing virtue."[145] At least in part, this gift-giving virtue involves understanding that there are no intrinsic values, no value in itself, as value is a gift that is given or conferred upon things, and that we are these givers and bestowers.[146]

When one walks along the paths around the lakes at Sils-Maria, if it is a bright, calm day the magnificent snow-peaked mountains are reflected in the water. What is given in nature, one might say, is indeed stunningly beautiful, but what is given is still always interpretable otherwise. Whether one makes the place a ski resort or leaves it completely alone for the deer and waterfowl to enjoy, it is still a value bestowed upon nature. The world that concerns us may be a fiction, but in order to stay true to the earth something about the givers and bestowers of value must change. The first clue to this transformation is suggested in the scene from the Prologue when, on his way down from his solitude in the mountains, Zarathustra encounters an old man in the forest. When the old man asks Zarathustra why he is coming down, Zarathustra responds, "I love human beings."[147] The old man responds that he does not love human beings; he loves God instead because human beings are too imperfect for him. He wants something back in return for his love. He hopes to get the greatest return on his investment in eternal life in the next world. Zarathustra responds that his love is a gift.[148] Here, again, the sun as an image of this bestowing love suggests the transformation of the bestowers of values. Throughout the text, the *golden sun* always gives its light without expecting a return. Becoming capable of this would seem to entail overcoming the exploitative will to power that reduces everything to a mere resource for extracting a return on an investment. As Parkes explains, this bestowing love leads to a new health, the great health, that wants "to embrace all things, so that it can bestow and contribute to the world with no egoistic thought of thanks or return."[149] Parkes draws the resonance with the teaching of *The Heart Sūtra*: "It is the same with the bodhisattva: the attainment of wisdom, which involves the realization of emptiness of the self through its interrelatedness with all things, naturally leads to an abundant generosity and a re-engagement with the world."[150] So how does the thought of eternal recurrence lead to the bestowing love? In the *Joyous Science* Nietzsche provides an important clue to *Zarathustra* when he appends the title "Incipit Tragoedia" to a preview of the book.[151] In *The Birth of Tragedy* Nietzsche suggests the highest aim of art, which he thought Greek tragedy had achieved, was its capacity to change us. *Zarathustra* is a tragedy because it aims to bring about the transfiguration of human beings. The thought of eternal recurrence would be the catalyst for this transformation.

In this respect the thought of eternal recurrence draws a comparison with the Zen kōan. With a kōan it is not enough to provide a rational explanation. Even if one could provide an explanation for Joshu's "Mu," for example, it wouldn't be enough to pass the test.

It wouldn't be enough if one could explain how Joshu's "Mu" is the perfect response to the question of whether or not a dog has Buddha Nature, since *mu* (無) can also mean "*emptiness*," the teaching of the *Heart Sutra* that everything is empty of inherent existence. As the thirteenth-century Chinese Zen Master Wumen (Japanese *Mumon*) puts it: "For the attainment of incomparable satori, one has to cast away his discriminating mind."[152] For the point of the kōan is not intellectual understanding, but rather the experience of *satori*, that sudden enlightenment, the profound transformation in the deepest depths below the surface consciousness of the discriminating mind. It is not enough to understand "Mu," as Mumon explained, "one must *be* 'Mu.'" It is not enough just to think about it, as Mumon had put it: "One must concentrate with your 360 bones and your 84,000 pores, making your whole body one great inquiry."[153] One would not pass the test until it is clear that the kōan has done its trick in becoming a catalyst for transformation.

Nietzsche first presents the thought of eternal recurrence in *The Joyous Science*:

> What if one day or night a demon came to you in your most solitary solitude and said to you: "This life, as you now live it and have lived it, you will have to live again, and innumerable times again, and there will be nothing new in it, but rather every pain and joy, every thought and sigh, and all the unutterably trivial or great things in your life will have to happen to you again, with everything in the same series and sequence—and likewise this spider and this moonlight between the trees, and likewise this moment and I myself. The eternal hourglass of existence will be turned over and over again, and you with it, speck of dust!"[154]

Nothing could seem to be worse than this fate. Most persons surely would want things to be different. Nietzsche continues, posing the question of the kōan: "If that thought took hold of you as you are, it would transform you and perhaps crush you; the question with regard to each and every thing, 'Do you want this again, innumerable times again?' would weigh upon your actions with the greatest weight! Or how well disposed would you have to become to yourself and to life, that you might *long for nothing more* than this final eternal confirmation and seal?"[155]

It is rather straightforward to see why the thought of eternal recurrence is so closely connected with the thought of *amor fati*, the love of fate, which Nietzsche expresses as a new year's resolution, writing at the beginning of 1882, just a few short months after the thought of eternal recurrence came to him at the rock: "I want to come to regard everything necessary as beautiful—so that I will become one of those who makes everything beautiful. *Amor fati*: from now on, let that be my love! I do not want to wage war against the ugly. I do not want to accuse anyone, I do not even want to accuse the accusers. May *averting my eyes* be my only negation! All in all, and on the whole, some day I hope to be an affirmer."[156]

Nietzsche's resolution to accept everything necessary as beautiful echoes the acceptance of fate in the Stoics, and resonates with something similar in Zen, which may be traced back to those stories in the *Zhuangzi* about characters with unusually powerful charismatic power (*de* 德) as a result of the way they have handled their circumstance or fate (*ming* 命). Parkes mentions a few of these characters in "The Wandering Dance," but my favorite is the humorous story of the ugliest man. He was ugly enough to astound the world, and yet everyone was drawn to him in an extraordinary way. He didn't have power to protect them, nor wealth to fill their bellies, but he had such powerful charisma because he didn't let the oscillations of fate upset the harmony of his spirit (qi 氣). This ability to "harmonize and delight" in the oscillations of fate and "never be at a loss for joy" enabled him to "make it be spring with everything."[157] Nietzsche's new year's resolution was to become just such a character. To be able to love fate and affirm eternal recurrence one would have to overcome the longing to be somewhere else than the present moment. One would have to overcome regret and the spirit of revenge. In "The Wandering Dance," Parkes highlights what may be the core of Nietzsche's philosophy, expressed in Zarathustra's words: "For that humanity be redeemed from revenge: that is for me the bridge to the highest hope and a rainbow after long storms."[158]

When Nishitani compares the thought of eternal recurrence with the standpoint of *śūnyatā*, he suggests that "we seem to be breathing the same pure mountain air that we felt in approaching the standpoint of Dōgen." Nishitani then cites the line in which Dōgen, in his first lecture upon returning from China, expressed something like the thought of *amor fati*: "I now while away my time accepting whatever may come."[159] Nevertheless, his final judgment in *Religion and Nothingness* is that Nietzsche's thought falls short of Zen, as Nietzsche's thought of eternal recurrence, he concludes, "does not make time to be truly time," and thus "cannot signify the point where something truly new can take place."[160] One wonders whether Nishitani may have forgotten what he had written earlier in the lecture on *amor fati* and eternal recurrence about what happens at the end of the passage in *Zarathustra* where the thought of eternal recurrence is expressed.

Zarathustra is a seafarer addressing sailors on a ship when he shares a vision and a riddle. It is important to note that he addresses only those bold searchers, attempters, and tempters, those who have, like Odysseus and his men, "embarked with cunning sails upon terrifying seas," those whose souls are lured by sirens' songs to founder in confounding depths.[161] Alluding to another Greek myth, he tells the riddle only to those who "do not want to grope along a thread with cowardly hand"—those who are not like Theseus who, after killing the Minotaur, needed a thread to find his way out of the labyrinth. Zarathustra tells the vision only to those willing to explore unexplored seas, taking up the temptations of dangerous thought experiments, not relying on the thread of Theseus, using the discriminating mind and its thread of sound argument to find one's way through the labyrinth.

The vision unfolds as a dream sequence with scenes suddenly shifting disconnectedly. After trudging through a desolate landscape with a dwarf, the spirit of gravity, sitting on his shoulder pouring leaden thoughts into his ear, Zarathustra confronts the dwarf in the gateway of the moment, calling up from his depths the thought of eternal recurrence. The problem of suffering that leads to the longing for another world, reducing this earth to a wasteland, is the problem of time and time's passing—wishing to be somewhere else than the present moment. In "The Wandering Dance," Parkes explains that the way to the overhuman, that rainbow after long storms, "involves abandoning the egoistic will that is impotent against the past and so wreaks revenge by branding its passing as deserved and all temporal experience as nugatory. To redeem the past by overcoming the 'spirit of revenge' is to learn to 'will backwards.'"[162] The thought of eternal recurrence forces one to face this moment, as the moment comes back again and again for all eternity. Here Parkes explains how the thought of eternal recurrence is a matter of seeing things as they are: "And if will to power is *what* everything is, eternal recurrence is how, the *way* all things are."[163] Parkes further explains that "to will the recurrence of a single good thing is to will the recurrence of everything bad" and this, Parkes emphasizes, suggests the "interdependence of all things."[164] Parkes then turns to what he describes as the "magnificently Dionysian culmination" of *Zarathustra* in the penultimate section of the book, in the passage where Zarathustra exclaims, "Did you ever say Yes to a single joy? Oh, my friends, then you said Yes to *all* woe as well. All things are chained together, entwined, in love—."[165]

The scene shifts to Zarathustra alone in the most desolate moonlight, not sure whether he is awake or dreaming; there is the sound of a nearby howling dog, and then that most horrible image—a young shepherd, writhing, convulsing, with a heavy black snake hanging out of his mouth. The snake is an obvious reference to the Ouroboros imagery, of a serpent biting its own tail, found in ancient Egypt and later in Gnosticism and alchemical texts, sometimes used as a symbol of the cyclical nature of time. In Zarathustra's dream vision the shepherd is choked up, nauseated by the thought of eternal recurrence. Zarathustra then challenges the bold seafarers to guess the riddle and interpret the vision. The shepherd finally heeds Zarathustra's call and bites through the snake, the thought of eternal recurrence, and he jumps up laughing: "No longer shepherd, no longer human—one transformed, illumined, who *laughed*!"[166]

In *The Self-Overcoming of Nihilism*, Nishitani calls attention to this laughter: "The most remarkable feature of Nietzsche's 'religion' may be the sound of *laughter* that echoes through it."[167] He compares Nietzsche's thought with Zen Buddhism, "the history of which," he notes, "also reverberates with laughter of various kinds."[168] In "The Wandering Dance," Parkes also emphasizes the importance of laughter. He points out that both *Zarathustra* and the *Zhuangzi* "are deeply humorous—each constituting perhaps the most amusing philosophy of its tradition—emphasizing laughter as an often necessary concomitant of insight into the way things are."[169] The importance of laughter in Zen is part of Zhuangzi's influence in Zen. Toward the end of *Beyond Good and Evil* Nietzsche proposes "an order of rank among philosophers depending on the rank of their laughter—all the way up to those capable of golden laughter."[170] Then there is that last mad letter, perhaps the last thing Nietzsche ever wrote, just a couple of days after he collapsed on the streets of Turin, where he explains that he is "condemned to while away the next eternity with bad jokes."[171] When one takes seriously Nietzsche's emphasis on laughter one starts to get the sneaking suspicion that the thought of eternal recurrence may be the bad joke with which Nietzsche is whiling away eternity. Gilles Deleuze once suggested that one really doesn't get Nietzsche if one doesn't get the jokes: "Those who read Nietzsche without laughing, without laughing often and a lot, and at times doubling up with laughter, might as well not be reading Nietzsche."[172]

One imagines Nietzsche setting out on a hike that day in August of 1881 on his first trip to Sils-Maria, likely his first hike along Lake Silvaplana. He already has the idea of a book set in the landscape of high mountains with his fictional Zarathustra coming down with his urgent message about staying true to the earth. As he makes his way along the lake, he is thinking about the sad tragic history of humanity, rooted in the longing for another world. He's thinking about the problem of suffering, wanting to be somewhere else than the present moment. With his poor eyesight he might not have noticed the rock in the distance, but he's thinking about this problem of time and time's passing. He's certainly aware of the Ouroboros imagery, and the ancient myths of eternal cycles of time. As he comes out of the woods and around a bend the rock suddenly looms up before him, and the idea of eternal recurrence hits him like a lightning-bolt. In *Ecce Homo*, Nietzsche refers to "the sacred spot where the first lightning-bolt of the thought of Zarathustra had flashed before me."[173] The lightning-bolt connects Zarathustra with Dionysus as the principal means of the god's power of transfiguration in Greek myth and tragedy is the lightning-bolt. Lightning also evokes Heraclitus who wrote, "A thunderbolt [steers] all things."[174] Struck by the thunderous lightning-bolt of the thought of eternal recurrence, Nietzsche stands before the pyramidal block of stone, and there he realizes that if the problem that reduces the earth to a wasteland is the longing for eternity in another world, then Zarathustra will be the teacher of eternal recurrence.

Nietzsche presents the thought of eternal recurrence as the heaviest weight, the most serious thought, the thought that "breaks the history of humanity in two."[175] Despite this seriousness, perhaps it is important to imagine Nietzsche roaring with laughter after the thought flashed before him at the rock. "So, you want eternity? You want to flee this earth and leave it behind?—Well try this eternity on!" It may be a bad joke to be sure, and the joke is then another joke, part of Nietzsche's deconstruction of the seriousness of philosophers naively believing they are awake when they are dreaming. But like a kōan it is a catalyst for a psychical transformation of human beings. Affirming eternal recurrence is just as impossible or absurd as the bodhisattva vow to return to life over and over in order to save each and every one of the numberless beings in the universe. But both the thought of eternal recurrence and the bodhisattva vow cut off the longing for another world, focusing our attention on the present moment, making possible that abundant generosity, which, Parkes suggests, opens "a radically new way of being" that "is profoundly relevant for ecological thinking."[176]

Notes

1   *From* Thus Spoke Zarathustra, *Translated by Graham Parkes © 2005. Reprinted by arrangement with Oxford University Press. Reproduced with permission of the Licensor through PLSclear.*

2   Wim Wenders, *Until the End of the World*, Warner Brothers, 1991.

3   Parkes's most recent book particularly emphasizes the relevance of Nietzsche's thought in this time of climate change. See Graham Parkes, *How to Think about the Climate Crisis: A Philosophical Guide to Saner Ways of Living* (London: Bloomsbury Academic, 2020a).
4   See Greg Garrard, *Ecocriticism* (London and New York: Routledge, 2004), 90.
5   Graham Parkes, "Nietzsche's Environmental Philosophy: A Trans-European Perspective," *Environmental Ethics* 27 (1) (2005): 77.
6   Parkes's translation from *The Antichrist* §14. Nietzsche's critique of Western philosophy, along with the related polemic against Christianity, bears some resemblance to the thesis by historian Lynn White Jr. that "Christianity bears a huge burden of guilt" for the ecological crisis (Lynn White Jr., "The Historical Roots of Our Ecologic Crisis," *Science* 155 (1967): 1206). White's paper became influential in the environmental movement after it came out in 1967, and it was quite controversial for its critique of Christianity. White emphasizes that "Christianity is the most anthropocentric religion the world has seen" and he traces the roots of the ecological crisis to the dualism of man and nature and the teleological view that "it is God's will that man exploit nature for his proper ends" (White, "The Historical Roots of Our Ecologic Crisis," 1205; Parkes, "Nietzsche's Environmental Philosophy," 85).
7   Parkes's translation from *On the Genealogy of Morals* III, §9. Parkes, "Nietzsche's Environmental Philosophy," 85.
8   Graham Parkes, "The Wandering Dance: *Chuang Tzu* and *Zarathustra*," *Philosophy East and West* 33 (3) (1983): 235.
9   An interesting feature of White's paper is that, toward the end, he praises the "beatniks" who "show a sound instinct in their affinity for Zen Buddhism, which conceives of the man-nature relationship as very nearly the mirror image of the Christian view" (White, "The Historical Roots of Our Ecologic Crisis," 1206). There has been a great deal of work exploring the relevance of Daoism and Zen in considering the ecological crisis in recent years. See, for example, John L. Culliney and David Jones, *The Fractal Self: Science, Philosophy, and the Evolution of Human Cooperation* (Honolulu: University of Hawai'i Press, 2017). For the affinity for Zen in the work of one of those "beatniks," see Jason M. Wirth, *Mountains and Rivers and the Great Earth: Reading Gary Snyder and Dōgen in an Age of Ecological Crisis* (Albany: State University of New York Press, 2017); *Nietzsche and Other Buddhas* (Bloomington, Indiana: Indiana University Press, 2019) also explores the relationship between Nietzsche's thought and Zen in *Nietzsche and Other Buddhas*.
10 Graham Parkes, "Staying Loyal to the Earth: Nietzsche as an Ecological Thinker," in *Nietzsche's Futures*, edited by John Lippitt (Basingstoke: Macmillan, 1999), 185.
11 Parkes, "The Wandering Dance," 235.
12 It is worth noting that the sharp separation between human beings and nature, which is such a distinctive feature of Western thought, does not arise in Chinese philosophy because of what Roger Ames has called the "assumed mutuality and collaterality" of the "three powers" of Heaven (*tian* 天), Earth (*di* 地), and human beings (*ren* 人) in Chinese cosmology (Roger T. Ames, "Roger T. Ames Responds," in *Appreciating the Chinese Difference: Engaging Roger T. Ames on Methods, Issues, and Roles*, edited by Jim Behuniak [Albany: State University of New York Press, 2018], 259). The notion of Heaven, as Parkes explains, did not "signify a transcendent realm beyond this world, as in the dualistic metaphysics of the Platonist or Christian traditions, since the three powers were always regarded as belonging together" (Graham Parkes, "The Art of Rulership in the Context of Heaven and Earth," in *Appreciating the*

Chinese Difference: Engaging Roger T. Ames on Methods, Issues, and Roles, edited by Jim Behuniak [Albany: State University of New York Press, 2018], 66).
13 Parkes, "The Art of Rulership in the Context of Heaven and Earth," 79.
14 Parkes, "The Art of Rulership in the Context of Heaven and Earth," 82.
15 Parkes, "The Wandering Dance," 236.
16 Parkes, "The Wandering Dance," 237.
17 Parkes, "The Art of Rulership in the Context of Heaven and Earth," 81.
18 Graham Parkes, "In the Light of Heaven before Sunrise: Zhuangzi and Nietzsche on Transperspectival Experience," in *Daoist Encounters with Phenomenology: Thinking Interculturally about Human Existence*, edited by David Chai (New York: Bloomsbury, 2020b), 61.
19 Friedrich Nietzsche, *Beyond Good and Evil*. Translated by Walter Kaufmann (New York: Vintage Books, 1966), 3.
20 Parkes's translation from *On the Genealogy of Morals* III, §12. Parkes, "In the Light of Heaven before Sunrise," 71.
21 Parkes, "In the Light of Heaven before Sunrise," 71.
22 Parkes, "In the Light of Heaven before Sunrise," 70.
23 Parkes, "The Wandering Dance," 242–3.
24 Parkes, "The Wandering Dance," 241.
25 Friedrich Nietzsche, *The Joyous Science*. Translated by R. Kevin Hill (New York: Penguin Classics, 2018), 73.
26 Zhuangzi, *Zhuangzi: Basic Writings*. Translated by Burton Watson (New York: Columbia University Press, 2003), 45.
27 Parkes, "The Wandering Dance," 242.
28 Parkes, "The Wandering Dance," 243.
29 Parkes, "The Wandering Dance," 243–4.
30 Parkes, "The Wandering Dance," 243.
31 Graham Parkes, "Human/Nature in Nietzsche and Taoism," in *Nature in Asian Traditions of Thought*, edited by J. Baird Callicott and Roger T. Ames (Albany: State University of New York Press, 1989), 86.
32 Parkes, "The Wandering Dance," 243–4.
33 Parkes, "The Wandering Dance," 244.
34 Parkes, "Staying Loyal to the Earth," 170.
35 Parkes, "Staying Loyal to the Earth," 169; Nietzsche, *The Joyous Science,* 122.
36 Parkes, "Staying Loyal to the Earth," 179.
37 Friedrich Nietzsche, *Thus Spoke Zarathustra: A Book for Everyone and Nobody*. Translated by Graham Parkes (Oxford: Oxford University Press, 2005), 143.
38 Parkes, "Staying Loyal to the Earth," 172.
39 Graham Parkes, "Nature and the Human 'Redivinized': Mahāyāna Buddhist Themes in *Thus Spoke Zarathustra*," in *Nietzsche and the Divine*, edited by John Lippit and Jim Urpeth (Manchester: Clinamen Press 2000), 192.
40 Parkes, "Nietzsche's Environmental Philosophy," 89.
41 Parkes, *How to Think about the Climate Crisis*, 178.
42 Zhuangzi, *Chuang-tzu: The Seven Inner Chapters and other writings from the book Chuang-tzu*. Translated by A. C. Graham (London: George Allen & Unwin, 1981), 148.
43 Zhuangzi, *Chuang-tzu,* 84.
44 Parkes, "Nature and the Human 'Redivinized,'" 192–3.
45 Parkes, "Nietzsche's Environmental Philosophy," 81.
46 Parkes, "Nietzsche's Environmental Philosophy," 87.

47 Parkes, "In the Light of Heaven before Sunrise," 70.
48 Zhuangzi, *Chuang-tzu*, 68.
49 Parkes, "In the Light of Heaven before Sunrise," 67.
50 Parkes, "In the Light of Heaven before Sunrise," 70–1.
51 This is the view, as Steven Vogel explains, of "nature" as "a stable world that precedes humans, ontologically prior to human activity and to the social structures (and the language) within which that activity takes place" (Steven Vogel, "Nature as Origin and Difference: On Environmental Philosophy and Continental Thought," *Philosophy Today*, SPEP Supplement, 1998, 170). As Vogel also explains, the poststructuralist project of deconstruction that begins with Derrida "is a project of taking that which appears to be original, foundational—in a word: natural—and revealing the complex processes of linguistic and social construction required to produce that appearance" (Vogel, "Nature as Origin and Difference," 170).
52 Vogel addresses this concern, noting that there has been some anxiety among environmental philosophers since there is this "vague sense that 'postmodernism,' by turning the whole world into a text, denies the very existence of nature and therefore the significance of attempts either to understand the dangers to which it is currently exposed or to argue for the need to protect it" (Vogel, "Nature as Origin and Difference," 169; Graham Parkes, "Zhuangzi and Nietzsche on the Human and Nature," *Environmental Philosophy* 10 [1] [2013]: 2).
53 Nietzsche, *The Joyous Science*, 13.
54 Heraclitus, "The Fragments," in *The Art and Thought of Heraclitus: A New Arrangement and Translation of the Fragments with Literary and Philosophical Commentary*, edited by Charles H. Kahn (Cambridge: Cambridge University Press, 1979), 33.
55 Nietzsche, *Beyond Good and Evil*, 2.
56 Parkes, "In the Light of Heaven before Sunrise," 78.
57 Nietzsche, *Beyond Good and Evil*, 46–7.
58 This suggestion that we should understand that the world that concerns us is a fiction anticipates Lyotard's famous characterization of the postmodern condition as an "incredulity toward metanarratives" (Jean-François Lyotard, *The Postmodern Condition: A Report on Knowledge*. Translated by Geoffrey Bennington and Brian Massumi [Minneapolis: University of Minnesota Press, 1984], xxiv–xxv). This is also what Derrida meant by the controversial phrase "*There is nothing outside of the text* [*il n'y a pas de hors-texte*]" (Jacques Derrida, *Of Grammatology*. Translated by Gayatri Chakravorty Spivak (Baltimore: Johns Hopkins University Press, 1974, 158), often misunderstood as the claim that there is nothing outside of language. What the phrase really says is that "there is no outside-text" or, in other words, there is no truth without veils, no access to a reality that is not already a product of interpretation.
59 Graham Parkes, "Renatured Humans on a Sacred Earth: The Power of Nietzsche's Ecological Thinking," in *A New Politics for Philosophy: Perspectives on Plato, Nietzsche, and Strauss*, edited by George A. Dunn and Mango Telli (Lanham, MD: Lexington Books, 2022), 230.
60 In this latest essay, Parkes draws from Lawrence Lampert in contending that Nietzsche does suggest an experience where we can "get out of our human corner" and see things as they really are. Parkes quotes Lampert in raising the crucial question and then provides the response: "'Could Nietzsche have left his own human corner and arrived at a view of nature free of humanization and in some fundamental

sense true to nature?' Lampert thinks that he could, and that the task for Nietzsche was then how to convey the significance of these experiences, 'experiences only beginning to be felt by others'" (Laurence Lampert, *Nietzsche and Modern Times: A Study of Bacon, Descartes, and Nietzsche* (New Haven, CT: Yale University Press, 1993, 335–6; Parkes, "Renatured Humans on a Sacred Earth," 232).

61 Friedrich Nietzsche, *Ecce Homo: How to Become What You Are*. Translated by Duncan Large (Oxford: Oxford University Press, 2007), 95.
62 Friedrich Nietzsche, *The Birth of Tragedy*. Translated by Walter Kaufmann (New York: Vintage Books, 1967a), 23.
63 Against this narrative, Nietzsche offers a different interpretation: "The 'kingdom of heaven' is a state of the heart—not something that is to come 'above the earth' or 'after death.' ... The 'kingdom of God' is nothing that one expects; it has no yesterday and no day after tomorrow, it will not come in 'a thousand years'—it is an experience of the heart; it is everywhere, it is nowhere" (Friedrich Nietzsche, *The Antichrist* in *The Portable Nietzsche*. Translated by Walter Kaufmann [New York: Penguin Books, 1977], 608).
64 Nietzsche, *The Birth of Tragedy*, 46; Friedrich Nietzsche, *Die Geburt der Tragödie aus dem Geiste der Musik* (Frankfurt am Main: Insel Verlag, 1987), 46.
65 John Sallis suggests this preview of Nietzsche's mature thought in *The Birth of Tragedy* in the "shimmering shining" which results when the Apollonian and Dionysian are brought together in Greek tragedy: "Tragedy both reveals and conceals the Dionysian abyss. And yet, such revealing and concealing are no longer simply binary opposites, nor is the disclosure thus to be thought as a mere mean between these opposites. In the determination of tragedy Nietzsche is under way to a thinking of disclosure that would differentiate it decisively from mere uncovering (limited by a symmetrical opposite). For it is a matter of a disclosure of the abyss, of that which withdraws from any presentation, of that which cannot as such be present (or absent, as long as absence is considered merely the complementary opposite of presence). It is a matter of a disclosure in which, nonetheless, the unpresentable is brought to shine in the distance as sublime" (John Sallis, *Crossings: Nietzsche and the Space of Tragedy* [Chicago: University of Chicago Press, 1991], 100).
66 Nietzsche, *The Joyous Science*, 133–4.
67 Nietzsche, *The Joyous Science*, 225.
68 Parkes, "The Wandering Dance," 244.
69 Parkes, "The Wandering Dance," 244.
70 Nietzsche, *The Joyous Science*, 272; Friedrich Nietzsche, *Die Fröhliche Wissenschaft* (Frankfurt am Main: Insel Verlag, 1982), 271.
71 Nietzsche, *The Joyous Science*, 133; *Die Fröhliche Wissenschaft*, 137.
72 Nietzsche, *The Joyous Science*, 226.
73 Nietzsche, *The Joyous Science*, 73.
74 Friedrich Nietzsche, *The Will to Power*. Translated by Walter Kaufmann and R. J. Hollingdale (New York: Vintage Books, 1968), 435.
75 Nietzsche, *The Will to Power*, 419, 452.
76 "Here, when the danger to his will is greatest, *art* approaches as a saving sorceress, expert at healing. She alone knows how to turn these nauseous thoughts about the horror or absurdity of existence into notions with which one can live" (Nietzsche, *The Birth of Tragedy*, 60). This is, at least in part, why Nietzsche suggests that the high point of Greek culture was not Socrates and Plato, but rather, Aeschylus and Sophocles. Socrates and Plato had a naively optimistic view that it was possible to

awaken from the dream and discover the truth about the nature of things, while Aeschylus and Sophocles had the courage to face the abysmal absurdity of existence.

77 Zhuangzi, *Chuang-tzu,* 148.
78 Zhuangzi, *Chuang-tzu,* 38–9.
79 Zhuangzi, *Zhuangzi: The Essential Writings: With Selections from Traditional Commentaries.* Translated by Brook Ziporyn (Indianapolis: Hackett Classics, 2009), 19.
80 Parkes, "The Wandering Dance," 242.
81 Parkes, "The Wandering Dance," 239.
82 Parkes, "The Wandering Dance," 239.
83 Whereas Confucius put a lot of emphasis on proper naming (*zhengming* 正名), the *Daodejing* emphasizes the nameless (*wuming* 無名): "Way-making (*dao* 道) that can be put into words is not really way-making,/ And naming (*ming*) that can assign fixed reference to things is not really naming./ The nameless (*wuming* 無名) is the fetal beginnings of everything that is happening" (Roger T. Ames and David L. Hall, *Dao De Jing: A Philosophical Translation* [New York, NY: Ballantine Books, 2010], 77). The yin emphasis of the Daodejing is suggested in the abundant yin imagery and the plethora of wu (無) terms—such as wuming (無名) "nameless," wushi (無事) "non-interfering," wuyu (無欲) "objectless desire," wuzhi (無知) "unprincipled knowing," and, of course, wuwei (無為) "non-coercive action." Needless to say, wu (無) can serve as a negation, but can also mean "emptiness" as when Laozi suggests that it is the emptiness of a clay vessel that makes it useful (Ames and Hall, *Dao De Jing*, 91). It is a decidedly yin term, and thus its frequent use in the text suggests this yin emphasis in the *Daodejing.*
84 Robin Wang draws attention to the yin emphasis in Daoism: "The spontaneous potency of the *Dao* is associated with the female body, which is a common metaphor for the *Dao* in the *Daodejing*. It reveals not just the importance of yin and its generative force, but also designates a yin origin that is hidden, implicit, or empty" (Robin R. Wang, *Yinyang: The Way of Heaven and Earth in Chinese Thought and Culture* [Cambridge: Cambridge University Press, 2012], 55). She also suggests that this yin emphasis in Daoism might be explained as a strategy similar to that employed in traditional Chinese medicine: "For example, in Chinese traditional medical diagnoses, too much yin in the body is a sickness of yang, and too much yang in the body is a sickness of yin. Changes in yin will affect yang, and vice versa. This mutual resonance is crucial to yinyang as a strategy because it entails that one can influence any element by addressing its opposite, which in practice most often takes the form of responding to yang through yin" (Wang, *Yinyang*, 10). Rather than emphasizing an experience of the presence of beings, the yin strategy, as Wang suggests, would seem to be the more modest approach of the awareness of what is hidden: "This attentiveness to the hidden background from which things originate and transform is an awareness of the yin side and is a common strategy of yinyang thought" (Wang, *Yinyang*, 17). "What sages rely on are the yin factors: yin emphasizes background and hidden structures. The yang specifies what is dominant, open, and in front" (Wang, *Yinyang*, 144).
85 Zhuangzi, *Chuang-tzu,* 147; Parkes, "The Wandering Dance," 239.
86 Parkes, "In the Light of Heaven before Sunrise," 70.
87 It is interesting to consider whether the Nietzschean poststructuralist position of cherishing the modesty of nature, and thus recognizing that the world that concerns us is always a fiction, is more consistent with the yin emphasis of Daoism. It may

seem outrageous to even consider Nietzsche's thought as yin, but as Wang explains in explicating yinyang theory, everything depends on context: "Because of this dependence on context, a single thing can be yin in one way and yang in another. [...] It is also this difference that enables yinyang as a strategy—to act successfully, we must sometimes be more yin and sometimes more yang, depending on the context" (Wang, *Yinyang*, 7–8). As she further explains: "Everything and every event can be seen as either yin or as yang, and then related with other things on this basis" (Wang, *Yinyang*, 20). The Italian philosopher Gianni Vattimo might be seen as articulating this yin strategy when he defends Nietzsche's nihilistic "weak thought" (*Il pensiero debole*) as a strategic countermovement in response to the history of Western thought: "I interpret 'nihilism' in the sense first given it by Nietzsche: the dissolution of any ultimate foundation, the understanding that in the history of philosophy, and of western culture in general, 'God is dead;' and 'the real world has become a fable'" (Gianni Vattimo, *Nihilism & Emancipation: Ethics, Politics, & Law*. Translated by William McCuig [New York: Columbia University Press, 2004], xxv).

88 Nietzsche, *The Antichrist* in *The Portable Nietzsche*, 586–7.
89 Nietzsche, *The Antichrist* in *The Portable Nietzsche*, 587.
90 Nietzsche, *Beyond Good and Evil*, 68.
91 Graham Parkes, "Nietzsche and East Asian Thought: Influences, Impacts, and Resonances," in *The Cambridge Companion to Nietzsche*, edited by Bernd Magnus and Kathleen M. Higgins (Cambridge: Cambridge University Press 1996), 373.
92 Bret W. Davis, "Zen after Zarathustra: The Problem of the Will in the Confrontation Between Nietzsche and Buddhism," *Journal of Nietzsche Studies* 28 (2004): 113.
93 Davis, "Zen after Zarathustra," 89.
94 Davis, "Zen after Zarathustra," 98.
95 Davis, "Zen after Zarathustra," 105.
96 Davis, "Zen after Zarathustra," 108.
97 Graham Parkes, "Will to Power as Interpretation," *Journal of Nietzsche Studies* 46 (1) (2014a): 42–3.
98 Graham Parkes, "Zarathustra and Asian Thought: A Few Final Words," *Journal of Nietzsche Studies* 46 (1) (2014b): 87.
99 Parkes, "Will to Power as Interpretation," 44.
100 Parkes, "Will to Power as Interpretation," 43.
101 Parkes, "Will to Power as Interpretation," 54.
102 Nietzsche, *The Will to Power*, 550.
103 Parkes, "Nietzsche's Environmental Philosophy," 84.
104 Parkes, "Will to Power as Interpretation," 44.
105 Nietzsche's conception that this play of forces that is the will to power is at once the whole universe, but also at play in human beings and the smallest organisms, suggests the fractal patterning which Culliney and Jones have called attention to in their work, *The Fractal Self*. They draw on the metaphor of Indra's Net from the *Avataṃsaka Sūtra* in which the universe is depicted as a net of jewels stretching infinitely in all directions, and that when one examines each jewel one finds "each of the many of them reflects the light of every other" (Culliney and Jones, *The Fractal Self*, 2). They go on to describe this fractal patterning in the emergence of the cosmos: "This fractally structured emergence subsequently enabled development of the cosmos' complex forms and behaviors in ways that we are just beginning to understand. Complexity in the cosmos organized itself from the bottom up and built,

across scale from nanometers to parsecs and through billions of years, worlds so wondrous that they intersect with dreams" (Culliney and Jones, *The Fractal Self*, 30).
106 Nietzsche, *Beyond Good and Evil*, 203.
107 Davis, "Zen after Zarathustra," 113.
108 Ronald Hayman, *Nietzsche: A Critical Life* (New York: Penguin Books, 1982), 320.
109 Nietzsche, *Beyond Good and Evil*, 201.
110 Nietzsche, *Beyond Good and Evil*, 203; Friedrich Nietzsche, *Jenseits von Gut und Böse* (Frankfurt am Main: Insel Verglag, 1984), 179.
111 Nietzsche, *Beyond Good and Evil*, 203.
112 Nietzsche, *Thus Spoke Zarathustra*, 66.
113 Nietzsche, *Thus Spoke Zarathustra*, 51.
114 Parkes, "Staying Loyal to the Earth," 185.
115 Parkes, "In the Light of Heaven before Sunrise," 72.
116 Parkes, "Will to Power as Interpretation," 51.
117 The Sanskrit expression for "seeing things as they are" is *yathābhūtaṃ darśanaṃ*. We can see the expression in *The Pali Canon*, in verse 203 of the *Dhammapada* where the Buddha explains that seeing things as they are leads to enlightenment: "Greediness is the worst of diseases; propensities are the greatest of sorrows. To him who has known this truly, *nirvāṇa* is the highest bliss (*jigacchā paramā rogā saṅkhārā paramā dukhā /etaṃ ñatvā yathābhūtaṃ nibbāṇaṃ paramaṃ sukham*)" (S. Radhakrishnan, *The Dhammapada* [Oxford: Oxford University Press, 1950], 126). The Pali term *saṅkhārā* (Skt: *saṃskāra*), rendered here as "propensities," is one of the five aggregates (Skt: *skandha*; Pali *khanda*) that make up the self in the Buddha's teaching of "no-self" (Skt: *anātman*; Pali: *anatta*). Here *saṅkhārā* refers to the "mental constructs" that shape the way all conditioned things show up for us.
118 In *The Fire Sermon*, the Buddha seems to suggest that to live is to burn. He goes through all the parts of the self, explaining how all is burning. The repeating refrain is when he suggests what we are burning with: "Burning with what? Burning with the fire of lust, with the fire of hate, with the fire of delusions; I say it is burning with birth, aging and death, with sorrows, with lamentations, with pains, with griefs, with despairs" (Walpola Rahula, *What the Buddha Taught: Revised and Expanded Edition with Texts from Suttas and Dhammapada* [New York: Grove Press, 1974], 95). In the PBS documentary *The Buddha: The Story of Siddhartha*, two contrasting interpretations of *The Fire Sermon* are presented, and in these two views the fundamental question concerning Buddhism may be brought to light. Max Moerman explains "We're on fire. We may not know it, but we're on fire and we have to put that fire out. We're burning with desire, burning with craving, everything about us is out of control." The poet W. S. Merwin offers a different take, suggesting that we have to find a way to turn the three poisons around to their opposites: "The Buddha goes on to talk about the three poisons, greed, anger, and ignorance, and how the three poisons are what is making the fire, and the way out of doing this is, not to deny the three poisons, but to recognize that if you turn them around, you come to their opposites; instead of greed you have generosity, instead of anger you have compassion, and instead of ignorance you have wisdom" (David Grubin, *The Buddha: The Story of Siddhartha* [PBS, 2010]).
119 This is expressed in Vasubhandu's classic summary of Yogācāra teaching in the *Thirty Verses*, where he explains how the metaphors of "self" and "nature" take place in the transformation of consciousness: "This transformation of consciousness (*vijñāna*) is a discrimination (*vikalpa*), and as it is discriminated, it does not exist [in-itself],

and so everything is perception-only (*vijñapti-mātra*)" (Stefan Anacker, *Seven Works of Vasubandhu: The Buddhist Psychological Doctor* [Delhi: Motilal Banarsidass, 1984], 187). This doctrine of *vijñapti-mātra* draws a comparison with the view Nietzsche already expressed in his early essay "Truth and Lie in a Nonmoral Sense," in which he explains that "the intellect unfolds its principal powers in dissimulation (*Verstellung*)" (Nietzsche 1979, 80). (The Sanskrit *vi* is equivalent to *dis* in English and *Ver* in German.) Nietzsche's point is that the intellect does not unfold its powers in simulation, copying reality; it is instead always adding, selecting, interpreting reality from particular perspectives. Even in this early text, Nietzsche suggests this process takes place, to some extent, below the surface of consciousness, and this anticipates his mature view that it is not the conscious ego that interprets, but the will to power in the unconscious depths.

120 Dan Lusthaus challenges the interpretation of *vijñapti-mātra* as a metaphysical idealism emphasizing that "no Indian Yogācāra text ever claims that the world is created by mind." He goes on to describe correct cognition as "the removal of those obstacles which prevent us from seeing causal conditions in the manner they actually become." He further explains that correct cognition is "euphemistically called *tathatā*, 'suchness,' which Yogācāra texts are quick to point out is not an actual thing, but only a word (*prajñapti-mātra*)." Nevertheless, Lusthaus concludes that "Yogācārins describe enlightenment as resulting from Overturning the Cognitive Basis (*āśraya-parāvṛtti*), i.e., overturning the conceptual projections transforms the basic mode of cognition from consciousness (*vi-jñāna*, dis-cernment) into *jñāna* (direct knowing). Direct knowing was defined as non-conceptual (*nirvikalpa-jñāna*), i.e., devoid of interpretative overlay" (Dan Lusthaus, *Buddhist Phenomenlogy: A Philosophical Investigation of Yogācāra Buddhism and the* Ch'eng Wei-shih lun [London and New York: RoutledgeCurzon, 2002], 534–7).

121 In the *Thirty Verses*, Vasubhandu explains that seeing everything in its suchness (*tathatā*) is nothing other than getting the wisdom of the *Prajñāpāramitā* teaching of *śūnyatā* that everything is empty of own-being (*svabhava*): "It is the ultimate truth of all events, as so it is 'Suchness' (*tathatā*)." Instead of suggesting the cessation of the process of *vijñapti-mātra*, however, the verse ends: "Since it is just so all the time, and it is just perception-only (*vijñapti-mātra*)" (Anacker, *Seven Works of Vasubandhu*, 187).

122 Graham Parkes, *Composing the Soul: Reaches of Nietzsche's Psychology* (Chicago: University of Chicago Press, 1994), 149–51.

123 This is Bret Davis's translation in his recent book *Zen Pathways* (Bret W. Davis, *Zen Pathways: An Introduction to the Philosophy and Practice of Zen Buddhism* [Oxford: Oxford University Press, 2022], 29).

124 Okamura explains that Dōgen is introducing "an example used in Yogācāra called 'the four views on one and the same water'" (Shokaku Okamura, *The Mountains and Waters Sūtra: A Practitioner's Guide to Dōgen's "Sansuikyō"* [Somerville, MA: Wisdom Publications, 2018]), 161; Eihei Dōgen, "Sansuikyō," in *The Mountains and Waters Sūtra: A Practitioner's Guide to Dōgen's "Sansuikyō…,"* edited by Shohaku Okumura, translated by Carl Bielefeldt (Somerville, MA: Wisdom Publications, 2018), 29.

125 Dōgen, "Sansuikyō," 29.

126 Dōgen, "Sansuikyō," 32.

127 Dōgen, "Sansuikyō," 30.

128 Okamura explains: "This is Dōgen's expression of emptiness, with no fixed and permanent self-nature. Everything is completely interdependent origination; nothing

is fixed. This is the reality of all beings according to Dōgen. Everything dwells in its Dharma position at this moment. But even though we dwell in this Dharma position, at the same time we are liberated from this position. We cannot stay here; in the next moment, we go somewhere else. This constant flowing, according to Dōgen, is the reality of our life." Okamura goes on to describe this as an incredibly liberating view: "It allows us to release our fixed concept of ourselves, our idea of human life, our point of view, and our system of values" (Okamura, *The Mountains and Waters* Sūtra, 168). Okamura goes on to explain: "Dōgen and the *Heart Sūtra* are saying nothing is fixed, and this is liberation" (Okamura, *The Mountains and Waters* Sūtra, 170).

129 Parkes, "The Wandering Dance," 247.
130 "Dōgen didn't like the term *kenshō*: it implies that our self (our body and mind, the five aggregates) is separate from nature and that our (non-physical) eyes can see it. In reality the nature cannot be seen; it cannot be the object of the subject, because the nature is ourselves. We cannot see ourselves; our eyes cannot see our eyes. There is no way we can see the nature; that is Dōgen's point" (Okamura, *The Mountains and Waters* Sūtra, 120).
131 Davis, *Zen Pathways*, 28.
132 Parkes, "The Wandering Dance," 243.
133 Davis, *Zen Pathways*, 28.
134 Parkes, "The Wandering Dance," 239.
135 Keiji Nishitani, *Religion and Nothingness*. Translated by Jan Van Bragt (Berkeley: University of California Press, 1982), 215.
136 Keiji Nishitani, *The Self-Overcoming of Nihilism*. Translated by Graham Parkes and Aihara Setsuko (Albany: State University of New York Press, 1990), 180.
137 Nishitani, *The Self-Overcoming of Nihilism*, xxi.
138 Nietzsche, *Ecce Homo*, 65.
139 Parkes, "Renatured Humans on a Sacred Earth," 224.
140 Nietzsche, *Ecce Homo*, 71.
141 Parkes, "Renatured Humans on a Sacred Earth," 226–7; Laurence Lampert, *Nietzsche's Teaching: An Interpretation of Thus Spoke Zarathustra* (New Haven, CT: Yale University Press, 1986), 176.
142 Parkes, "Renatured Humans on a Sacred Earth," 239, 242.
143 Parkes, "Nietzsche's Environmental Philosophy," 89.
144 Nietzsche, *The Joyous Science*, 194; *Die Fröhliche Wissenschaft*, 189.
145 Nietzsche, *Thus Spoke Zarathustra*, 65.
146 Throughout *Zarathustra* Nietzsche plays with the fact that both *geben* and *schenken* can mean to "give," "present," "bestow," or even "confer." *Geschenk* can be rendered as "gift" or "present," and thus when Zarathustra explains at the beginning of the Prologue that the reason he has come down from the mountain is to bring human beings "*ein Geschenk*," Parkes renders this as "a present" whereas Kaufmann uses "a gift." Parkes translates *schenkende Tugend* as "bestowing virtue" and Kaufmann uses "gift-giving virtue." In the passage from *The Joyful Science* above when Nietzsche explains that there is no value in itself because value "has been given [*gegeben*] and bestowed [*geschenkt*] upon it," Nietzsche's text goes on to say "*und* wir *waren diese Gebenden und Schenkenden*" (Nietzsche, *Die Fröhliche Wissenschaft*, 189) that might more literally be rendered "and *we* are these givers and bestowers."
147 Nietzsche, *Thus Spoke Zarathustra*, 10.
148 This theme of the gift is the thread running through Derrida's reflections in *The Politics of Friendship*. Toward the end of the text, Derrida turns to the section "On

the Friend" in which Zarathustra says, not once but thrice, that "woman is not yet capable of friendship" (Nietzsche, *Thus Spoke Zarathustra*, 50). But, as Derrida points out, Zarathustra goes on to say that this is also true for men: "Confirming what has just been pronounced on women, Zarathustra suddenly *turns towards* men—he apostrophizes them, accusing them, in sum, of being in the same predicament. Woman was not man, a man free and capable of friendship, and not only of love. Well now, neither is man a man. Not yet. And why not? Because he is not generous enough, because he does not know how to give enough to the other. To attain to this infinite gift, failing which there is no friendship, one must know how to give to the enemy. And of this, neither woman nor man (up until now) is capable" (Jacques Derrida, *Politics of Friendship*. Translated by George Collins [London and New York: Verso, 1997], 283). Derrida goes on to point out the irony of the resonance of Zarathustra's teaching of this gift of friendship with the message of Jesus: "For is not what has just been repeated, doubled, parodied, perverted and assumed also the Gospel message?" (Derrida, *Politics of Friendship*, 284). The problem—and this Derrida suggests is Nietzsche's critique of Christianity—is that the Gospel message of love still conceived love as an investment rather than a gift. This is the reason for Derrida's rueful reflections on the future of democracy, as the key to democracy, it turns out, is also this gift-giving love. It seems the problem at the heart of democracy is also the challenge of remaining loyal to the earth: can human beings become capable of this gift?

149 Parkes, "Zarathustra and Asian Thought," 87.
150 Parkes, "Nature and the Human 'Redivinized,'" 183.
151 Nietzsche, *The Joyous Science*, 221.
152 Zenkei Shibayama, *Zen Comments on the Mumonkan* (San Francisco: Harper & Row, 1974), 19.
153 Shibayama, *Zen Comments on the Mumonkan*, 19.
154 Nietzsche, *The Joyous Science*, 220–1.
155 Nietzsche, *The Joyous Science*, 220–1.
156 Nietzsche, *The Joyous Science*, 177.
157 Zhuangzi 2003, 114–5.
158 Nietzsche, *Thus Spoke Zarathustra*, 86.
159 Nishitani, *Religion and Nothingness*, 215.
160 Nishitani, *Religion and Nothingness*, 215–16.
161 Nietzsche, *Thus Spoke Zarathustra*, 134.
162 Parkes, "The Wandering Dance," 247.
163 Parkes, "The Wandering Dance," 247.
164 Parkes, "The Wandering Dance," 248.
165 Nietzsche, *Thus Spoke Zarathustra*, 283; Parkes, "The Wandering Dance," 248.
166 Nietzsche, *Thus Spoke Zarathustra*, 138.
167 Nishitani, *The Self-Overcoming of Nihilism*, 66.
168 Nishitani, *The Self-Overcoming of Nihilism*, 66.
169 Parkes, "The Wandering Dance," 236.
170 Nietzsche, *Beyond Good and Evil*, 232.
171 Hayman, *Nietzsche: A Critical Life*, 335.
172 Nietzsche, *The Antichrist* in *The Portable Nietzsche*, 18.
173 Nietzsche, *Ecce Homo*, 70.
174 Heraclitus, "The Fragments," 83.
175 Nietzsche, *Ecce Homo*, 94.
176 Parkes, "Nietzsche's Environmental Philosophy," 81.

References

Ames, Roger T. 2018. "Roger T. Ames Responds." In *Appreciating the Chinese Difference: Engaging Roger T. Ames on Methods, Issues, and Roles*, edited by Jim Behuniak, 259–62. Albany: State University of New York Press.

Ames, Roger T. and Hall, David L. 2010. *Dao De Jing: A Philosophical Translation*. New York: Ballantine Books.

Anacker, Stefan. 1984. *Seven Works of Vasubandhu: The Buddhist Psychological Doctor*. Delhi: Motilal Banarsidass.

Culliney, John L. and Jones, David. 2017. *The Fractal Self: Science, Philosophy, and the Evolution of Human Cooperation*. Honolulu: University of Hawai'i Press.

Davis, Bret W. 2004. "Zen after Zarathustra: The Problem of the Will in the Confrontation Between Nietzsche and Buddhism." *Journal of Nietzsche Studies* 28: 89–138.

Davis, Bret W. 2022. *Zen Pathways: An Introduction to the Philosophy and Practice of Zen Buddhism*. Oxford: Oxford University Press.

Deleuze, Gilles. 1977. "Nomad Thought." *Nietzsche's Return Semiotext(e)* 3 (1): 12–21.

Derrida, Jacques. 1974. *Of Grammatology*. Translated by Gayatri Chakravorty Spivak. Baltimore: Johns Hopkins University Press.

Derrida, Jacques. 1997. *Politics of Friendship*. Translated by George Collins. London and New York: Verso.

Dōgen, Eihei. 2018. "Sansuikyō." In *The Mountains and Waters Sūtra: A Practitioner's Guide to Dōgen's "Sansuikyō,"* edited by Shohaku Okumura, translated by Carl Bielefeldt, 23–36. Somerville, MA: Wisdom Publications.

Garrard, Greg. 2004. *Ecocriticism*. London and New York: Routledge.

Grubin, David. 2010. *The Buddha: The Story of Siddhartha*. Alexandria: PBS.

Hayman, Ronald. 1982. *Nietzsche: A Critical Life*. New York: Penguin Books.

Heraclitus. 1979. "The Fragments." In *The Art and Thought of Heraclitus: A New Arrangement and Translation of the Fragments with Literary and Philosophical Commentary*, edited by Charles H. Kahn, 27–86. Cambridge: Cambridge University Press.

Jones, David. 2005. "Crossing Currents: The Over-flowing/Flowing-over Soul in *Zarathustra* and *Zhuangzi*." *Dao: A Journal of Comparative Philosophy* 4 (2): 235–51.

Lampert, Laurence. 1986. *Nietzsche's Teaching: An Interpretation of Thus Spoke Zarathustra*. New Haven, CT: Yale University Press.

Lampert, Laurence. 1993. *Nietzsche and Modern Times: A Study of Bacon, Descartes, and Nietzsche*. New Haven, CT: Yale University Press.

Lusthaus, Dan. 2002. *Buddhist Phenomenlogy: A Philosophical Investigation of Yogācāra Buddhism and the Ch'eng Wei-shih lun*. London and New York: RoutledgeCurzon.

Lyotard, Jean-François. 1984. *The Postmodern Condition: A Report on Knowledge*. Translated by Geoffrey Bennington and Brian Massumi. Minneapolis: University of Minnesota Press.

Nietzsche, Friedrich. 1966. *Beyond Good and Evil*. Translated by Walter Kaufmann. New York: Vintage Books.

Nietzsche, Friedrich. 1967a. *The Birth of Tragedy*. Translated by Walter Kaufmann. New York: Vintage Books.

Nietzsche, Friedrich. 1967b. *On the Genealogy of Morals* and *Ecce Homo: How One Becomes What One Is*. Translated by Walter Kaufmann. New York: Vintage Books.

Nietzsche, Friedrich. 1968a. *Also sprach Zarathustra: Ein Buch für Alle und Keine*n. Berlin: Walter de Gruyter & Co.

Nietzsche, Friedrich. 1968b. *The Will to Power*. Translated by Walter Kaufmann and R.J. Hollingdale. New York: Vintage Books.
Nietzsche, Friedrich. 1977. *The Antichrist* in *The Portable Nietzsche*. Translated by Walter Kaufmann. New York: Penguin Books.
Nietzsche, Friedrich. 1979. "On Truth and Lies in a Nonmoral Sense." In *Philosophy and Truth: Selections from Nietzsche's Notebooks of the Early 1870's*, edited by Daniel Breazeale, 79-97. Atlantic Highlands, NJ: Humanities Press International, Inc.
Nietzsche, Friedrich. 1982. *Die Fröhliche Wissenschaft*. Frankfurt am Main: Insel Verglag.
Nietzsche, Friedrich. 1984. *Jenseits von Gut und Böse*. Frankfurt am Main: Insel Verlag.
Nietzsche, Friedrich. 1987. *Die Geburt der Tragödie aus dem Geiste der Musik*. Frankfurt am Main: Insel Verglag.
Nietzsche, Friedrich. 1996. *On the Genealogy of Morals*. Translated by Douglas Smith. Oxford: Oxford University Press.
Nietzsche, Friedrich. 2005. *Thus Spoke Zarathustra: A Book for Everyone and Nobody*. Translated by Graham Parkes. Oxford: Oxford University Press.
Nietzsche, Friedrich. 2007. *Ecce Homo: How to Become What You Are*. Translated by Duncan Large. Oxford: Oxford University Press.
Nietzsche, Friedrich. 2018. *The Joyous Science*. Translated by Randolph Kevin Hill. New York: Penguin Classics.
Nishitani, Keiji. 1982. *Religion and Nothingness*. Translated by Jan Van Bragt. Berkeley: University of California Press.
Nishitani, Keiji. 1990. *The Self-Overcoming of Nihilism*. Translated by Graham Parkes and Aihara Setsuko. Albany: State University of New York Press.
Okamura, Shokaku. 2018. *The Mountains and Waters Sūtra: A Practitioner's Guide to Dōgen's "Sansuikyō."* Somerville, MA: Wisdom Publications.
Parkes, Graham. 1983. "The Wandering Dance: *Chuang Tzu* and *Zarathustra*." *Philosophy East and West* 33 (3): 235-50.
Parkes, Graham. 1989. "Human/Nature in Nietzsche and Taoism." In *Nature in Asian Traditions of Thought*, edited by John Baird Callicott and Roger T. Ames, 79-97. Albany: State University of New York Press.
Parkes, Graham. 1994. *Composing the Soul: Reaches of Nietzsche's Psychology*. Chicago: University of Chicago Press.
Parkes, Graham. 1996. "Nietzsche and East Asian Thought: Influences, Impacts, and Resonances." In *The Cambridge Companion to Nietzsche*, edited by Bernd Magnus and Kathleen M. Higgins, 356-83. Cambridge: Cambridge University Press.
Parkes, Graham. 1999. "Staying Loyal to the Earth: Nietzsche as an Ecological Thinker." In *Nietzsche's Futures*, edited by John Lippitt, 167-88. Basingstoke: Macmillan.
Parkes, Graham. 2000. "Nature and the Human 'Redivinized': Mahāyāna Buddhist Themes in *Thus Spoke Zarathustra*." In *Nietzsche and the Divine*, edited by John Lippit and Jim Urpeth, 181-99. Manchester: Clinamen Press.
Parkes, Graham. 2005. "Nietzsche's Environmental Philosophy: A Trans-European Perspective." *Environmental Ethics* 27 (1): 77-91.
Parkes, Graham. 2013. "Zhuangzi and Nietzsche on the Human and Nature." *Environmental Philosophy* 10 (1): 1-24.
Parkes, Graham. 2014a. "Will to Power as Interpretation." *Journal of Nietzsche Studies* 46 (1): 42-61.
Parkes, Graham. 2014b. "Zarathustra and Asian Thought: A Few Final Words." *Journal of Nietzsche Studies* 46 (1): 82-8.

Parkes, Graham. 2018. "The Art of Rulership in the Context of Heaven and Earth." In *Appreciating the Chinese Difference: Engaging Roger T. Ames on Methods, Issues, and Roles*, edited by Jim Behuniak, 65–90. Albany: State University of New York Press.
Parkes, Graham. 2020a. *How to Think about the Climate Crisis: A Philosophical Guide to Saner Ways of Living*. London: Bloomsbury Academic.
Parkes, Graham. 2020b. "In the Light of Heaven before Sunrise: Zhuangzi and Nietzsche on Transperspectival Experience." In *Daoist Encounters with Phenomenology: Thinking Interculturally about Human Existence*, edited by David Chai, 61–84. New York: Bloomsbury.
Parkes, Graham. 2022. "Renatured Humans on a Sacred Earth: The Power of Nietzsche's Ecological Thinking." In *A New Politics for Philosophy: Perspectives on Plato, Nietzsche, and Strauss*, edited by George A. Dunn and Mango Telli, 223–47. Lanham, MD: Lexington Books.
Radhakrishnan, Sarvepelli. 1950. *The Dhammapada*. Oxford: Oxford University Press.
Rahula, Walpola. 1974. *What the Buddha Taught: Revised and Expanded Edition with Texts from Suttas and Dhammapada*. New York: Grove Press.
Sallis, John. 1991. *Crossings: Nietzsche and the Space of Tragedy*. Chicago: University of Chicago Press.
Shibayama, Zenkei. 1974. *Zen Comments on the Mumonkan*. San Francisco: Harper & Row.
Vattimo, Gianni. 2004. *Nihilism & Emancipation: Ethics, Politics, & Law*. Translated by William McCuig. New York: Columbia University Press.
Vogel, Steven. 1998. "Nature as Origin and Difference: On Environmental Philosophy and Continental Thought." *Philosophy Today*, SPEP Supplement, 169–81.
Wang, Robin R. 2012. *Yinyang: The Way of Heaven and Earth in Chinese Thought and Culture*. Cambridge: Cambridge University Press.
White Jr., Lynn. 1967. "The Historical Roots of Our Ecologic Crisis." *Science* 155: 1203–7.
Zhuangzi. 1981. *Chuang-tzu: The Seven Inner Chapters and Other Writings from the Book Chuang-tzu*. Translated by Angus C. Graham. London: George Allen & Unwin.
Zhuangzi. 2003. *Zhuangzi: Basic Writings*. Translated by Burton Watson. New York: Columbia University Press.
Zhuangzi. 2009. *Zhuangzi: The Essential Writings: With Selections from Traditional Commentaries*. Translated by Brook Ziporyn. Indianapolis: Hackett Classics.

12

Downbeats and Upbeats in the Meters of Nietzsche, Zhuangzi, and Parkes

David Jones

The Imaginal Ground of Water

If Nietzsche had ever opened the *Zhuangzi*, he may have forsaken his quest to become the first *tragic philosopher*, at least in the way he envisioned the movement of the will to life through the terror of the re-evaluation of all values until becoming "*oneself* the eternal joy of becoming, beyond all terror and pity—that joy which included even joy in destroying."[1] Nietzsche looked to his Ancients for signs of a predecessor of this *tragic wisdom* and even concedes in *Ecce Homo*, although later renounces his concession, that perhaps the eternal recurrence of the same "*might* in the end have been taught already by Heraclitus."[2] If Nietzsche had ever entered the *Zhuangzi* text, he might have found an all-too-Chinese Heraclitus.

In its first chapter, Nietzsche would have encountered a taste of his own type of perspectival philosophy beset with bestial images not so unlike his own. But in the *Zhuangzi*, unlike his *Zarathustra*, these beasts would change forms and become their Other:

> There is a fish in the Northern Oblivion named Kun, and this Kun is quite huge, spanning who knows how many how many thousands of miles. He transforms into a bird named Peng, and this Peng has quite a back on him, stretching who knows how many thousands of miles.[3]

The very first undertaking in the *Zhuangzi* is to remove the reader's solid ground and replace this ground with a fluid buoyancy of a floating through. The ground of the Kun is the fluid, changeable sea and the Peng's is the flowing sky constantly transformed by light, wind, color, stars, and darkness. Nietzsche's "eternal joy of becoming" does not seem to yield him the same protean nature of shape-shifting that we see in the *Zhuangzi*. Surely, Nietzsche's will to power is his Proteus, allowing him to move "beyond all terror and pity" to the "eternal joy of becoming" that "included even joy in destroying," especially the joy of destroying the self.

But a cicada and fledgling dove laugh at the Peng's transformation; they see things differently. It's only the Kun's location that has changed; its vast size and the inherent limitations imposed by its massive girth remain with the Peng, even though he has transcended the earthly realm of the sea by entering the celestial realm of sky. The cicada and fledgling dove alight on an elm or sandalwood to rest and end up just plummeting to the ground, prompting them to ask: "What's all this about ascending ninety thousand miles and heading south?"[4] Their small perspectives of self, reality, and world limit them, just as the massive girth does to the Peng's need for wind "to be piled up thickly enough … to support [its] enormous wings" and its inability to see into the particularities of differences in "the vast distance, going on and on without end."[5]

The temporal counterpart to these relative spatial limitations between the cicada and fledgling dove and the Peng is soon realized in the *Zhuangzi* for "the morning mushroom knows nothing of the noontide; the winter cicada knows nothing of spring and autumn. In southern Chu there is a tree called Mingling, for which five hundred years is as a single spring, and another five hundred years as a single autumn" and "in ancient times, there was even one massive tree whose spring and autumn were each eight thousand years long."[6] These nested modalities and perspectives of time are inclusive of the human experience as well and the comparative relativity of lived time in the figure of Pengzu, legendary for having lived several hundred years who "everyone tries to match," is shunned: "Pathetic, isn't it?"[7] But there is a distinctive possible perspective reserved for the assiduously astute members of this shunned human species.

The anecdotal discussions in the *Qiwulun,* 齊物論 chapter ("Equalizing Assessments of Things") of the *Zhuangzi* are perhaps its best examples of this perspectivism:

> "This" is also a "that." "THAT" posits a "this" and a "that"—a right and a wrong—of its own. But "THIS" also posits a "this" and a "that"—a right and a wrong—of its own. So is there really any "that" versus "this," any right versus wrong? Or is there really no "that" versus "this"? When "this" and "that"—right and wrong—are no longer coupled as opposites—that is called the Course [*Dao*] as Axis, the axis of all courses.[8]

Only from the perspective of a "this" is there a "that," but when one assumes the perspective of a "that" this "this" becomes a "that" and the "this" becomes a "that" *ad infinitum*. Brook Ziporyn points out in a footnote that the text literally says this uncoupling of opposites "is called the Course-Axis," or the *Dao*-Axis, and this could "imply either the axis of the Course [*Dao*] or the Course [*Dao*] *is* an axis."[9] Such going to and fro, backward and forward, side-to-side *ad infinitum* can quickly become a constant movement *ad nauseam* if one is psychologically unprepared for the challenges of the "perspectival life," that is, although we are restricted to our own perspective, which is our "this," we are nevertheless aware of the existence of other possible perspectives, the "that"—hence we naturally realize a "this" and a "that" and a "that" and a "this" in our present "this" perspective. This is all so unless we engage in the rarefaction of the present "this" perspective and turn it into a besotted, solipsistic, other-denying perspective that disavows the manifest sense of the "this" perspective of other perspectives. Even the zealous Zarathustra, the advocate of the circle, who has had his *Grund* pulled from under him in "The Convalescent," tells his most abysmal thought:

> Hail to me! You are coming—I hear you! My abyss is *speaking*, my ultimate depth I have turned out into light!
> Hail to me! Come! Give me your hand!——ha! Let go! Haha!——Disgust, disgust, disgust!———woe is me!

Zarathustra rests for seven days and arising on the seventh his animals speak to him:

> "O Zarathustra" … Step out from your cave: the world awaits you like a garden. The wind is playing with heavy fragrances that would come to you; and all streams would like to follow you, …
> "All things are yearning for you, … All things would be your physicians!
> …. Like yeast-soured dough you lay there, then your soul rose and swelled over its brims.—"[10]

Surfacing from his cave is Zarathustra's emerging into the world—its many other beings, animate and inanimate alike. This emergence is his healing, his freedom from the boring style of the "*first décadent*," Plato, the "big words and fine attitudes" of the Socratic virtues, and the "*ressentiment against*" life so cherished by Christianity.[11] But this is only the lion's roar of *Nein* in the "Three Transformations," which is its reactive rejection and passive nihilism that predicates the life affirming and actively nihilistic "sacred Yea-saying" of the child; for the predatory lion is the necessary evolving stage for a "forgetting, a beginning anew, a play, a self-propelling wheel, a first movement, sacred Yea-saying."[12] This rebirth into the world—back into nature—requires a transformation in (and to) Zarathustra's self; for Zarathustra's self needs to swell and overflow its encapsulated container of exclusivity that keeps him a disparately different individual from other species, one without connection, nor intimacy, with the natural processes of the natural world and its abundant species. Unlike the decree of God in Genesis 1:28,[13] Zarathustra now understands, "For me—how could there be an outside-me? There is no outside!"[14]

And as Zarathustra climbs from his mountain cave, from his last remaining and confining perspective of a "this" self, and as his soul swells over its rims and rushes into the world, his animals say to him, "for those who think as we do all things are already dancing."[15] All the "these" ("this-es") and "those" ("that-s") of the world are no longer static; now they dance as Zarathustra's animals do. The world itself becomes a dance of ineluctable forces of activity playing themselves out in individuated "this" forms.

For Zhuangzi, "when "this" and "that"—right and wrong—are no longer coupled as opposites—that is called the Course [*Dao*] as Axis, the axis of all courses" and "when this axis finds its place at the center, it responds to all the endless things it confronts, thwarted by none. For it has an endless supply of 'rights' and an endless supply of 'wrongs.' Thus, I say nothing compares to the Illumination [*ming*, 明] of the Obvious."[16] Nietzsche would have been able to understand the *shengren* (聖人), the Daoist sage, as one who has this clarity (*ming*, 明) of understanding, as one who places herself at the axis of *dao* (道), where the sage "responds to all the endless things [she] confronts, thwarted by none." Graham Parkes has focused some of his work on this clarity as seeing things as they really are in the Daoist and Zen experience. He writes in "In the Light of Heaven before Sunrise: Zhuangzi and Nietzsche on Transperspectival Experience" that "Nietzsche elaborates the idea of knowing things as they are in themselves, rather than as human awareness construes them, in a remarkable series of notebook entries from 1881."[17] He then goes on to quote Nietzsche directly: "The task: to see things as they are! The means: to be able to see with a hundred eyes, from many persons!" To accomplish this task requires practice: "What is needed is *practice* in seeing with *other* eyes: practice in seeing apart from human relations, and thus seeing factually [*sachlich*]! To cure human beings from their delusions of grandeur." We might be tempted to equate "the idea of knowing things as they are in themselves" with the unknowable reality of "the thing-in-itself" of some two-fold theory of appearance and reality that Nietzsche criticized in such statements as:

> The fictitious world of subject, substance, "reason," etc. is needed—: there is in us power to order, simplify, falsify, artificially distinguish. "Truth" is the will to be master over the multiplicity of sensations:—to classify phenomena into definite categories. In this way start from a belief in the "in-itself" of things (we take phenomena as *real*).[18]

> But here one realizes that this hypothesis of beings is the source of all world-defamation (—the "better world," the "true world," the "world beyond," the "thing-in-itself.").[19]

For Nietzsche (and Parkes) there are no things-in-themselves, but there is a way—or there are ways—of seeing the world and its ever-changing natures that give rise to the emergent patterning of nature itself—that is, with "*other* eyes." Parkes explains this further, "We can adopt a multiplicity of perspectives because we consist in a multiplicity of drives, which often manifest themselves as inner persons. But in order to 'think oneself away out of humanity' it is necessary to acknowledge as well the vast inscape of the human soul, which consists in land and sea, rocks and waves, wind and stars."[20] This clarity, this seeing and understanding things as they really are—insubstantial, flowing, along with the unfolding of the course of everything, including rocks, is first gleaned by Daoists.

For at the axis of *dao*, the sage is the umbilicus of *yin/yang* and *dao* becomes the contextual field of a perspectival "this" and "these" and "that" and "those." No longer do the values of rightness and wrongness exercise control in their domain as the passage continues; rather, they become only interpretations of the perspectives of right and wrong:

> Something is affirmative because someone affirms it. Something is negative because someone negates it. Courses [Daos] are formed [*cheng* 成] by someone walking them. Things are so by being called so. Whence thus and so? From thus and so being affirmed of them. Whence not thus and so? From thus and so being negated

of them. Each thing necessarily has some place from which it can be affirmed as thus and so, and some place from which it can be affirmed as acceptable.[21]

As Nietzsche puts it in *Beyond Good and Evil*, "there are no moral phenomena at all, but only a moral interpretation of phenomena" and as he puts it in the *Notebooks*, "Against positivism, which halts at phenomena—'There are only *facts*'—I would say: No, facts [it] is precisely what there is not, only interpretations. We cannot establish any fact 'in itself': perhaps it is folly to want to do such a thing."[22] We can see a possible Daoist reading of Nietzsche's statement here by considering the character *cheng* 成. *Cheng*, translated above as "formed," also means taking shape, completion, accomplishment, fullness, formation, and even maturity, success, and perfection.[23] The walking of the Courses is one of making things so, just so, or natural, that is, *ziran* 自然, where the uncoupling of opposition is a necessary step in the dance of seeing the mutuality of all that is. The "walking of the Courses" is really "a dancing with them," treating the other courses with an equanimity, as partners in a cosmic dance where self and other finally dance together. Nietzsche too arrived at this thought in the *Joyous Science*: "Our first question about the value … of a human being, or a musical composition … : Can they walk? Even more, can they dance?"[24] Seeing things as they are with the clarity of the sage is to poise the self as a participant in the dance of all things.

Concerning morality, and in concert with Nietzsche, the Daoist position holds that once such questions of "this and that" arise, the natural balance coursing of *yin/yang* has been disrupted already. In Buddhist terms in the *Saṁuyutta Nikāya*, 2.28, "(1) 'When this is, that is; (2) this arising, that arises; (3) when this is not, that is not; (4) this ceasing, that ceases.'"[25] Interdependent Arising, *pratītya samutpāda*, identifies that everything is impermanent, constantly changing, interconnected, and mutually conditioning. We too see this in the aesthetic and metaphysical dimensions of the uncoupling of "this/that" and "that/this" and "these/those" and "those/these."

> So no thing is not right, no thing is not acceptable. For whatever we may define as a beam as opposed to a pillar, as a leper to the great beauty Xishi, or whatever might be [from some perspective] strange, grotesque, uncanny, or deceptive, there is some course that opens them into one another, connecting them to form a oneness. Whenever fragmentation is going on, formation, completion, is also going on. Whenever formation is going on, destruction is also going on.[26]

In this Shiva-like dance we learn that "all things are neither formed nor destroyed, for these two also open into each other, connecting to form a oneness."[27] Hence, according to Zhuangzi,

> it is only someone who really gets all the way through them that can see how the two sides open into each other to form a oneness. Such a person would not define rightness in any one particular way but would instead entrust it to the everyday function [of each being]. Their everyday function is what works for them, and "working" just means this opening up into each other, their way of connecting. Opening to form a connection just means getting what you get: go as far as whatever

you happen to get to, and leave it at that. To be doing this without knowing it, and not because you have defined it as right, is called "the Course [Dao]."[28]

But as we begin to listen to the tenor of the apparent harmonious intonations of Nietzsche and Zhuangzi, we begin to hear their first sounds of dissonance. The challenge for the Daoist sage is to forget morality. As Xu You asks Yierzi in "The Great Source as Teacher" chapter of the *Zhuangzi*, "'How did Yao instruct you?'" and explains that he was told "to submit wholeheartedly to Humanity and Responsibility and to speak clearly of right and wrong." And then is rebuked: "Then what on earth did you come here for? Yao has already tattooed your face with Humanity and Responsibility and de-nosed you with right and wrong. How can you ever roam in the far-flung and unconstrained paths of wild, unbound twirling and tumbling?"[29] Nietzsche's thinking on the Dionysian should come to mind when we read this passage and the response Yierzi gives to Xu You's challenging question—"Nonetheless, I'd like to roam around its outskirts"—and it should also remind us of the same type of repudiation Nietzsche had for traditional Western morality. But the "outskirts" will not completely do, for they are insufficient in bringing the self into the movement of the world, for just like Zarathustra, "how could there be an outside-me? There is no outside!"[30]

Morality for Zhuangzi is viewed as punishment, a de-nosing of the face, a tattooing of one's face that will remain forever in its occlusion of the "unconstrained paths of the wild," of the Dionysian. Morality restricts one to an exclusively human perspective; it locks one into a single direction and prohibits the "Wandering Far and Unfettered" (the title of chapter 1) from "opening up into each other, [which is the] ... way of connecting." The "roam in the far-flung and unconstrained paths of wild, unbound, twirling and tumbling" is the possibility of *dao*, the possibility of possibility, and the fluid way of the sage.

This fluidity is illustrated in another anecdote in "The Great Source as Teacher" chapter. Yan Hui said to Confucius: "I am making progress" and was asked what he meant. "I have forgotten Humanity and Responsibility" was the answer. "That's good, but you're still not there." Another day he saw Confucius: Again, "What do you mean?" asks Confucius. And the answer from Yan Hui is "I have forgotten ritual [*li*] and music [*yue*]." But, sadly for Yan Hui, he was still not there yet. He returns yet another day and again announces that he is "making progress." "I just sit and forget."[31] Yan Hui has arrived at the "Opening to form a connection" and has seen "how the two sides open up into each other to form a oneness," for he is no longer defining "rightness in any one particular way but [is] instead entrust[ing] ... this opening up into each other, [this] way of connecting."[32]

Confucius is jolted by this proclamation of just sitting and forgetting and asks what he means. Yan Hui, who is Confucius's favorite disciple, explains: "It's a dropping away of my limbs and torso, a chasing off of my sensory acuity, which disperses my physical form and ousts my understanding [*zhi*] until I am the same as the Transforming Openness."[33] *Zhi* (知) is indeed an understanding but can also be seen as realizing with a performative aspect—a performative realizing awareness, and not just a cognitive act.[34] And this "Transforming Openness" is *dao* (道), the world's way. Confucius then responds, as many teachers do, with a question: "The same as it [the Transforming

Awareness]? But then you are free of all preference! But then you are free of all constancy! You truly are a worthy man!" and begs him to accept him as his disciple.³⁵ Yan Hui is the same as the *transforming openness* because he is free from the singular perspective of self.

To go along with no memory of the good (*yi*, 義) and norms (*li* 禮) is to eradicate the conscious rule of the heroic ego (*das Ich*) and to be responsively natural (*ziran*, 自然). The "I" as the czar of the psyche is always future or past oriented; its raison d'etre is to provide a ground for the continuity and constancy of the self through time, and responsibility is the *telos* of this self. This ground and its *telos* always establish themselves as that which is past and as that which will be but fail to ground themselves in the present. This self is always in the "has-been" or the "yet-to-be" and can never find itself in the moment of the now. By giving itself to memory and projecting itself into the future, the "I" wills remembering as an activity that forces itself into viewing all life as essential and substantial. And the viewing of all life as essential and substantial uncouples "this and that" and all becomes a "that." Nietzsche would have understood Zhuangzi's active forgetting of the self for he also finds the joy and need of "active forgetting" when he writes regarding promises:

> That this problem has been solved to a large extent must seem all the more remarkable to anyone who appreciates the strength of the opposing force, that of *forgetfulness*. Forgetting is no *vis inertiae* as the superficial imagine; it is rather an active and in the strictest sense positive faculty of repression, that is responsible for the fact that what we experience and absorb enters our consciousness as little while we are digesting it …. [The] purpose of active forgetfulness … is like a doorkeeper, a preserver of psychic order, repose, and etiquette: so that it will be immediately obvious how there could be no happiness, no cheerfulness, no hope, no pride, no *present*, without forgetfulness.³⁶

Without this active force, there can be no present—the self is always forced to abide and subsist in a time that does not exist, and this is the time of morality and promises, the times of futures and pasts; this is the time of "this-es" and "that-s."

It is only through active forgetting that the soul can rise and swell over its brim into the present and back into the world; and the soul can only do this when morality (the *yi* 義 of the Confucians for Zhuangzi) has been forgotten and when we are reminded that existence is not really "an imperfect tense that never becomes present."³⁷ The idea, as Nietzsche enjoins us to consider in *Beyond Good and Evil*, is "to translate man back into nature; to become master over the many vain and overly enthusiastic interpretations and connotations that have so far been scrawled and painted over that eternal basic text of *homo natura*; to see to it that man henceforth stands before man as even today, hardened in the discipline of science, he stands before the *rest* of nature."³⁸

For both Zhuangzi and Nietzsche, active forgetting is to will away the monarchy of the "I" and this is also given expression in the child's forgetting the camel's burdens of tradition in "On the Three Transformations" in *Zarathustra*. As Alphonso Lingis emphasizes,

What … is … noble is the ability to forget: not merely to *forgive* one's hurts and humiliations, one's impotencies, but what is more to *forget* them, to be able to pass over the past to welcome the rushes of what comes in the present. That is the secret power of the noble life: the life that arises innocent before each moment, each event, each person, as though the past had no claim and no law, as though all the ghosts and phantoms of the past dissipated before the light of the present.[39]

The "life that arises innocent before each moment" is only attained when the control of consciousness over the psyche is abdicated—only then can the psyche's natural "repose" enter the eternal succession of becoming, "the light of the present." To reach this "repose" one must, like Yan Hui, let go of one's conscious rule of the soul: first to let go of the rites (*li*, 禮) and music (*yue*, 樂), that is, traditional value of the inherited Confucian tradition, and then to let go of the constricting perspectives of Humanity and Responsibility, that is, morality (*yi*, 義), and finally to let the self itself slip away and become stretched out into and across the utter insubstantiality and impermanence of all that is—*dao*, or *Abgrund*, the opening to world. And as the self overflows into the insubstantiality of the world, it enters time for the first time and becomes a part of time and its eternal cyclical transformations; it enters the moment. By entering the moment, self becomes soul. The voice of Heraclitus, Nietzsche's favorite philosopher, is within our ears here as well, for:

> One cannot step twice into the same river,
> > nor can one grasp any mortal being (*ousia*) a second time in succession,
> > but swift and piercing it changes.
> > scattering and it gathers again,
> > again, and later, but at once it forms and dissolves,
> > and approaches and lets go.[40]

And for the self to overflow into the world's insubstantiality and enter time for the first time, it too must become a part of time and its cyclical transformations. The self enters the moment and by entering the moment, it becomes soul, and has composed itself as soul.

The Emptying Soul

The eternal dimension of the temporal was very much a project for both Zhuangzi and Nietzsche. As the animals say to Zarathustra who had just awakened on the seventh day from his dead sleep and has emerged from the confines of his cave in *The Convalescent*:

> Everything goes, everything comes back; eternally rolls the wheel of Being. Everything dies, everything blooms again, eternally runs the year of Being.

> Everything breaks, everything is joined anew; eternally is built the same house of Being. Everything separates, everything greets itself again; eternally true to itself remains the ring of Being.

In every now Being begins; around every here rolls the ball of there. The centre is everywhere. Crooked is the path of eternity.—⁴¹

Nietzsche's decentered approach to the world's unfolding resonates with Zhuangzi's vision of *this* and *that* and placing oneself at the axis of *dao*, which is the fluid perspectival perspective that knows itself as a perspective amongst innumerable (and sometimes competing) perspectives that are constitutive of everything (*wanwu* 萬物). This perspective that knows itself as yet only another perspective among the presence of numerous other perspectives is what I have referred to some years back as the *perspectival perspective*.⁴² This perspectival perspective is what allows the entrance to the "Transforming Openness" of *dao* as axis or the access of all courses, and "when this axis finds it place in the center, it responds to all the endless things it confronts, thwarted by none."⁴³ The *perspectival perspective* is the fluid perspective that flows with all other perspectives in the natural world—the "human has been translated back into nature" and has become a being that cannot be "grasp[ed] [by] any mortal being (*ousia*) a second time in succession" because "at once it forms and dissolves, and approaches and lets go." This fluid perspective is the domain of the sage and the overhuman who has overcome the fixed perspective of self that focuses primarily on itself and understanding all things accordingly. For the sage, the self is no longer a self; it is now a soul that belongs to and moves with the rhythmic oscillations of the World Soul.⁴⁴

It is the focused perspective, the perspective of a singular consciousness—the Apollonian over the Dionysian—that prohibits other contending perspectives their equal claim to world, reality, and to life; it is the absolute dominance of the ego's conscious ruling perspective over the non-conscious forces that lays claim to their lineage. Did Nietzsche's favorite philosopher also not realize this too when he pronounced that "strife is the father of all and king of all things?" In *The Joyful Science*, Nietzsche makes this point straightforwardly in two sections:

> For the longest time, conscious thought was considered thought itself. Only now does the truth dawn on us that by far the greatest part of our spirit's [*gestigen*] activity remains unconscious and unfelt … *Conscious* thinking, especially of that of the philosopher, is the least vigorous and therefore also the relatively mildest and calmest form of thinking; and thus precisely philosophers are most apt to be led away about the nature of knowledge.⁴⁵

And earlier in the text:

> Thus consciousness is tyrannized—not least our pride in it. One thinks that it constitutes the *kernel* of man; what is abiding, eternal, ultimate, and most original in him. One takes consciousness for a determinate magnitude. One denies its growth and intermittences. One takes it for the "unity of the organism."⁴⁶

What happens to us unconsciously and without feeling (*uns unbewusst, ungefühlt verläuft*) lies at the root of why Nietzsche gives us a genealogy and not a critique—a genealogy gives us a genesis of the very forces at work in a critique, it allows us to view and re-view from the inside, and see from the inside out and outside in.[47] Such a willingness to be self-reflective is one of life's greatest challenges and Nietzsche, as Freud noted, was the paradigmatic example of engaging in *knowing oneself*.[48] Nietzsche notes this challenge in *The Joyous Science*, "'Everybody' is farthest away—from himself'; all who try the reins knows this to their chagrin, and the maxim 'know thyself' addressed to human beings by a god is almost malicious."[49] In "The Child with the Mirror," which begins the second part of *Zarathustra*, we see a more vivid example of this challenge. Zarathustra has returned to his cave in the mountains and realizes his enemies have grown powerful and have distorted his teaching. But this realization, which has come to him in a frightening dream, is an omen that brings him happiness, for like all buddhas and bodhisattvas before him, he seeks to end the disease and its suffering that has plagued humankind from its beginning, even though for Nietzsche suffering plays a more positive role than it does for the Daoists.

Parkes lucidly sees this point when he writes that "if the saint [early on in *Zarathustra*] seems an unlikely member of a Nietzschean trinity, one must bear in mind that this character is of a distinctly Dionysian-Schopenhauerian cast, insofar as his 'I' has been dissolved in such a way that his 'suffering life' is experienced as 'the deepest feeling of belonging, sympathy, and unity with all living things.'"[50] Zarathustra's "deepest feeling, sympathy, and unity of all living things" is his "Wild Wisdom." Zarathustra will descend once again from his cave with a gift to bestow, and this gift is his love:

> My impatient love overflows in torrents, downwards, toward rising and setting. Out of silent mountains and thunderstorms of pain my soul rushes into the valleys. ... And may the river of my love plunge through impassable places! How should a river not at last find its way to the sea! Indeed a lake is within me, solitary and self-contained; but the river of my love draws it off—down to the sea![51]

The self's perspective of consciousness views the vista of world as *this* and *that*, *it* and *other*, as a world that has Being—a world that awaits discovery by some externally unrelated self. To allow this perspective principal control is not to be in-the-world in a lived sense, in a lifeworld; it means to be absent and without presence in any world. Eternity becomes only that which can be thought of, remembered like the Platonic Ideas, or desired for through its unacknowledged attachment to some sense of an immortal soul—this sense of attachment pervades Chinese philosophy and provided the opportunity for the syncretization of Buddhism. Both the Daoists and Nietzsche know this Eternity to be a non-place, a non-republic to which to emigrate;[52] they both realize morality is the necessary passport that allows entrance to Plato's Good, the Christian paradisiacal heaven, or even the immanent moral universe of the idealistic Confucians. This transcendence to the eternal in whatever form is what Zarathustra and Yan Hui will to forget. They engage in active forgetting because their souls wish to empty into the world; they wish to give the Eternal back to the world, and to give eternity back to time's "gathering and scattering," "forming and dissolving," "separating and greeting again."

Forgetting self is fundamental in Nietzsche's project of Dionysian overflowing: "I love him whose soul is overfull, so that he forgets himself, and all things are in him: thus all things become his going-under."[53] In *Prologue* 1, we are told by the advocate of the circle, the advocate of life, that "bless the cup that wants to overflow, that the water may flow from it golden and carry everywhere the reflection of your [the star's] delight [*Wonne*]! Behold! This cup wants to become empty again, and Zarathustra wants to become human again."[54] Zarathustra's emptying is a learning how to "sit and forget" the self, but his emptying process requires an existential devotion to the horror and terror of the *Angst* experience of having no self, which is a necessary condition for the teacher of eternal re-coming,[55] without which there is no dancing over abysses. Such an experience, however, is all but missing in Yan Hui's forgetting. And here lies a different step in their dances, and a different and subtle syncopation in the downbeat of their metrical compositions.

Although Nietzsche's Zarathustra will be challenged and be required to become his soul's "commander and legislator" by nurturing his *Angst*, this commandeering is tempered by his childlike reintegration into the world. This reintegration is a consequence of his emptying. His soul overflows its container with intoxicating bliss (*Wonne*).[56] The word *Wonne* carries connotations of overflowing, intoxication, and bliss, which are not often brought out in translations. To "carry everywhere the reflection of intoxicating-overflowing-bliss" conjures oracles of earthly mythological images, such as the sun's fiery light flowing and melting beneath the horizon and with its setting, its daily death, is time's passing away by mixing the two opposing forces of fire and water into an alchemical mead that arouses one to drink its honey-colored nectar until the cup is empty. To drink this Dionysian draught until one's interior buzzes with the amber-colored blood of the gods, until the viscera of the soul throbs with divine blood and induces inhuman convulsions, is Zarathustra's thirst, not Yan Hui's. Not until this cup is emptied can the human transform into the child (*Floss*)[57] of the "Three Transformations" who is an "Innocence ... and forgetting, a beginning anew, a play, a self-propelling wheel, a first movement, a sacred Yea-saying."[58] And as "the sun is new each day" for Heraclitus, the world is also newly created each day for the child Zarathustra. But for Zarathustra to become human again, he must become a child; he must will his *katharsis* and purge all that he holds onto in the depths of his soul. As Parkes writes:

> The individual is like a deep well, sunk down through the stratum upon stratum of dark earth, the deeper the older, and fed by innumerable underground streams. For the person without history, "that which at every moment we experience of this continual glowing" will be superficial, an awareness of only the topmost levels.[59]

All must be retched from the soul's darkest crypt, even Zarathustra's last holding-on-to, his most abysmal thought, must also come up; this over-rich nectar of the gods leads him to the Overhuman: "Get up, abyss-deep thought, out of my depths! I am your cock and morning-dawning, you sleepy worm: up! up! My voice shall surely crow you awake! ... And once you are awake you shall stay awake eternally."[60] Zhuangzi's Yan

Hui is also awakened, but there is no suffocating snake of suffering robbing him of his soul's breath. Yan Hui's awakening entails overcoming the suffering and frustration that stems from his bondage, his attachment, and craving for a self of "merit and fame." The Daoist sage's awakening is realized from the release of the anthropocentric perspective (hence the many references to biota in the *Zhuangzi*), that is, a freedom from the perspective that impoverishes and challenges the sage's style and fluidity that places a focus on a goal beyond the natural world. The sage meets this challenge in her contextual turn, but for Nietzsche the depth beckons him from the underworld.

Depth

Nietzsche and Zhuangzi focus their attention on a central ontological problem, which is rooted in the psychological: how to get to the world of the here and now and how to become who we are. So far, it has been suggested their solutions to this problematic are similar with hints of difference. At least this is the imaginal ground with respect to the themes of forgetting morality and self that are fomented by some form of an idealized good. However, the *Angst* experience so appreciably apparent in Nietzsche's *Zarathustra* is all but missing in Zhuangzi's characters. Yan Hui's matter of fact response of "sitting and forgetting" does not carry any of the heavy existential crisis experienced by Zarathustra. The quality and style of reaching the here/now and becoming who we are diverge as two streams do after momentarily mixing their waters and moving along their own ways. Their differences in style are found in the texture and nature of their responses to the experiences of confronting the self in their self-overcoming. In comparing their responses requires a short excursion into the theme of nothingness, which will reveal where they cross, how they flow, and how they part in their respective directions and styles. An expression of the Daoist sense of nothingness is found in the *Daodejing*:

> Way-making [*dao*] being empty [*chong*] [*zhong*]
> You make use of it
> But do not fill it up.
>
> So abysmally deep—
> It seems the predecessor of everything that is happening (*wanwu*) ....
>
> So cavernously deep—
> It only seems to persist.
>
> I do not know whose progeny it is;
> It prefigures the ancestral gods.[61]

Ames and Hall read *chong* (沖) "to surge up" as *zhong* (盅) "empty" because of the "processive and fluid character of experience [that] precludes the possibility of either beginning or final closure by providing within it an ongoing space for self-renewal. Within the rhythms of life, the swinging gateway opens and novelty emerges spontaneously to revitalize the world …. Whatever is most enduring is ultimately overtaken in the ceaseless transformation of things."[62] It is through *chong/zhong* that *dao*-making executes the chance necessity of the conditional reality of the world where abiding permanence of substantiality is illusory. This emptiness is not conceived as the opposite of something—either something or nothing—because it is more ontological. *Chong/zhong* frees objects of the world from their rigid forms so that their true characters can disclose themselves, that is, *chong/zhong* is the essential reality of nothingness, or emptiness. When the Daoist sage places herself at the umbilicus of *yin* and *yang*, she too discloses her true nature, that is, nothing. This is because her attunement (和) with the way of the world, with all that happens (*wanwu*, 萬物), is perfectly blended allowing the sage to do "things noncoercively [*wuwei*, 無為]. And yet nothing goes undone."[63] The sage needs only to realize that her cup (self) is already empty. In the *Zhuangzi* "the Consummate Person has no fixed identity, the Spirit Man has no particular merit, the Sage has no one name."[64] The sage is of no fixed identity, no particular merit, and without any one name because she has merged with the world's flowing by giving up the distinction between inside/outside, takes no merit for deeds because her ego no longer exists, and has no singular name because she is now named according to the ten-thousand things (*wanwu*, 無為), that is, to the multiplicity that she and the world to which she belongs continue to become. With this comes a nobility shared by few and a complete immersion into and emergence out of the natural world, for

> the Consummate Person is miraculous, beyond understanding! The lakes may burst into flames around him. But this can't make him feel it is too hot. The rivers may freeze over, but this can't make him feel it is too cold. Ferocious thunder may crumble the mountains, the winds may shake the seas, but this cannot make him feel startled. Such a person chariots on the clouds and winds, piggybacks on the sun and moon, and wanders beyond the four seas. Even death and life can do nothing to change him—much less the transitions between benefit and harm![65]

When the self stops seeing itself as a relational condition in the world, it is only an ego, "the seat of delusion and suffering of frustration and bondage," as Herbert Fingarette realized in his landmark article "The Problem of the Self in the *Analects*."[66] Like his counterpart sage, Zarathustra's cup also needs emptied, but the emptying process of overflowing induces a retching suffering for him. Zarathustra's suffering has a different puissance than the sage's suffering, for in the process of overcoming attachment, the sage's forgetting self is less existentially driven.

Although there are numerous passages in the *Zhuangzi* that indicate the necessity of disabusing oneself of the delusional ego, they tend to strike a different feeling tone from Zarathustra's tenor. The response to the challenge outlined above is dissimilar to, and perhaps even incompatible with, Nietzsche's. Like the Daoists, Nietzsche too was aware of the dangers of defining self as ego when he writes:

> A thought comes when "it" wishes, and not when I "wish," so that it is a falsification of the facts … to say that the subject "I" is the condition of the predicate "think." *It*

thinks; but that this "it" is precisely the famous old "ego" is, to put it mildly, only a supposition, an assertion, and assuredly not an "immediate certainty."[67]

The quest for certainty, beginning with Plato, was a reaction to the fear of becoming. Plato's fear regulates Heraclitus to the sensible realm in his Theory of Ideas and finds expression in the *Theaetetus*'s call to the attention of the doctrine of flux.[68] However, in the *Theaetetus*, Socrates's engagement is with the later Heraclitean version of the Master's philosophy—Plato had learned his "Heraclitus" only indirectly through Cratylus. Instead of "Upon those who step into the same rivers, different and again different water flow …" quoted above, Plato received "You cannot step into the same river twice," which according to Cratylus meant "You cannot even step into the same river once!" The version that came to Plato was an extreme version of flux, one that forfeited Heraclitus's appeal to the necessity of *Logos* that brought order and knowability to the process of becoming. Heraclitus was misperceived as a dangerous threat because he was thought of as providing a basis for Protagoras's Sophism of relativism and skepticism. Hypothesized Being became life's *pharmakon*—a cure for the deadly disease of change, a cure for this deadly disease of life; and Nietzsche was the first Western philosopher to see this. The history of Western philosophy has moved along with misinterpretations, even errors, and the consequences have been deleterious, especially in our relationship to the natural world. Our problems abound and intensify in their proliferations—Nietzsche was the first to really see this and its root cause, and Parkes has devoted his life trying to get us to come to grasp the terms of its meaning and being.

The microcosmic expression of the fear of becoming was manifested in the development of the ego. Such ego development belongs to a metaphysics of Being:

> It is *this* [basic presupposition of the metaphysics of language or reason] which sees everywhere deed and doer; this which believes in will as cause in general; this which believes in the "ego" [*Ich*], in the ego as being [*Sein*], in the ego as substance, and which *projects* its belief in the ego-substance [*Ich/Substanz*] on to all things—only thus does it *create* the concept "thing" …. Being is everywhere thought in, *foisted on*, as cause; it is only from the conception of "ego" that there follows, derivatively, the concept "being" [*Sein*].[69]

Not only does Nietzsche deny this delusional ego as self, but even more radically, he denies the substantial self altogether, along with the substantiality of all the things we know through our language. Analogously, throughout the history of Chinese philosophy philosophers would constantly need to "rectify names" (zhengming 正名) to address these issues as well. Nietzsche does not conscript any "technical" terms such as "ego" for service—nor would the Chinese thinkers, given their philosophical orientation, need special terms for this reified self—but simply uses the German word for "I."[70] This choice indicates an even more radical move on his part because he denies the self as a being or the being of self. There is no self; the self is merely a fiction: "'the subject' is the fiction that many similar states in us are the effect of one substratum: but it is we who first created the 'similarity' of these states; our adjusting them and making them similar is the fact, not their similarity (—which ought rather to be denied—)" for the subject is only "the term for our belief in a unity underlying all the different impulses of the highest feeling of reality."[71] What is left for Nietzsche to affirm is the self as event, a struggle of forces, a becoming, an innocence before each dawning moment where the "phantasms of the past" pale in the flame of the present. Up to this point the resonances between Nietzsche and Zhuangzi still resound, but as we move deeper, their silences sing more audibly present, and as the music changes, so does the dance.

Zarathustra's innocence before the dawning moment is not a childlike innocence of non-differentiation like Zhuangzi's Yan Hui; it is rather an innocence that has been informed by the lion's roar, a roar of force, of one force commanding and imposing its perspective on other forces within the soul's chaos. The roaring force of the lion is more than a just a heteronomous *Nein* to external dominating forces; it too is a "no" to the fictitious self, to the self as constructed by those very external forces that rule the soul—a self that fails to see itself as external to itself until it can ultimately deny itself; and as such as acting only in accord with one's desires rather than one's moral duty in Kantian moral philosophy. The lion's roar is more than this because it is also a "no" to *das Ich*'s control over the dominion of soul and its Dionysian play of becoming a pervious soul where the dam between inner/outer collapses from the weight of the nature's flowing. Only with this "no" can we realize "our body is nothing but one social gathering of many souls."[72]

This sense for a multiple soul is not without its Chinese counterpart. Chinese philosophers, especially Confucius, make the point we are the roles we live and that there is no self beyond these lived roles—even the dead ancestors remain alive as long as we remember them.[73] Its overlapping relations of intimacy with others past, present, and in the future is what defines this sense of self.[74] At any given time, the self is a gathering of others. But for Nietzsche, as these forces gather and scatter, different and new souls are constantly being born and dying, and in the chaos a will emerges for more, for more power: "Every living thing does everything it can not to preserve itself but to become *more*" and "what man wants, what every smallest part of a living organism wants, is an increase of power."[75] The willing that wills more is the primal potency beneath all foundations and forms; it is chaos and Nietzsche calls this groundless chaos will to power. Lingis has made this point: "The Will to Power is an abyss (*Abgrund*), the groundless chaos beneath all grounds, all the foundations and it leaves the whole order of essences groundless."[76] The will to power is an abyss, "a no-ground behind every ground."[77] The will to power is the potency or power (*Macht*) that is necessary to create (*machen*) things of existence; it is the force that forms worlds. But this force must not be misconstrued as a unified force that makes other forces possible because this too would fall into the metaphysical trap—this was Nietzsche's complaint against the Atomists and the Daoists' warning to those more idealistic Confucians who held the universe to be a metaphysically moral one. The will to power is the possibility of power to create, to give birth to new forms that cry out for more. Each created form reveals its difference from other forms and affirms the manifold existence of the "myriad things" (*wanwu*, 萬物). As Nietzsche says, "Everything simple is merely imaginary, is not 'true.' But whatever is real, whatever is true, is neither one nor even reducible to one."[78] This will to power is a diverse ground, a ground without ground. This groundless ground is the dance of Dionysus for "one must still have chaos within, in order to give birth to a dancing star."[79] And to see this deep star shine, Zarathustra must go under. Although Chinese philosophers had their own forces in *yin*, *yang*, and *qi*, the temper and timbre of the difference between Nietzsche and Zhuangzi's philosophy is disclosed in Zarathustra's willing and willingness to go under and Zhuangzi's desire to stay afloat and move with *dao*'s current, the balanced tension of *yin/yang* and *qi* that flows through all things.

Deeper Depth

> You will not discover the utmost verge of the soul
>   even if you travel its every path
>   so deep a logos it has—Heraclitus

Before he can bring himself to empty his cup, his inherited encapsulated soul, and before he can re-soul the desert of world long left dry since the times of Thales, Zarathustra must go under. Zarathustra's movement, his direction downward before moving outward, is the only way *outward* for him. This is the terrain of his, not Yan Hui's, psychological and ontological topography. Zarathustra wishes to go down as far as he is capable because he must accept the Oracle of Delphi's legacy of "Knowing Himself." The images in *Zarathustra* are ones of depth: "… a man's disposition [*Gemüth*] is deep, its torrent [*Strom*] rushes in subterranean caves";[80] " … I must descend [*steigen*] into the depths [*Tiefe*]: just as you do in the evening when you go behind the sea and still bring light to the underworld, you overrich star!";[81] "I love him whose soul is deep even in being wounded, and who can perish even from the smallest experience can perish: thus he goes gladly over the bridge";[82] "slow is the experience of all deep wells: long must they wait before they know just *what* has fallen into their depths."[83] And there are many more.

Zarathustra descends from his cave on the mountain's summit down to the village; this is his *going-under* (*untergehen*): "I want to descend," he tells his evening star. His going-under is a descent into a profound abyss; but first and foremost, it is a descent into his own abyss and this expedition into his abyss is his descent into the world of the human. He knows that one must first crack one's own abyss to become the tragic philosopher before Dionysian depth can be penetrated. Nietzsche places the word "*Tiefe*" (depth) into the mouth of Zarathustra in the above quote, but this word carries connotations that go beyond direction and location. In fact, the term has nothing to do with direction or location as indicated by "*steigen*." *Steigen* means to ascend, soar, go up, mount, climb, but in context can only mean to descend—to rise up one must first go down—because it would make little literal sense to say that "I must ascend to the depth."[84] But this is precisely what Zarathustra tells his star. Zarathustra gives us a clue, a sign, that tells us his mission has little to do with logic and its human made laws of contradiction[85] but has more to do with those mythological quests that are repeated, again and again, first by his Greek forefathers, such as Odysseus, Herakles, Perseus, and Theseus. Zarathustra knows that only gods and heroes can journey into *the deep*, into the underworld.

Zarathustra's descent into the deep gives him power (*Strom*), a strong current (*Strom*), and makes him flood (*Strom*) like a torrent (*Strom*), which enables him to soar to the heights of gods and heroes. His river rushes and roars (*rauschen*) with the intoxication (*Rausch*) of the gods. Zarathustra's overcoming drowning in the Dionysian depths of his internal river is qualitatively different from the Swimmer's experience in the *Zhuangzi* whose sole purpose is to just stay afloat as we see in the next section. Much of *Zarathustra* is about Zarathustra's descent to the world of the human, back to the marketplace, and into the depths of his soul—and the Ox Herding pictures of Zen come to mind. As he explores the depths of soul, as he travels its every path, his soul responds as though a hydraulic ram and with this hydraulic activity his cup overflows into the world—as he goes under, he empties into and re-souls the world, a world left dry since the time of Thales's successors.[86] For the hydraulic system to begin its displacement of soul, Zarathustra must descend deep and rush and roar into his underground caves; his spring is deep, and long must he wait until he finally realizes what fell into its most unfathomable depth.

In his underground caves he finds his cast of characters: the Spirit of Gravity, the Dwarf, Soothsayer, Scholars, Priests, Tarantulas, the Spirit of Revenge, the Fire-hound, and so forth—all splinter characters of his multiple-psyche as Parkes calls them.[87] But he must wait long for the deepest, most profound and archaic fragment of his soul to be spewed forth—the fragment Zarathustra had glimpsed as a shade on his journey to the underworld through the eternally smoking fire-spewing mountain not far from his blessed isles where he learned that "the greatest events—those are not our loudest, but our stillest hours."[88]

Waters Deep

Daoists have long held water as being an image for *dao*. Chapter 8 of the *Laozi* begins this way:

> The highest efficacy is like water,
> It is because water benefits everything (*wanwu*)
> Yet vies to dwell in places loathed by the crowd
>
> That it comes nearest to proper way-making.
>
> In dwelling, the question is the right place.
> In thinking and feeling. It is how deeply.[89]

Although "proper way-making," or the appropriate way of making the *dao*, is associated with water and its malleable power, depth is only alluded to in "thinking and feeling" and finding the "right place," or natural place.

Water is also an important image in the *Zhuangzi*, but its feeling-tone has a much different tenor to it from the above images relating to Zarathustra. Water is always on its way in the *Zhuangzi*; it is fluidity and motion that are emphasized, not the hidden secrets of the water's depth. The sage's authentic response to the challenge of water is to "get with the flow" of *dao*:

> Confucius was viewing the Lu waterfall, which plummets several hundred feet, whitening the waters for forty miles around, impassable to fishes and turtles. And yet he saw an old man swimming there in the torrent. Thinking the man had attempted suicide due to some suffering in his life, Confucius sent his disciples to run along the bank and try to pull him up. But the old man emerged several hundred paces downstream, walking along the bank singing, his hair streaming down his back. Confucius hurried after him and said, "I thought you were a ghost, but now I see you are a man! Do you have a course [*dao*] that allows you to tread upon the waters?"[90]

Although the "ghostly" swimmer dives from the heights of the waterfall into the churning depths, he does not stay under long; he surfaces quickly and swims several hundred paces downstream with the current. The swimmer wishes only to swim, to move with the world's current and by swimming along this way, the man is insubstantial in the world's non-differentiating flow; he moves effortlessly (*wuwei*, 無為) in his fractal connection[91] with the world's becoming; he has disciplined himself "in order to be oneself the eternal joy of becoming," as Nietzsche would say. But the old swimmer of Lu wishes not to stay under and enjoy the underground caves; he has no desire to journey to his or any other underworld—this is not his quest, nor his passion. Instead, he surfaces without much delay and strolls along the banks and sings a song. In response to Confucius's question, he replies: "No, I have no course. I got my start in the given, developing via my own inborn nature, and reached completion through fate. I enter into the navels of whirlpools and emerge in the surging eddies. I just follow the course of the water itself, without making any private one of my own. This is how I tread the waters."[92] Chapter 16 of the *Laozi* illustrates this to be a theme in Daoism: "Extend your utmost emptiness as far as you can / And do you best to preserve your equilibrium (*jing*)."[93]

The *shengren* (sage) does go under into the "navels of whirlpools" and "surging eddies" as he knifes through their depth but seems little concerned with the psychological requirement displayed by Zarathustra. From the inflows, the *shengren* soon floats out with the outflows. The swirling undercurrents seem to pose little interest. They do not tantalize him, nor do they cause him seasickness as they do for Zarathustra. There is no nausea of the swirling depths of soul, nor its torrential chaos. For the sage, there is no quest into the underworld, for he "just follow[s] the course of the water itself, without making any private one of [his] own" and "[doing his] best to preserve [his] equilibrium." The emphasis of this passage is to "follow the course of the water" and to stay afloat. Following the way of the water is to be attuned to *dao*. As the *Daodejing* states: "the highest efficacy is like water"[94] and "Way-making (*dao*) is an easy-flowing stream/ Which can run in any direction."[95] In a sense, Confucius was right to see the swimmer as a ghost because he is suspended in the *zhong* (emptiness 盅) of all things and moving with its flow (*chong* 沖). Of note, however, he does not see the swimmer's quest as mythological or psychological, like we must regard Zarathustra's, but one of self-cultivation.

Being afloat on water is to be suspended in the *chong/zhong* of all things; it is to find one's equilibrium. The identity of the water is characterized by its nothingness—since it has no form, but con-forms, it is always in motion and has no locus, it is always here, then there. The *shengren*'s response to the challenge of water is to get with the flow—to become attuned to the nothingness of its becoming (*zhong* 盅) and its surging forth (*chong*, 沖). The surging forth is its emergence, but the emergence is also a reversion: "In the process of things emerging together (*wanwu*) / We can witness their reversion. / Things proliferate, / And each again returns to its root. / Returning to the root is called equilibrium."[96] This is not without parallel in Nietzsche since he too is an advocate of becoming: "The affirmation of passing away *and destroying*, which is the decisive feature of a Dionysian philosophy; saying Yes to opposition and war; *becoming*, along with a radical repudiation of the very concept of *being*—all this is clearly more closely related to me than anything else thought to date."[97]

The Daoist sage's response, however, appears primarily ontological, that is, to borrow Heidegger's emphasis on the participation of human beings in the emergence of Being, and secondarily psychological. Nietzsche, on the other hand, is foremost psychological and then ontological. Again, Heidegger is helpful here: "Being as a whole reveals itself as *physis*, 'nature,' which here does not yet mean a particular sphere of beings but rather beings as such as a whole, specifically in the sense of emerging presence."[98] For Nietzsche, who visited these realms before Heidegger, to get to this "emerging presence" entails a vital and necessary *untergehen*—this is his prerequisite. Not only is this going-under the process of Zarathustra's becoming, but it is also his necessary point of departure for his whole excursion into self-overcoming. Not only do Nietzsche's and Zhuangzi's preferences for departure indicate a major divergence in the currents of the two philosophies of becoming, but they also suggest a different feeling for the nature of reversion, or returning, in Eternal Re-Coming.

Confucius asks the old swimmer what he meant by "getting your start in the given, developing via your own inborn nature, and reaching completion through fate" and receives this response: "I was born on the land and thus I feel securely at home on the land. That's the given. I grew up with the water and thus feel securely home in the water. That's my own nature. And I am thus and so but without knowing how or why I am thus and so. That's fate."[99] For the Daoist sage, not knowing and the acceptance of their fate is enough in and of itself because their destiny is connected with *dao*; their destiny is the flowing of *dao*. The challenge for the *shengren* is to bring body and soul (their nature) into the current of *dao* and to respond naturally (*ziran* 自然, literally "being so of itself").[100]

A different current, however, carries Nietzsche's *amor fati*. His current rushes as a torrent through a narrow canyon after early spring melt. One must love one's fate in order to become what one is and the suffering and nausea of the deepest depths is a necessary condition for this becoming; fate brings chaos to soul as a necessary condition for the soul's subsequent bliss of overflowing into the world, of ensouling the world. Zarathustra, the prophet of the *Übermensch* and eternal recurrence, can never be satisfied with an "it is so without me knowing why it is so" because to know is to suffer and this suffering must be affirmed. It too must be a joy! "My formula for greatness in a human being is *amor fati*: that one wills nothing to be otherwise, not forward, not backward, not in all eternity. Not merely to suffer what is necessary, … but to *love* it."[101]

The *Übermensch* must even love the chaos of his soul. He must love the suffering and nausea that the eternal return brings because it is necessary. Zarathustra's suffering must be known and cultivated; he will not be satisfied with the Swimmer of Lü-liang's destiny. In Daoism, the joy is reunion with world through self-overcoming, the over-coming of this/that, it/other and this reunion is realized without Zarathustra's existential and psychologically driven need to deal with those wretched demons that need retched from the soul's depths. The overhuman needs to know and experience suffering and nausea because these are the requisite routes back into the world.

Deep Suffering

> It is by disease that health is pleasant—Heraclitus

As we have seen, suffering is less prominent in the *Zhuangzi* and is not so central, and seemingly less crucial to the Daoist project. This is not to suggest that suffering was not recognized in the Chinese worldview and in Daoist philosophy, for there was plenty of suffering to go around. Buddhism's Four Noble Truths resonated sufficiently with Chinese culture and Daoism played a significant role in Buddhism being accepted during the Tang Period. Nevertheless, the existential role of suffering was in large part missing from Daoist philosophy. Parkes was sensitive to this difference of tone that is being intensified and deepened here. At the end of his original article in Part One, he concludes in the penultimate paragraph that "this difference in mental temperature— the *Zhuangzi*'s cool harmony as against Zarathustra's friction-generated heat—is less a result of residual egoism on the part of the overhuman than of a disparity in the degrees of encapsulation of the self that is to be overcome." For Zarathustra, he must wait long to discover what fell into his deepest spring; he must wait to find what has crawled into his throat and down to his soul as he lies sleeping—"Up, abysmal thought, out of my depth! … Let go! … Nausea, nausea, nausea—woe unto me!" And what will crawl from Zarathustra's soul will be both beautiful and disgusting. Once again, the animals' speech in "The Convalescent":

> —"O Zarathustra," said his animals in reply, "for those who think as we do all things are already dancing: they come and shake hands and laugh and flee—and come back again.
>
> Everything goes, everything comes back; eternally rolls the wheel of Being. Everything dies, everything blooms again, eternally runs the year of Being.
>
> Everything breaks, everything is joined anew; eternally is built the same house of Being. Everything separates, everything greets itself again; eternally true to itself remains the ring of Being.

> In every now Being begins; around every here rolls the ball of there. The centre is everywhere. Crooked is the path of eternity.—[102]

In this passage the exquisite nature of the eternal return is portrayed. For oneself to be centered one must be with the world's center, which is everywhere; it is all-over. The soul must overflow into the world where the "center is all over, everywhere." Such a decentered eternal world is also central in the *Zhuangzi* where decentering the human perspective of "It/Other" or "This/ That" (animals do not share this dualistic perspective)[103] is the first level of disciplining for the sage. But Zarathustra's snake is at hand (throat).

Zarathustra's animals attempt to cheer him after a seven-day dead sleep in his underworld: "My abyss is *speaking*, my ultimate [*letzte*] depth [*Tiefe*] I have turned out into the light."[104] Zarathustra's eagle has placed two sacrificial lambs at his feet, which have been robbed from lesser shepherds. His animals tell him to step out of his cave because the "world awaits you like a garden" and the "wind is playing with heavy fragrances"[105] that wish to tickle his nostrils. But all is not roses in Zarathustra's Eden. He lovingly calls the animals of his psyche politic "buffoons" (*Schalks-Narren*) and "barrel organs" (*Drehorgeln*) because he knows there is another side to the eternal return, a darker masked side.[106] He reminds them "how that monster crawled down [his] throat and choked [him] and how he bit off its head and vomited it out."[107] This suffocation is Zarathustra's encounter with his authentic becoming; he must come to grips with his *Angst* if he is ever to accept the eternal return. "*Angst*" reveals this experience through its Latin root *ango* and Greek root *agcho*, which mean to press together, strangle, to choke, to occlude the soul (*psuchē*) of its breath (*psuchē*). Zarathustra's final depth (*letzte Tiefe*), his most deep-seated *Angst* has been brought to light—and the headless *Schlange* is disgusting, covered with vomit, and it has made Zarathustra sickened in revulsion for it had traveled deeply into his very being. And what choked Zarathustra was his "great loathing for the human being" for "the human being recurs eternally! The small human being recurs eternally" and Zarathustra realizes that "the greatest" and "the smallest human being" are "all-too-similar to each other" for they are "all-too-human."[108] This is what had traveled into his deepest being for the longest time. Although Zhuangzi too is no fan of the humanity of the Confucians and despite their other apparent affinities, Zarathustra's final depth represents an abysmal divergence between the two traditions.

I Searched Myself—Heraclitus

Zarathustra must learn and is unconsciously driven to discover the disgusting nature of his last monster. In "The Convalescent," Zarathustra apprises his animals of the cruelty of human beings and that the "human being is the cruellest beast toward itself"[109] and finally realizes that it was "the great loathing for the human being" that had crawled down his throat and choked him. It is true that Zhuangzi too was disgusted with the smallness of man and suggested there was "disillusion, not to say disgust, with human life as it is ordinarily lived"[110] and that "the man who has freed himself from conventional standards of judgment can no longer be made to suffer."[111] Even Confucius shares this disgust with the *xiaoren* (小人), the small person: "Exemplary persons (*junzi* 君子) understand what is appropriate (*yi* 義); petty persons understand what is of personal advantage (*li* 利)."[112] However, for Zhuangzi, the suffering of disgust is to be overcome for it is merely illusion, not cultivated as it must be for Zarathustra, because for Zarathustra not only does the greatest man recur, but the smallest recurs as well. Zarathustra, the prophet of the overhuman and the teacher of the eternal return, must now become a semiologist of suffering; he must become the hidden god of Delphi, Dionysus, who "neither speaks nor keeps secret but reveals with a sign" (Heraclitus, Fragment 14), and offer the world a sign of suffering so it too can know the necessity of suffering: "And the eternal return even of the smallest!—that was my disgust with all being-there! Alas, Nausea! Nausea! Nausea!——" But his animals will not let him go, and say to him:

> "Speak no further, you convalescent!"—... "But go out where the world awaits you like a garden.
>
> "Go out to the roses and bees and the flocks of doves! But especially to the songbirds, that you may learn from them how to *sing*!
>
> "For singing is for convalescents: ...
>
> "Better still, dear convalescent, first fashion yourself a lyre, a new lyre!
>
> "For do you not see, O Zarathustra! For your new songs you will need new lyres.
>
> "Sing and foam over, O Zarathustra. With new songs you must heal your soul: that you may bear your enormous fate, which has been no human's fate up to now!
>
> "For your animals know well, O Zarathustra, who you are and must become: behold, *you are the teacher of the eternal recoming*—that is now *your* fate! ..."[113]

To fulfill his destiny, Zarathustra must love his fate; he must embrace the returning moment that has announced that he is the teacher of the eternal return, the teacher of the eternal re-coming of the greatest and smallest, the greatest and smallest both within and without. His soul must be healed with the *pharmakon*—the cure and poison—of new songs, songs of the child Zarathustra that will overflow (*brause über*) his soul's brim and become an exemplification of his overflowing Dionysian intoxication. For Zarathustra, his choice is clear—either to live authentically or to die inauthentically, for he is fated to be the teacher of eternal re-coming.

In the *Zhuangzi*, death is treated as just another event to be accepted as part of the process as seen the story of Nie Que[114] and Wang Ni, chapter 2 when Wang Ni says of the Consummate Person "Even death and life can do nothing to change him" for "such a person chariots on the clouds and winds, piggybacks on the sun and moon, and wanders beyond the four seas."[115] Such a wandering is flowing over and with the skies and seas, and death and life are like this. In the sections on Zhuangzi's Wife's Death and Zhuangzi's encounter with the Hollow Skull in chapter 18 we can hear the resonances with Nietzsche's Zarathustra: Zhuangzi finds the Skull's acceptance of death incredulous and poses the hypothetical of returning the Skull's "body to life" and receives this response from the skull "with a deep frown knitted in it brows": "How could I refuse the joy of a king on his throne, to suffer again the toils of humankind."[116] When his wife died, Zhuangzi was found by Huizi, known for his paradoxes, in his offering of condolences "squatting with his knees out, drumming on a pot singing." Huizi finds his actions shameful and receives this reply: "Mingled together in the amorphous, something altered, and there was the energy; by alteration in the energy there was the shape, by alteration of the shape there was the life. Now once more altered she has gone over to death. This is to be companion with spring and autumn, summer and winter, in the procession of the four seasons … I with my sobbing knew better than to bewail her."[117] And with his own pending death, Zhuangzi's disciples wished to give him a lavish funeral and are checked by their Master: "I will have heaven and earth as my coffin and crypt, the sun and moon for my paired jades, the stars and constellations for my round and oblong gems, all creatures [*wanwu* 萬物] for my tomb gifts and pallbearers. What could possibly be added?"[118] For Zhuangzi, there is clearly a choice, to bring oneself into the flowing of the natural unfolding of all that is, but for Zarathustra this choice is less benign, less inactive, and far less acquiescent.

For Zarathustra there is a wrenching choice that must be affirmed; he teaches that one should die at the right time: "Many die too late, and some die too early. The teaching sounds strange: 'Die at the right time!' Die at the right time: thus teaches Zarathustra. Of course, if one never lives at the right time, how could one ever die at the right time? Would that such a one never been born!"[119] Zarathustra, the teacher of the overhuman and eternal re-coming, also teaches about death, or how to die authentically. In the penultimate passage of "The Convalescent," Zarathustra's animals speak for him as though they too are part of his soul and proclaim the end of Zarathustra's going-under, the end of his setting and perishing (*Untergehen*), and the beginning of his re-birth: "Loving and going-under [*Untergehen*]: that has rhymed for eternities. The will to love: that means being willing to die too."[120] There is no further province to be explored within; the cup, the container of his soul, has been shattered and its insides emptied outside. Zarathustra's *Untergehen* is his passive nihilistic affirmation, his lion's roar that creates the possibility of the active nihilistic affirmation of the child's "yes." There is no negative affirmation of the lion's roar in the *Zhuangzi*, but rather a positive affirmation navigating us back into the flow.

The animals of Zarathustra's polycentric psyche refuse to perish at this time and speak thus:

> And if you desired to die now, oh Zarathustra: behold, we also know, how you would speak there to thee:—but your animals entreat you, that you do not die yet!
> You would speak … without trembling … !—
> "Now I die and fade away, … and in no time I am nothing. Souls are as mortal as bodies.
> But the knot of original-events [*Ursachen*], turns again, in which I am entwined [*verschlungen*],—it will create me again! I myself belong to the original-events of the eternal return.

> I come again, with this sun, with this earth, with this eagle, and with this serpent [*Schlange*]—not to a new life or a better life or a similar life:
> —I come eternally again to this same and selfsame life, within the greatest and also the smallest, that I again teach the doctrine of eternal return of all things,—
> —that I speak again the word of Great Earth and of Human's-Noon, that I again to human beings proclaim the Overhuman.[121]

Zarathustra's death is not a literal death; his going-under is like the light of the setting sun that is brought to the underworld each night. The reciprocal relationship between life and death is established as an integral turning of all things in the becoming of the world, in the being of becoming. Zarathustra has aligned himself with the being of becoming and has become entwined or (inter)woven into the multifarious fabric of the world. The Daoist sage is entwined in the knot of original events by entering the axis of *dao*, or center of the cusp of *yin* and *yang*. But the *shengren*'s blending into the tapestry of life requires acceptance of the natural course of all things, not the affirmation of willing that this life as now lived will repeat itself infinite times and that one must learn to die at the right time. The Daoist sage will return as well, but as a leaf, the tree, the insect, … because "death and life alter nothing in himself."[122]

In true Dionysian fashion, Zarathustra's animals are always at hand, the eagle of the sky and the serpent of the earth. The serpent, the creature closest to the earth that touches the earth with all areas of her under-body as she slithers toward her destination, is an ancient image of that wisdom. Zarathustra, the redeemer of the earth, needs to bite off the head of wisdom that removed soul from world with its love of the transcendent and metaphysical. The "*schlungen*" of *verschlungen* is vividly cognate with *Schlange*, serpent. Zarathustra is (re)threaded into the fabric of existence, but in order to be entwined in this eternal worldly web of forces he must bite off his internal snake's head before his soul can finally flow out into the flowing of the world.

The *shengren*'s trip seems more benign; the sage has no visible internal monsters impeding her return back into the flow. Zarathustra's internal battle, however, leaves him bonded eternally to the primeval forces (*Ursachen*) in a different way. Zarathustra realizes this bond as the heaviest of all weights; his soul is now stretched back through time and allied with the forces that make life possible. Thus ends his going-under. Zhuangzi's characters, however, seem to know all of this without experiencing the suffering of the snake's head and the taste of its severed and disembodied wisdom. The *shengren*'s journey lacks Zarathustra's turmoil for the sage seems to float effortlessly (*wuwei* 無為) to the center once she has realized her true nature, her "being just so of herself" (*ziran* 自然). The terminus of their trips back into the world is the same, but Zarathustra and Zhuangzi's ways diverge and are driven by distinct currents of overflowing and flowing-over.

Zarathustra's animals finish their speech to Zarathustra and his internal dialogue is stilled. They wait for his response, but Zarathustra remains silent. He lay with his eyes closed, but is not asleep "for he was just then conversing with his soul." In "On the Great Yearning" he tells his soul that there is no soul more loving, vast (*umfangender*) and voluminous (*umfänglicher*);[123] it is "over-rich and heavy" and "waiting out of overflow" asks: "Where would future and past be closer together than in you?"[124] Zarathustra, the advocate of suffering, has now truly become the advocate of the circle—past and future find their eternal sense in the encircling (*umfangender*) of the timeless rhythms of soul. Zarathustra concludes his conversation with his soul by imparting that he has now accorded his totality to her care. Zarathustra spoke thus:

> O my soul, now I have given you all and even my deepest [*Letztes*], and all my hands have become empty for you:—*that I called you to sing*, behold, that was my deepest!
> That I called you to sing, speak now, speak: *which* of us should now—return thanks?—Better still: sing for me, sing, O my soul! And allow me to return thanks!—[125]

There is *nothing* in Daoism, just like this.—

Notes

I am grateful to Peter Groff for his helpful comments on an earlier version of this essay. A much earlier version appeared as "Crossing Currents: The Over-flowing/Flowing-over Soul in *Zarathustra* and *Zhuangzi*" in *Dao: Journal of Comparative Philosophy* Volume IV, No. 2 (June 2005). I am grateful to its editor, Huang Yong, for permission to recast the original article in its divergent form.

1. Friedrich Nietzsche, *The Portable Nietzsche*. Translated by Walter Kaufmann (New York: The Viking Press, 1968b), 563.
2. Friedrich Nietzsche, *On the Genealogy of Morals* and *Ecce Homo*. Translated by Walter Kaufmann and R. J. Hollingdale (New York: Vintage Books, 1969), 274.
3. Zhuangzi, *Zhuangzi: The Essential Writings*. Translated by Brook Ziporyn (Indianapolis: Hackett Publishing, 2009), 3.
4. Zhuangzi, *Zhuangzi: The Essential Writings*, 4.
5. Zhuangzi, *Zhuangzi: The Essential Writings*, 4.
6. Zhuangzi, *Zhuangzi: The Essential Writings*, 4.
7. Zhuangzi, *Zhuangzi: The Essential Writings*, 4.
8. Zhuangzi, *Zhuangzi: The Essential Writings*, 12.
9. Zhuangzi, *Zhuangzi: The Essential Writings*, 12.
10. Friedrich Nietzsche, *Thus Spoke Zarathustra*. Translated by Graham Parkes (New York: Oxford University Press, 2005), 189.
11. See Friedrich Nietzsche, *Twilight of the Idols and the AntiChrist*. Translated by R. J. Hollingdale (Baltimore: Penguin Books, 1968a), 106, 108, and 110 respectively. *Twilight of the Idols*, "What I Owe the Ancients."
12. Nietzsche, *Thus Spoke Zarathustra*, 24.
13. God blessed them and said to them, "Be fruitful and increase in number; fill the earth and subdue it. Rule over the fish in the sea and the birds in the sky and over every living creature that moves on the ground." *New International Version*.
14. Nietzsche, *Thus Spoke Zarathustra*, 190.
15. Nietzsche, *Thus Spoke Zarathustra*, 190.
16. Zhuangzi, *Zhuangzi: The Essential Writings*, 12.

17 David Chai (ed.), *Daoist Encounters with Phenomenology: Thinking Intellectually about Human Existence* (London: Bloomsbury, 2020), 70.
18 Friedrich Nietzsche, *The Will to Power*. Translated by Walter Kaufmann and R. J. Hollingdale. Edited by Walter Kaufmann (New York: Vintage Books, 1967), 280.
19 Nietzsche, *The Will to Power*, 377.
20 Chai, *Daoist Encounters with Phenomenology*, 70.
21 I use the Ziporyn translation, which is becoming the translation of choice for the *Zhuangzi*, whenever possible. At other times I will resort to A. C. Graham's fine translation. Zhuangzi, *Zhuangzi: The Essential Writings*, 13.
22 Friedrich Nietzsche, *Beyond Good and Evil*. Translated by Walter Kaufmann (New York: Vintage Books, 1966), 85; Nietzsche, *The Will to Power*, 266.
23 Zhuangzi, *Zhuangzi: The Essential Writings*, 213.
24 Friedrich Nietzsche, *The Gay Science*. Translated by Walter Kaufmann (New York: Vintage Books, 1974), 322.
25 John M. Koller, *Asian Philosophies* (Upper Saddle River, NJ: Prentice Hall, 2002), 167.
26 Zhuangzi, *Zhuangzi: The Essential Writings*, 13.
27 Zhuangzi, *Zhuangzi: The Essential Writings*, 13.
28 Zhuangzi, *Zhuangzi: The Essential Writings*, 13.
29 Zhuangzi, *Zhuangzi: The Essential Writings*, 48.
30 Nietzsche, *Thus Spoke Zarathustra*, 190.
31 Zhuangzi, *Zhuangzi: The Essential Writings*, 49.
32 Zhuangzi, *Zhuangzi: The Essential Writings*, 13.
33 We see, at least prima facie, a difference here with Nietzsche's thinking. The dispersion of one's physical form is lauded as a good by Zhuangzi, perhaps following the Daoist predilection to fly with the wind, but for Nietzsche there is a correlate between the body and wisdom. Parkes has made significance of the body in Nietzsche's philosophy, especially as the locus of value along with the earth (Parkes, *Composing the Soul*, 132; see 331, for the relation of soul and body; 361, for the attunement of bodily energies and flowing with the flows; 376 for the disorder of the energies). Zhuangzi, *Zhuangzi: The Essential Writings*, 49.
34 Roger Ames and Henry Rosemont have this to say about *zhi* in the context of the *Analects*: "'To realize' has the same strong epistemic connotations as 'to know' or 'knowledge' in English. You may say you believe whatever you like, but you only *know*, or *realize* something, if that something is indeed the case. In addition, it underscores the performative, perlocutionary meaning of *zhi*: the need to author a situation and 'make it real'" (Roger T. Ames and Henry Rosemont, Jr., *The Analects of Confucius: A Philosophical Translation* [New York: Ballantine Books, 1998], 55). In terms of Zhuangzi's and Nietzsche's project, we must realize our "knowing" is always driven by being in a perspective and this realization is an awakening to being a participant in the world and its processes.
35 Zhuangzi, *Zhuangzi: The Essential Writings*, 49.
36 Also see *Beyond Good and Evil* section 230 (Nietzsche, *Beyond Good and Evil*, 159–62; Nietzsche, *On the Genealogy of Morals* and *Ecce Homo*, 57).
37 Friedrich Nietzsche, *The Use and Abuse of History*. Translated by Adrian Collins (Indianapolis: Bobbs-Merrill, 1957), 6.
38 Nietzsche, *Beyond Good and Evil*, 161.
39 Alphonso Lingis, "The Will to Power," in *The New Nietzsche: Contemporary Styles of Interpretation*, edited by David B. Allison (New York: Dell Publishing, 1977), 54.
40 Fragment DK B91. All Heraclitus translations are mine.

41 Nietzsche, *Thus Spoke Zarathustra*, 190.
42 See David Jones, "The Over-flowing/Flowing-over Soul in Zarathustra and Zhuangzi," *Dao: A Journal of Comparative Philosophy* 4 (2) (2005): 235–51. As mentioned above, Parkes has coined the term "transperspectival" and in the *Zhuangzi* context he refers to it in the following manner: "It's a matter of broadening our perspectives, expanding the 'natural light' (*lumen naturale*) of human understanding to become co-extensive with what Zhuangzi calls the 'broad light of Heaven.' Viewing things in this light lets us see them impartially, since the light of Heaven is above all indifferent to what it illuminates" (Chai, 68). It is unclear to me how this is different from what I have been calling the "perspectival perspective" (as far back as my PhD thesis, "Ariadne's Complaint—Becoming of Being in Herakleitos, Chuang Tzu, and Nietzsche"), but he may, of course, have something more profound in mind. For me, the "perspectival perspective" remains a perspective, one that is more comprehensive by nature.
43 Zhuangzi, *Zhuangzi: The Essential Writings*, 12.
44 For more on this see Bradley Douglas Park's excellent chapter, "Being in our Right Mind: The Rhythmic Body and the Emergence of the Social Mind."
45 The phrase "*der allergrösste Teil unseres geistigen Wirkens*," here translated by Kaufman as "the greatest part of our spirit's activity," could also be translated as "the greatest part of our spiritual work" or "the greatest part of our intellectual work." Nietzsche, *The Gay Science*, 262.
46 See also section 354: "The problem of consciousness ... confronts us only when we begin to comprehend how we could dispense with it; ... *For what purpose*, then, any consciousness at all when it is in the main *superfluous*?" (Nietzsche, *The Gay Science*, 297, 85).
47 See Deleuze 1962, 87–9.
48 "... [Freud] several times said of Nietzsche that he has a more penetrating knowledge of himself than any other man who ever lived or was likely to ever live" (Ernest Jones, *Sigmund Freud: Life and Work*, Volume 2 (London: The Hogarth Press, 1958), 385). Ernest Jones also relates that in the second month of Freud's own self-analysis that "he was learning the truth of Nietzsche's maxim: 'One's own self is well hidden from oneself: of all mines of treasure one's own is the last to be dug up'" (Jones, *Sigmund Freud*, 322).
49 Nietzsche, *The Gay Science*, 263.
50 Graham Parkes, *Composing the Soul: Reaches of Nietzsche's Psychology* (Chicago: University of Chicago Press, 1994), 214.
51 Nietzsche, *Thus Spoke Zarathustra*, 72.
52 Nietzsche writes in the notebooks that "*Radical nihilism* is the conviction of an absolute untenability of existence when it comes to the highest values one recognizes; plus the realization that we lack the least right to posit a beyond or an in-itself of things that might be 'divine' or morality incarnate" (Nietzsche, *The Will to Power*, 9).
53 Nietzsche, *Thus Spoke Zarathustra*, 14.
54 Nietzsche, *Thus Spoke Zarathustra*, 9.
55 Parkes has started translating *ewige Wiederkunft* as "eternal recoming" rather than "eternal recurrence." However, he does retain "eternal return" for *ewige Wiederkehr*. I follow his lead here but use a hyphen.
56 *Wonne* does mean "delight" as it is often translated, and Parkes also translates it this way. However, it is worth pointing out that the word carries stronger feelings of joy, rapture, ecstasy, bliss, to be enraptured, and to stand ecstatically outside oneself.

It is often used in compounds: *Wonnebebend* (trembling with delight), *Wonnegefühl* (feeling bliss), *Wonnetrunken* (intoxicated bliss), and *Wonnefloss* (colloquial for a baby).

57 *Floss* also means a float, buoy, raft and can mean flowing water from the verb *fliessen*, to flow, run, melt, and pass away. The child of the "Three Transformations" anticipates the death of the man Zarathustra as he is carried into the current of a surging forth world, and to his transformation. One is reminded of the Buddhist "Parable of the Raft" where we are prompted to realize that a raft is for crossing over, but not for holding onto once one has crossed.
58 Nietzsche, *Thus Spoke Zarathustra*, 24.
59 Parkes, *Composing the Soul*, 124.
60 Nietzsche, *Thus Spoke Zarathustra*, 188.
61 Ames and Hall, *Daodejing*, 87.
62 Ames and Hall, *Daodejing*, 83.
63 As Robin Wang points out, "although there is only one use of the term yinyang in the *Daodejing*, the text has a persistent orientation to the paradoxical interdependence of opposites in the world" (Wang 2012, 30). (Roger T. Ames and David L. Hall, *Daodejing: "Making This Life Significant"* [New York: Ballantine Books, 2003], 151).
64 Watson points out in a note that these are "not three different categories but three names for the same thing." See Zhuangzi, *The Complete Works of Chuang Tzu*. Translated by Burton Watson (New York: Columbia University Press, 1964), 26; Zhuangzi, *Zhuangzi: The Essential Writings*, 45.
65 Zhuangzi, *Zhuangzi: The Essential Writings*, 18.
66 Herbert Fingarette, "The Problem of the Self in the *Analects*," *Philosophy East and West* 29 (2) (1979): 129.
67 Nietzsche, *Beyond Good and Evil*, 24.
68 See lines 179c–183c (Plato 92–100).
69 Nietzsche, *Twilight of the Idols and the AntiChrist*, 38.
70 Freud follows Nietzsche's lead and uses "I" and not "ego." Bruno Bettelheim has an excellent discussion of this in his *Freud and Man's Soul*: "To mistranslate *Ich* as "ego" is to transform it into jargon that no longer conveys the personal commitment we make when we say "I" or "me"—not to mention our subconscious memories of the deep emotional experience we had when, in infancy, we discovered ourselves as we learned to say "I"" (Bruno Bettelheim, *Freud and Man's Soul* (New York: Vintage Books, 1982), 53.
71 Nietzsche, *The Will to Power*, 268–9.
72 Nietzsche, *Beyond Good and Evil*, 27.
73 Henry Rosemont Jr. consistently makes this point in his *Against Individualism*.
74 See Thomas P. Kasulis, *Intimacy or Integrity: Philosophy and Cultural Difference*.
75 Nietzsche, *The Will to Power*, 367, 373.
76 Lingis, "The Will to Power," 38.
77 Nietzsche, *Beyond Good and Evil*, 289.
78 Nietzsche, *The Will to Power*, 291.
79 Nietzsche, *Thus Spoke Zarathustra*, 15.
80 Nietzsche, *Thus Spoke Zarathustra*, 58.
81 Nietzsche, *Thus Spoke Zarathustra*, 9.
82 Nietzsche, *Thus Spoke Zarathustra*, 14.
83 Nietzsche, *Thus Spoke Zarathustra*, 46.
84 *Langenscheidt's New College German Dictionary*, Berlin and Munich, 1973.

85 Ames and Hall make a similar point in their commentary on chapter 15 of the *Laozi*: "Persons who have been most successful at making their way in the world have been fully immersed in the process itself, assuming for themselves the profound and complex character of the experience that they have forged." See Ames and Hall, *Daodejing*, 98.
86 This hydrodynamic activity of *überfliessen* (overflowing) and *untergehen* is crucial for Nietzsche. A "cup" that wants to overflow must first go down. This Dionysian overflow is essential for the play of self-overcoming through going-over (*ubergehen*) and *untergehen* in Prologue, 4 of *Zarathustra*.
87 For more on the multiple psyche in Nietzsche, see Graham Parkes, "Ordering the Psychic Polytic: Choices of Inner Regime for Plato and Nietzsche," *Journal of Nietzsche Studies* 2 (Autumn) (1991): 53–77; and Parkes, *Composing the Soul*, 346–62.
88 Nietzsche, *Thus Spoke Zarathustra*, 114.
89 Ames and Hall, *Daodejing*, 87.
90 Zhuangzi, *Zhuangzi: The Essential Writings*, 81.
91 For a discussion of this fractal connection, see John L. Culliney and David Jones, "The Fractal Self and the Organization of Nature: The Daoist Sage and Chaos Theory," *Zygon: Journal of Science and Religion* 4 (1999): 643–54.
92 Zhuangzi, *Zhuangzi: The Essential Writings*, 81.
93 Ames and Hall, *Daodejing*, 99.
94 Ames and Hall, *Daodejing*, 89.
95 Ames and Hall, *Daodejing*, 130.
96 Ames and Hall, *Daodejing*, 99.
97 Nietzsche, *On the Genealogy of Morals* and *Ecce Homo*, 273.
98 Martin Heidegger, "On the Essence of Truth." Translated by John Sallis. In *Martin Heidegger: Basic Writings*, edited by David Farrell Krell (New York: Harper and Row Publisher, 1977), 129.
99 Zhuangzi, *Zhuangzi: The Essential Writings*, 81.
100 A. C. Graham, *The Book of Lieh-tzŭ: A Classic of Tao* (New York: Columbia University Press, 1990), 2.
101 Friedrich Nietzsche, *Friedrich Nietzsche: Werke in Drei Bänden*. Edited by Karl Schlechta (München: Carl Hanser Verlag, 1965), 258.
102 My alternative translation, not an improvement on the Parkes translation, but brings out the earlier theme of seeing things as themselves a bit more clearly.

> —"Oh Zarathustra, thereupon the animals said, for those who think as we do, all things themselves are dancing: they come and reach out their hands and laugh and run away—and come back.
> All goes, All comes back; the wheel of being rolls eternally. All dies, All blooms on once more, eternally runs the year of being.
> All breaks, All becomes joined anew; eternally the same house of being is built. All departs, All greet themselves again; eternally the ring of being remains true to itself. In each Now being begins; near each Here rolls the sphere There. The center is all-over. Crooked is the path of eternity."
>
> Nietzsche, *Thus Spoke Zarathustra*, 190.

103 Thomas Metzinger argues this point from the "cognitive agency" of the "cognitive subject." Based upon the "phenomenal self-model" (PSM), he states that "most animals are conscious to one degree or another, but their PSM is not the same as ours. Our evolved type of conscious self-model is unique to the human brain, in that

by representing the process of representation itself, we can catch ourselves … in the act of knowing" (5). See also 120.

104 Not only does the German "*letzte*" mean final, but also has the connotations of deep, which further emphasizes the noun "*Tiefe*." In "The Convalescent," Nietzsche wants to make a point about the relation of depth and suffering.

105 Nietzsche, *Thus Spoke Zarathustra*, 189.

106 Parkes's translation of *Schalks-Narren* "buffoons" is certainly accurate. My preferred translation, however, is "roguish fools" to pick up on the centrality of aesthetic phenomena being indispensable for making existence bearable. As Nietzsche writes in section 107 of *The Joyous Science*:

> At times we need a rest from ourselves by looking upon, by looking *down* upon, ourselves and, from an artistic distance, laughing *over* ourselves or weeping *over* ourselves. We must discover the *hero* no less than the *fool* in our passion for knowledge; we must occasionally find pleasure in our folly, or we cannot continue to find pleasure in our wisdom. Precisely because we are at bottom grave and serious human beings—really, more weights than human beings—nothing does us as much good as a *fool's cap*: we need it in relation to ourselves—we need all exuberant, floating, dancing, mocking, childish, and blissful art lest we lose the *freedom above things* that our ideal demands of us. (Nietzsche, 1974, 164)

Although Nietzsche uses a different word in section 6 of the "Prologue," this plays off on the sense of the jester or clown (*Possenreißer*), as well as being "overburdened with … wisdom" in the Prologue's opening. The *Possenreißer* too can be viewed as a splinter psyche of Zarathustra.

107 Nietzsche, *Thus Spoke Zarathustra*, 190; translation slightly modified.
108 Nietzsche, *Thus Spoke Zarathustra*, 191.
109 Nietzsche, *Thus Spoke Zarathustra*, 191.
110 Crell, *Chinese Thought*, 99.
111 Zhuangzi, *The Complete Works*, xi.
112 Ames and Rosemont, *The Analects*, 92.
113 My interpolation of "recoming" for "recurrence." Nietzsche, *Thus Spoke Zarathustra*, 192.
114 A. C. Graham translates the name Nie Que as "Gaptooth." This is in line with most of the characters in the *Zhuangzi* as being marginalized by Confucian society but brought into the center of the discourse.
115 Zhuangzi, *Zhuangzi: The Essential Writings*, 18.
116 Zhuangzi, *Chuang Tzu: The Inner Chapters*. Translated by A. C. Graham (London: George Allen & Unwin, 1981), 125.
117 Zhuangzi, *Chuang Tzu: The Inner Chapters*, 124.
118 Zhuangzi, *Zhuangzi: The Essential Writings*, 117.
119 Nietzsche, *Thus Spoke Zarathustra*, 62.
120 Nietzsche, *Thus Spoke Zarathustra*, 107.
121 To get at some of the themes being addressed here, I have somewhat departed from Parkes's translation, which is the best translation of *Thus Spoke Zarathustra*. See Nietzsche, *Thus Spoke Zarathustra*, 193.
122 Zhuangzi, *Chuang Tzu: The Inner Chapters*, 58.

123 Both German words are very close in meaning. *Umfangender* also has the sense of encircling which plays very nicely with the thought of eternal re-coming. *Umfänglich* adds a sense of spatiality, size, and extension. This may hint at the closeness of space and time in Nietzsche's thought of eternal re-coming.
124 Nietzsche, *Thus Spoke Zarathustra*, 194–5.
125 Parkes translation slightly amended (Nietzsche, *Thus Spoke Zarathustra*, 196). Parkes translates *Letztes* as "ultimate," and this is clearly accurate. I have chosen "deepest" since *Letztes* can also mean "lowest" or "bottom" as well as "last" or "final." I do so to emphasize the deeper psychological motifs throughout *Zarathustra* than can be found in the *Zhuangzi*.

References

Ames, Roger T. and Hall, David L. 2003. *Daodejing: "Making This Life Significant."* New York: Ballantine Books.
Ames, Roger T. and Rosemont, Henry, Jr. 1998. *The Analects of Confucius: A Philosophical Translation*. New York: Ballantine Books.
Bettelheim, Bruno. 1982. *Freud and Man's Soul*. New York: Vintage Books.
Creel, H. G. 1953. *Chinese Thought from Confucius to Mao Tsê-tung*. Chicago: University of Chicago Press.
Culliney, John L. and Jones, David. 1999. "The Fractal Self and the Organization of Nature: The Daoist Sage and Chaos Theory." *Zygon: Journal of Science and Religion* 4: 643–54.
Culliney, John L. and Jones, David. 2017. *The Fractal Self: Science, Philosophy, and the Evolution of Human Cooperation*. Honolulu: University of Hawaii Press.
Deleuze, Gilles. 1962. *Nietzsche and Philosophy*. Translated by Hugh Tomlinson. New York: Columbia University Press.
Fingarette, Herbert. 1979. "The Problem of the Self in the *Analects*." *Philosophy East West* 29 (2): 129–40.
Graham, Angus C. 1990. *The Book of Lieh-tzŭ: A Classic of Tao*. New York: Columbia University Press.
Heidegger, Martin. 1977. "On the Essence of Truth." Translated by John Sallis. In *Martin Heidegger: Basic Writings*, edited by David Farrell Krell, 113–41. New York: Harper and Row Publisher.
Jones, David. 2005. "The Over-flowing/Flowing-over Soul in Zarathustra and Zhuangzi." *Dao: A Journal of Comparative Philosophy* 4 (2): 235–51.
Jones, Ernest. 1958. *Sigmund Freud: Life and Work*, Vol. 2. London: The Hogarth Press.
Kasulis, Thomas P. 2002. *Intimacy or Integrity: Philosophy and Cultural Difference*. Honolulu: University of Hawaii Press.
Koller, John M. 2002. *Asian Philosophies*. Upper Saddle River, NJ: Prentice Hall.
Laozi. 2003. *Daodejing: A Philosophical Translation*. Translated by Roger T. Ames and David L. Hall. New York: Ballantine Books.
Lingis, Alphonso. 1977. "The Will to Power." In *The New Nietzsche: Contemporary Styles of Interpretation*, edited by David B. Allison, 37–63. New York: Dell Publishing.
Metzinger, Thomas. 2009. *The Ego Tunnel*. New York: Basic Books.
Nietzsche, Friedrich. 1957. *The Use and Abuse of History*. Translated by Adrian Collins. Indianapolis: Bobbs-Merrill.

Nietzsche, Friedrich. 1965. *Friedrich Nietzsche: Werke in Drei Bänden*. Edited by Karl Schlechta. München: Carl Hanser Verlag.

Nietzsche, Friedrich. 1966. *Beyond Good and Evil*. Translated by Walter Kaufmann. New York: Vintage Books.

Nietzsche, Friedrich. 1967. *The Will to Power*. Translated by Walter Kaufmann and R. J. Hollingdale. Edited by Walter Kaufmann. New York: Vintage Books.

Nietzsche, Friedrich. 1968a. *Twilight of the Idols and the AntiChrist*. Translated by R. J. Hollingdale. Baltimore: Penguin Books.

Nietzsche, Friedrich. 1968b. *The Portable Nietzsche*. Translated by Walter Kaufmann. New York: The Viking Press.

Nietzsche, Friedrich. 1969. *On the Genealogy of Morals* and *Ecce Homo*. Translated by Walter Kaufmann and R. J. Hollingdale. New York: Vintage Books.

Nietzsche, Friedrich. 1974. *The Gay Science*. Translated by Walter Kaufmann. New York: Vintage Books.

Nietzsche, Friedrich. 2005. *Thus Spoke Zarathustra*. Translated by Graham Parkes. New York: Oxford University Press.

Parkes, Graham. 1991. "Ordering the Psychic Polytic: Choices of Inner Regime for Plato and Nietzsche." *Journal of Nietzsche Studies* 2 (Autumn): 53–77.

Parkes, Graham. 1994. *Composing the Soul: Reaches of Nietzsche's Psychology*. Chicago: University of Chicago Press.

Plato. 1957. *Theaetetus*. Translated by Francis MacDonald Cornford. Indianapolis: Bobbs-Merrrill.

Rosemont, Henry Jr. 2015. *Against Individualism: A Confucian Rethinking of the Foundations of Morality, Politics, Family, and Religion*. Lanham, MD: Lexington Books.

Wang, Robin R. 2012. *Yinyang: The Ways of Heaven and Earth in Chinese Thought and Culture*. New York: Cambridge University Press.

Zhuangzi. 1964. *The Complete Works of Chuang Tzu*. Translated by Burton Watson. New York: Columbia University Press.

Zhuangzi. 1981. *Chuang Tzu: The Inner Chapters*. Translated by Angus C. Graham. London: George Allen & Unwin.

Zhuangzi. 2009. *Zhuangzi: The Essential Writings*. Translated by Brook Ziporyn. Indianapolis: Hackett Publishing.

Part Five

Keeping on Dancing

13

Gratitude of a Wondering Wanderer

Graham Parkes

One repays a teacher poorly if one always remains only a student.
... "This--is just my way:--where is yours?"

Nietzsche's Zarathustra[1]

Given limitations of space and time, let me just say, simply: *Thanks!* Gratitude for all this, to friends and colleagues who have given us such wonderful things here. And they are an occasion for wonder, itself a prime occasion for the practice of philosophy.[2] So let me round out this rich collection of perspectives with some reflections on the course of my engagement with the practice, which combined philosophical wondering with geographical wandering.

My introduction to philosophy came when I was a schoolboy at Glasgow Academy, thanks to a reading group of would-be intellectuals who called themselves The Humanists. The first book we read after I was invited to join was *The Rebel* by Albert Camus. I was all for rebelling, but since I couldn't understand most of the assigned reading I felt rather stupid, and not much enlightened by the subsequent discussion. If this was philosophy—I hadn't a clue, but a member of the group suggested it was—I didn't think it was for me.

The text for the following month, however, was *Thus Spoke Zarathustra* by Friedrich Nietzsche. I was baffled by that one, too, but the difference was a feeling that something important was going on, even though I couldn't tell what it was. On the back cover of the book it said that Nietzsche was a German philosopher, but I had no idea what philosophers did (except write opaque books like *The Rebel*). A friend later explained that a philosopher was "someone who gets paid to sit around and think," which struck me as very nice work if you could get it. At any rate, even though I'd hardly got it at all, *Zarathustra* had struck a chord, and I resolved to look into the matter further.

I had the opportunity a few years later, after getting a scholarship in modern languages at Oxford and then switching to philosophy and psychology. In late 1967 a series of powerful psychedelic experiences, mostly in the meadows bordering the tree-lined river, had revealed a world that filled me with wonder. (This, by the way, was where the dance began.) I thought perhaps philosophy would offer ways of understanding what was going on with these multiple realities, but Oxford Philosophy was concerned with other matters. A friend recommended looking into Indian and Tibetan thought, which turned out to be deeply instructive on the topic of worlds and places in between.

Despite lacking a clear idea of what I wanted to do in life, I was awarded a fellowship in philosophy by the University of California at Berkeley, which prompted a move there in 1970—over five thousand miles to the west, and into quite a different culture. The San Francisco Bay Area was the place then for a remarkable diversity of philosophies and religions: you could find everything from Advaita Vedānta to Zen, in universities and institutes, centers and temples. There was a rich profusion of cultures and arts, and all of this surrounded by places of staggering natural beauty: the Pacific coast, redwood groves, the high mountains of the Sierra Nevada—and relatively unspoiled in those days.

The desire to understand this feeling of awe for the natural world led to a study of East Asian philosophies, and I was fortunate to meet two more advanced students, from China and Korea, who introduced me to Daoism and Mahāyāna Buddhism respectively. (Such an encounter would have been unlikely if I'd stayed in the UK.) But since the terms of my fellowship wouldn't permit the extra time it would take to learn the classical language for a PhD in Chinese Buddhism, I ended up writing on Heidegger and depth psychology instead.

A year after finishing the doctorate, I managed to get a position at the University of Hawai'i, a stroke of very good fortune in view of my interest in Chinese thought. But when I arrived at the Philosophy Department, there were already three colleagues teaching Chinese philosophy, so would I be interested in switching to Japanese instead? As an accommodating young academic without tenure, I agreed and shifted the language learning accordingly. (Another path taken without deliberation: something happens and you acquiesce.)

The natural world in Hawai'i is a thing of wonder—a tropical outcome of copious rain falling and ocean waves breaking on volcanic rock over hundreds of millennia. But already in the early 1980s the place was being destroyed by "development," and a few people were getting very rich from the desecration. This was all the more sad to witness because the early Hawaiians had managed to maintain a good balance with the natural ecosystems that sustained them, from mountain-top down to seashore and bay. And now the invaders were destroying their nature as well as their culture based on those ecosystems. A familiar story under capitalism: environmental destruction as a source of financial profit.

As an adjunct professor in a new department for "Ecology and Health," I came to understand ecology in greater depth, and when climate change became an issue in the late 1980s, I started following the climate sciences—all the while discovering corresponding philosophies of nature in the East Asian traditions. Deriving from writing traditions, these latter were more sophisticated than the Hawaiian ideas based on their oral tradition, but the two views of human/nature were perfectly compatible. If modern Western thinking moves away from anthropocentrism at all, it tends to stop with biocentrism; but the East Asian idea of the world as a field of *qi* or *ki* 氣 energies, like the Polynesian idea of *mana*, undermines any hard distinction between animate and inanimate. On this kind of ecocentric or polycentric view, rocks and unworked stone (a common topic in the essays preceding) are revered as "kernels" of Earth's central energies.

I'll skip over my time at the National University of Ireland from 2008—except to mention the magnificent rock formations along the coast of County Cork. After seven years there, I went on to teach as a visiting professor in Prague, Tokyo, Singapore, and Shanghai, where I tried out my ideas about how to tackle the climate crisis with a variety of student audiences—and finally ended up in Vienna, where I had taught briefly around the turn of the century and still had colleagues in intercultural philosophy.

If there's anything faintly remarkable about that career, it's the almost total lack of planning on the level of the literal wandering. The move to Berkeley (first and best thing I ever did) was motivated by ill-informed impulse and intended only five years in the United States rather than forty. The move to Hawai'i resulted from a fluke in the advertisement and application process: I'd applied for a position in philosophy of law (!), but the Department happened to be looking for someone in Continental European philosophy as well, after someone's unexpected resignation.

I applied for the job in Ireland because they were looking for a professor who could "internationalize" the philosophy program, and I assumed that my competence in comparative and intercultural philosophy would impress. It didn't: the search committee chose somebody else—but they turned the offer down, and so it came to me. It was only much later that I learned that "internationalising the program" had nothing to do with content: it was a matter of attracting foreign students and having them pay high fees.

I'd like to see the circuitous path from Glasgow to Vienna as an instance of Zhuangzi's "free wandering without a destination" (*xiaoyaoyou* 逍遙遊), though it's probably more a case of clueless drifting, verging on some kind of spontaneity. But the ways from acid to zazen, Nietzsche to Zhuangzi, Heidegger to Dōgen, the arts to Earth systems—those make more sense, and in retrospect you can even see some method to it.

But it's also true that after so much wandering you can end up wondering "Was it all worthwhile?" After all, the judgment has often been: "Parkes is something of a dilettante. ... He doesn't do *real* philosophy. ... The man spreads himself too thin." So let's apply the ultimate test: if I had to live this life again, "as I live and have lived it ... and everything in the same order and sequence," would I be able to say Yes to that? (Nietzsche's thought of "eternal recoming.")[3] Or would I rather have me staying in one field, and becoming really good in it?[4] Or spending less time writing essays and teaching students, and more time publishing books so as to make a name for myself? Now, in the light of the essays in this volume, I can at least say *No* to those last two questions—and *Yes* to having wandered through diverse fields, and talked and read, and talked about writing, with my students.

Though he always wished for a few good students, Nietzsche himself went with the books, remarking with amazement how after its publication a book can "live on like a being endowed with spirit and soul." (Go *Wandering Dance*!) But he goes on to remark, anticipating the thought of recoming, how "*every* action of a human being" can become "the occasion for further actions, decisions, thoughts," so that "everything that happens is inextricably knotted together with everything that will happen." He calls this the "*immortality* of movement": "whatever has moved is, like an insect in amber, embedded and eternalised in the total association of all that is."[5] So, if your pedagogical efforts manage to move students and especially if they're now teaching students of their own—you never know how far and widely the pulsing of the mesh will permeate the matrix.

I have known on occasion the joy in seeing ideas that I've entertained being taken up by former students and colleagues and developed in new ways and contexts—but never on the scale of this particular book. So my heartfelt thanks, again, to my colleagues Roger, Yuriko-san, Kathy, Peter, and Jason, and former students David, Meilin, Andrew, Leah, Brad, and Tim, for making it all seem worthwhile. Tim has set the record for the most thorough engagement with the wandering dance idea, and that has set me wondering again. And without David—well, nothing. Thanks, finally, to Setsuko Aihara for the painting on the book's cover, a lustrous elaboration of a vague image that came to me of the wanderer dancing.

Let me finish by saying to all my friends here, as Brian Eno memorably sang to China his "china" (rhyming slang for "friend") fifty years ago, words that I've often sung to myself—in gratitude and mild surprise—over the years: *"I've wandered around and you're still here."*[6]

Yes, but the last word has to be *dance*.

Notes

1   Nietzsche, *Thus Spoke Zarathustra*, 1.22, "On the Bestowing Virtue," 3.11, "On the Spirit of Heaviness."
2   Plato, *Theaetetus* 155d, Aristotle, *Metaphysics A*, 982b11.
3   I've come to think that "eternal recoming" is a better translation for *ewige Wiederkunft* than "eternal recurrence" (while "eternal return" is fine for *ewige Wiederkehr*). Eternal *recoming* has a nice assonance with Nietzsche's favored *becoming*, while conveying the connotation of the ordinary meaning of *die Wiederkunft*: the Second Coming (of the Lord), or Parousia. But while Jesus is going to come back a second time, on Judgment Day, Zarathustra is coming again a third, fourth, and innumerable times more.
4   Nietzsche, *The Joyful Science*, aphorism 341.
5   Nietzsche, *Things Human, All-too-Human*, aphorism 208.
6   Brian Eno, "China My China" from *Taking Tiger Mountain* (1974); at 2':32'.

Contributors

Roger T. Ames is Humanities Chair Professor at Peking University, Co-Chair of the Academic Advisory Committee of the Peking University Berggruen Research Center, and Professor Emeritus of Philosophy at the University of Hawai'i. He is former editor of *Philosophy East & West* and founding editor of *China Review International*. Ames has authored several interpretative studies of Chinese philosophy and culture: *Thinking through Confucius* (1987), *Anticipating China* (1995), *Thinking from the Han* (1998), and *Democracy of the Dead* (1999) (all with D. L. Hall), *Confucian Role Ethics: A Vocabulary* (2011), and most recently *Human Becomings: Theorizing Persons for Confucian Role Ethics* (2021). His publications also include translations of Chinese classics: *Sun-tzu: The Art of Warfare* (1993); *Sun Pin: The Art of Warfare* (1996) (with D. C. Lau); the *Confucian Analects* (1998) and the *Classic of Family Reverence*: *The* Xiaojing (2009) (both with H. Rosemont), *Focusing the Familiar:* The Zhongyong (2001), and *The Daodejing* (with D.L. Hall) (2003). Almost all of his publications are now available in Chinese translation, including his philosophical translations of Chinese canonical texts. He has most recently been engaged in compiling the new *Sourcebook in Classical Confucian Philosophy* (2022) and its companion volume, *A Conceptual Lexicon for Classical Confucian Philosophy* (2021).

Meilin Chinn is Associate Professor of Philosophy at Santa Clara University whose work centers on Aesthetics and Chinese Philosophy. Her published writing includes essays on music and its relationship with truth ("Only Music Cannot Be Faked," *Dao: A Journal of Comparative Philosophy* (2017)), meaning ("Music With and Without Images," *Journal of Chinese Philosophy* (2020)), and the self ("Persona: Resounding Selves and Empty Music," *On the True Sense of Art: A Critical Companion to the Transfigurements of John Sallis* (2016)), as well as the aesthetics of race ("Race Magic and the Yellow Peril," *Journal of Aesthetics and Art Criticism* (2019)), and environmental philosophy ("Sensing the Wind: The Timely Music of Nature's Memory, Journal of Environmental Philosophy (2013)"). She is also the editor of a three-volume collection titled *Asian Aesthetics: Primary Sources*, underway with Bloomsbury Press.

Timothy J. Freeman is Associate Professor of Philosophy at the University of Hawai'i Hilo. His main areas of research are comparative philosophy, continental philosophy, philosophy of art, environmental philosophy, and political philosophy. He is author of "The Shimmering Shining: The Promise of Art in Heidegger and Nietzsche" appearing in the journal *Comparative and Continental Philosophy (2013)* and most recently, "Living on the Edge of a Volcano: Reflections on Nietzsche's Philosophy and Albert Saijo's Zensational Rhapsody" in the *Journal of World Philosophies* in the summer of 2023. He is also a ceramic artist, and his award-winning pit-fired ceramic vessels evoking the volcano landscape where he lives on the island of Hawai'i have appeared in numerous state-wide juried and group exhibitions including both the *Hawaii's Modern Masters: 10th Annual Celebration of the Arts* in Honolulu, and the exhibition *Minimenta: La grand exposition des petits formats* at the Galerie Goutte in Paris, France, in 2015. He received his doctorate from the University of Hawai'i.

Peter D. Hershock is Director of the Asian Studies Development Program and Coordinator of the Humane AI Initiative at the East-West Center in Honolulu. His philosophical work makes use of Buddhist conceptual resources to address contemporary issues of global concern. He has authored or edited more than a dozen books on Buddhism, Asian philosophy, and contemporary issues, including *Reinventing the Wheel: A Buddhist Response to the Information Age* (1999); *Buddhism in the Public Sphere: Reorienting Global Interdependence* (2006); *Valuing Diversity: Buddhist Reflection on Realizing a More Equitable Global Future* (2012); and *Buddhism and Intelligent Technology: Toward a More Humane Future* (2021). His current project is *Consciousness Mattering: Buddhism on the Creative Entanglement of Consciousness, Evolution and Ethics*.

Kathleen Higgins is Professor of Philosophy at the University of Texas at Austin. Her main areas of research are philosophy of music, aesthetics, continental philosophy, and philosophy of emotion. She is author of *The Music between Us: Is Music a Universal Language?* (2012), *The Music of Our Lives* (2011), *Nietzsche's "Zarathustra"* (2010), *Comic Relief: Nietzsche's Gay Science* (2000), and co-author (with Robert C. Solomon) of books on Nietzsche and the history of philosophy. She has edited or co-edited other books on such topics as Nietzsche, German Idealism, aesthetics, ethics, erotic love, non-Western philosophy, and the philosophy of Robert C. Solomon. She has been a Professorial Research Fellow at the University of Vienna, a Visiting Fellow at the Faculty of Psychology and Educational Sciences of Katholieke Universiteit Leuven, Visiting Fellow of the Australian National University Philosophy Department and Canberra School of Music, and a Resident Scholar at the Rockefeller Foundation's Bellagio Study and Conference Center. She has also been a frequent Visiting Professor at the University of Auckland. She is a former President of the American Society for Aesthetics.

David Jones is a University Foundation Distinguished Professor and Professor of Philosophy at Kennesaw State University and is an Affiliate Professor of Philosophy at the University of Hawai'i at Hilo. He has been a Visiting Professor of Confucian Classics at Emory and Visiting Scholar at the Institute for Advanced Studies in Humanities and Social Science at National Taiwan University. In addition, he is Editor of the journal *Comparative and Continental Philosophy*. His most recent books are *The Philosophy of Creative Solitudes* (2020) and *The Fractal Self: Science, Philosophy, and the Evolution of Human Cooperation* with John L. Culliney (2017). As President Emeritus of the Southeast Regional of the Association of Asian Studies and the Comparative and Continental Philosophy Circle, he was the East-West Center's Distinguished Alumnus in 2004–5. He received his doctorate from the University of Hawai'i.

Leah Kalmanson is Associate Professor and Bhagwan Adinath Professor of Jain Studies in the Department of Philosophy and Religion at the University of North Texas. She received her PhD from the Department of Philosophy at the University of Hawaiʻi at Mānoa in 2010. She is the author of *Cross-Cultural Existentialism: On the Meaning of Life in Asian and Western Thought* (2020) and co-author with Monika Kirloskar-Steinbach of *A Practical Guide to World Philosophies: Selves, Worlds, and Ways of Knowing* (2021). Her essays appear in journals including *Comparative and Continental Philosophy, Continental Philosophy Review, Frontiers of Philosophy in China, Hypatia, Journal of World Philosophies, Philosophy East and West, Pragmatism Today, Shofar*, and *Studies in Chinese Religions*, as well as the digital magazine Aeon.

Bradley Douglas Park is Professor of Philosophy at St. Mary's College of Maryland. His research interests center around the embodied mind, phenomenology, and East Asian philosophy. Outside of academia, he has been a student of traditional Appalachian music, particularly clawhammer banjo, and has performed at the Washington Folk Festival and at the West Virginia State Folk Festival.

Graham Parkes is a Professorial Research Fellow in philosophy at the University of Vienna, and his main interests are in Continental European philosophy, Chinese and Japanese thought, and philosophies of art and nature. His latest book is *How to Think about the Climate Crisis: A Philosophical Guide to Saner Ways of Living* (2021). After much wondering and wandering, he is striving to prolong "the earthly dance" into the age of arthritis.

Yuriko Saito is Professor Emerita of Philosophy at the Rhode Island School of Design, USA, and Editor of *Contemporary Aesthetics*, an open-access, peer-reviewed journal. She has published widely on everyday aesthetics, Japanese aesthetics, and environmental aesthetics. Her *Everyday Aesthetics* (2008) and *Aesthetics of the Familiar: Everyday Life and World-Making* (2017) were published by Oxford University Press. The latter was awarded the 2018 Outstanding Monograph Prize by the American Society for Aesthetics.

Andrew K. Whitehead specializes in East-West comparative philosophy, particularly Japanese and Chinese philosophies, and the German and French traditions of phenomenology and existentialism. He is the President of the Académie du Midi Philosophical Association, an Associate Editor of the journal *Comparative and Continental Philosophy*, an Executive Officer of the Comparative and Continental Philosophy Circle, and a guest professor for the Higher Institute of Philosophy (HIW) at KU Leuven in Belgium. He is the author of several articles and a number of co-edited books, including *Critique, Subversion, and Chinese Philosophy: Socio-Political, Conceptual, and Methodological Challenges* (2020) and *Imagination: Cross-Cultural Philosophical Analyses* (2018), both of which are published with Bloomsbury Academic.

Jason M. Wirth is Professor of philosophy at Seattle University and works and teaches in the areas of Continental Philosophy, Buddhist Philosophy, Aesthetics, and Environmental Philosophy. His recent books include *Nietzsche and Other Buddhas: Philosophy after Comparative Philosophy* (2019), *Mountains, Rivers, and the Great Earth: Reading Gary Snyder and Dōgen in an Age of Ecological Crisis* (2017), a monograph on Milan Kundera (*Commiserating with Devastated Things*, 2015), *Schelling's Practice of the Wild* (2015), and the co-edited volume (with Bret Davis and Brian Schroeder), *Japanese and Continental Philosophy: Conversations with the Kyoto School* (2011). He is the associate editor and book review editor of the journal *Comparative and Continental Philosophy*. He is currently completing a manuscript on the cinema of Terrence Malick as well as a work of ecological philosophy called *Turtle Island Anarchy*.

Subject Index

abyss 22, 25, 26, 149, 187–8, 205, 217, 225, 229–30, 235
   Abgrund 22, 188, 222, 229
   Dionysian abyss 20
   groundless ground 230
Aesthetics vii, 5, 51, 53, 69, 73–5, 77, 80–4, 87, 97–8, 100, 112, 121–2, 253–6
   aesthetic justification of life 75, 80
   aesthetic phenomenon 70–1
   aesthetic pleasure 119–20
   aesthetic turn 53
   Daoist aesthetic sensibility 32
   Nietzsche's aesthetics 79–80
   Parkes's work in 87
   rock aesthetics 138
   wabi aesthetics 73–5
ālāya-vijñāna (storehouse consciousness) 193
alchemy 59–60, 65, 132
amor fati 195–6, 198, 198, 233–4
anātman 190–1, 193, 208
Angst 26, 114, 225–6, 235
anima mundi 52–5, 59–61, 63
animism 8, 52, 137, 139
Anitya (impermanence)103, 190, 193
Anthropocene 4, 32, 192
anthropocentrism 15–16, 55–6, 77–8, 180–1, 189, 191, 250
Apollonian 20, 150–2, 187, 201, 223
   Apollonian phenomenology 167
Art 5–6, 36, 46, 48, 50, 53–4, 60, 69, 70–1, 73–5, 77–82, 84–5, 87–8, 90–100, 121–2, 165, 171–2, 174, 202–4, 212, 214, 244, 250–1, 253–5
   art of shared relational appreciation and enhancement 107
   artistic forms 165
   artwork 85, 87, 90–5
   improvisational art 106
   life as a work of art 75–7, 186–8

   "The Origin of the Work of Art" 87–8, 90–1, 97–8, 100
   philosopher as artist 186–8
   romantic art 91, 95
attunement (*Gestimmtsein*, 和) 89, 167, 170, 240
Ausbeutung (exploitation) 192

becoming 3–4, 7, 24, 37, 39, 55–6, 58, 60, 79, 91, 128–9, 134, 184, 191, 196–8, 215, 228, 229, 232–3, 238, 241, 252
   becoming aware 91
   becoming a Buddha 131
   becoming entrained 163
   becoming *homo natura* 56, 58, 60
   becoming meaningful 104
   becoming possessed by rhythm 150–151
   becoming present 102
   becoming what one is 191
   becomings 141, 253
being-in-the-world 8, 89, 152, 153
Bodhisattva 9, 104, 108, 197, 201, 224
body vii, 8, 15, 23, 28, 33, 37, 45, 50–1, 54–5, 59–60, 65, 103–5, 107, 129–30, 139, 142, 149–159, 162–4, 166–7, 169, 172–3, 180, 186, 191, 194, 198, 206, 229, 233, 237–8, 240
   as a field of resonance 55
   body and mind 41, 82, 105, 167, 210
   body and soul 233, 240
   body-mind 8, 77–8, 144, 150–1, 154
   body's *eigenmode* 155
   flesh 55, 63, 150, 152, 155
   lived bodyheartminding 41, 42
   musculoskeletal body 155, 167
   practiced body 154
   rhythmic body 8, 152, 155, 163, 167, 241
Book of Changes (易經大傳), *Yijing* 15, 17, 43, 45

Buddhism 9, 23, 47, 108, 130, 133, 135, 149, 181, 190–1, 193–5, 207–9, 212, 224, 234, 250, 254
　Chan 103, 107, 129, 131, 133
　Huayan Buddhism 24
　Mahāyāna 78, 109, 127, 133–4, 190–1, 193, 195, 201, 250
　Pure Land 130–1, 133, 135
　Zen viii, 8, 9, 72, 74, 77, 79, 82–3, 103, 107, 117, 131, 133–4, 135, 138, 148, 179–81, 184, 190–2, 194–200, 202, 207–10, 212, 214, 218, 231, 250, 253
Busshō 仏性 (Buddha Nature) 78–9, 134, 198

catharsis (*katharsis*) 225
cheng 成 (formed) 218–19
chong 沖 (to surge up) 226–7, 232, 233
　as *zhong* 盅 "empty" 226–7, 232, 233
Christianity, Christian 15, 17, 96, 129, 180, 187, 190, 202, 211, 217, 224
　as "Platonism for the people" 129
Climate 2, 4, 15, 46, 48, 62–3, 67, 101–2, 107, 109, 125, 182, 192, 202–3, 214, 250–1, 255
　climate change 6, 10, 102, 105, 179–80, 184, 202, 250
Confucianism 15, 17, 138
　Confucian role ethics 4, 33, 253
creativity 4, 26, 32, 53, 58, 107, 172, 174
　closed creativity and open creativity contrasted 105
　co-creativity 4, 32
　procreativity 37–8

da 大 (great, deep) 181, 189
dance vii, vii, 1–3, 33–4, 40, 46, 55, 65, 67, 90, 101, 128–9, 133, 135, 146, 148–51, 158, 164, 165, 167, 183, 188, 217, 219, 225, 229, 230, 249, 251, 252, 255
　dancer 21–2, 40, 56, 59, 149–51, 165, 183
　Daoist dance vii, 4, 31, 34
　The Great Dance 3, 8–9, 20
　　cosmic dance 33
　　wandering dance vii, 2–3, 9, 24, 26–7, 58, 60, 65, 67, 101, 107–9, 180–3, 186, 188–90, 195, 199–200, 202–3, 205–6, 210–1, 213, 251–2
dao 道 (way, course, way-making) 6, 10, 22–5, 27, 32, 33, 38–40, 45, 59, 61, 63, 95, 97, 105–6, 181, 206, 212, 216–18, 220, 222–3, 226–7, 230–3, 238, 241
　as *Abgrund* 222
Daodejing 4, 31–3, 35–41, 44, 46–7, 181, 206, 226, 232–43, 245, 253
Daoism 4–5, 8–9, 32–3, 97, 138, 149, 168, 174, 180–1, 184, 189, 195, 202, 206, 232, 234, 239, 250
Das Übermass (Excess) 187
Dasein (being-there) 88–90
de 德 (power) 24, 28, 33, 61, 65, 145, 199
death 1, 3, 18–20, 23, 25, 28, 59, 60, 71–2, 74, 114, 150, 187, 205, 208, 225, 227, 236–8, 242
　death of God 26, 187–8
　free death 71
　timely death 72, 237–8
decadence 190–1
depth 5, 7, 16, 73, 102, 126, 128–30, 150, 193–4, 198–9, 209, 217, 225–6, 230–5, 244, 250
　and going-under (*untergehen*) 18, 230, 233, 237, 243
depth psychology 4, 16, 59, 134, 193, 250
dharma 78, 132–5, 194, 210
Dharmakāya (reality-body) 103–4
Dionysus, Dionysian 6, 8, 20, 25, 95–7, 150–2, 167, 187–8, 195, 200–1, 205, 220, 223–5, 229–31, 233, 236, 238, 243
　Dionysian poets 167
disclosure 6, 88, 91–4, 96, 205
dream, dreaming 6, 19–21, 27, 71, 111, 132, 139, 179, 182–3, 188–9, 199–201, 206, 208, 224
　lucid dreamer 182–3, 188–9

eagle (and serpent) in *Zarathustra* 16, 24, 235, 238
ecology, ecological vii, 2, 4, 7, 31, 40–5, 50–1, 53, 59, 62, 66–7, 82, 84, 130–1, 138, 142, 144–7, 149,

Subject Index

169, 172–4, 179–80, 182–3, 188, 201–2, 204, 213–14, 250, 256
ecological crises 9, 49, 61, 125, 137, 142, 181, 192, 202, 214, 256
sonic/aesthetic ecology 6, 106
ego 18, 23–4, 26, 54, 56, 58, 60, 77–8, 103–4, 108–9, 129, 134, 191–2, 195, 197, 200, 209, 223, 227–8, 234, 242, 245
   das Ich 221, 229
   heroic, executive ego 221, 151
   monarchy of the "I" 107, 221
élan vital 139, 141, 143
embodiment 14, 65, 91, 152, 158, 168
   embodied practice 6, 101, 104, 105
emptiness 7, 37, 54, 57, 103, 105–6, 127, 131–2, 189, 195, 197–8, 206, 209, 227, 232
   empty (*kū* 空, xu 虛, *śūnyatā*, *zhong* 盅), emptying 7, 54, 56–8, 64, 127, 185, 189, 195, 222, 224–7, 230, 232, 239, 253
   epistemic emptiness 105
   and nothingness 226–7, 233
entrainment 8, 150–3, 159–60, 163–5, 167, 169, 174
   interbodily entrainment 153, 155, 167
   rhythmic entrainment 155, 159, 163
environmental ethics 35, 43, 44, 82, 84, 184, 202, 213
   Daoist environmental role ethics 41, 43
eternal recurrence, eternal re-coming 9, 16, 19, 22, 24–5, 57–61, 132, 181, 190, 195–201, 215, 233, 241, 252
   eternal return 25, 59, 131, 196, 234–8, 241, 252
ethics 4, 6, 32, 36, 44, 46, 49, 94, 104–7, 117, 207, 214, 253–4
   Confucian ethics 36
   Confucian role ethics 4, 33, 253
   Daoist role ethics 4, 33, 41, 43–4
   environmental ethics 4, 36, 41, 43, 44, 50, 82, 84, 184, 202, 213
   global ethics 32, 104
exemplary persons (*junzi* 君子) 236
experience 5, 7, 9–10, 18, 20, 23, 34, 35, 37, 40–1, 43, 51–2, 54, 56, 59, 62, 69, 77, 87–8, 91, 93–5, 99, 103, 112–14, 116, 120, 130, 132, 134, 141–2, 151, 160–1, 165, 170–2, 175, 180–90, 193, 195, 196, 198, 200, 204–6, 216, 218, 221, 224–6, 230–1, 234, 242–3, 249
   Angst experience 225–6, 235
   perspectiveless experience 194
   transperspective experience 183, 203, 214

family (*jia* 家) 35, 36–8, 44–5, 81, 126, 144, 164, 246, 253
fasting of the heart-mind 185, 189
fengshui 49–53, 55, 59–63, 67, 137, 141–4, 147–8
film 6–7, 34, 77, 87, 111–22, 179
   analytic divisions of 112
   and philosophical knowledge 115, 117–18
"focus-field" (*dedao* 德道) approach 33, 41–3, 46
Forgetting 221

garden 7, 34, 72, 79, 126, 128, 130, 133, 138, 142–3
   Kubota Garden 7, 126, 127, 135
   Ryōanji 7, 126, 128
gaze 161, 162, 165, 168, 170–1, 173–4
   gaze-cuing 160, 168
   gaze-following 160, 168
   in Merleau-Ponty 161
Genjōkōan 194–5
ghosts (gui 鬼) 78, 142, 104, 222
   and spirits (shen 神) 142, 145
global 4, 6, 10, 32, 44, 47, 48, 101–5, 107, 150, 180, 254
Gnosticism 200
god, gods 70, 91, 134, 140, 147–8, 151, 185, 187–8, 190, 197, 201–2, 205
   Death of God 26, 57, 188, 205, 217, 224–6, 230, 231, 236, 239
   God of the Eleatics 57
Grund (ground) 217
   groundless ground 230

harmony 10, 22, 25, 26, 97, 199, 234
   attunement 89, 167, 170, 240
   attunement (*Gestimmtsein*) 89
   attunement (和) 227

health, healthy 15, 17, 21–2, 50, 75, 138–9, 142–4, 150–1, 197, 234, 250
heart-mind (*xin* 心) 40, 41, 54, 143–5, 147, 185, 189, 194
   fasting of the heart (*xinzhai* 心齋) 54–5, 185, 189
heaven, heavens (*tian* 天)16–17, 22–4, 27, 36, 55–6, 58, 64–5, 67, 131–2, 138, 182–4, 189–90, 202–4, 206, 208, 214, 218, 224, 241
   all-under-Heaven (*tianxia* 天下) 44
   heaven(s) and earth 17, 27, 38, 46, 48, 51, 52, 54–6, 60, 61, 65, 103, 181, 184–5, 189–90, 202–3, 206, 214, 237, 246, 181, 184–5, 189, 214, 237, 246
   in Nietzsche 184–5, 187, 190, 196, 202–3, 205–6
   principles and patterns of the heavens (*tian li* 天理) 62
   tiandao 天道 36
   under heaven's canopy (*kanyu* 堪輿) 52, 63
holism, holistic 33, 41, 45, 156, 157
   holistic aesthetic ordering, value 4, 32, 39, 45
homo natura 4, 49, 56, 58–61, 221

image(s), imagery vii, 6, 9, 14–17, 21–3, 26–8, 33–4, 37–8, 40–1, 50, 53–6, 64–66, 73, 101, 111, 113–14, 117–22, 128–9, 132, 141, 179, 181, 183, 187–9, 195, 197, 200–2, 206, 225, 230–1, 238, 253
   image of wandering dance 101, 252
   imagery of garden cultivation 76
   musical imagery 89, 93, 99
improvisation 6, 101–2, 105–7, 155–6
intentionality 153, 154–5
   motor-intentionality 153–4
   original intentionality 154
interdependence 16, 17, 25, 47, 103, 127, 131, 190, 193, 195, 200, 242, 254
   of opposites 17–18, 21
irony 118–20, 211

jiao note 52

kenshō (見性) "seeing things as they are" 194–5, 210
kōan 197–8, 201
   Kōan of Eternal Recurrence 196

labyrinth 199
landscape(s) 1, 4, 49–50, 73, 79, 83, 133, 135, 144, 146, 148, 180, 199, 201, 253
language 2, 5–6, 13–14, 19, 26–7, 31, 33–5, 42, 47, 54–5, 57, 87, 92–6, 98
   poetic language 6–7, 92–3, 98–100, 111, 116–22, 126, 128, 159, 165, 167, 186, 204, 228, 249, 250, 254
laughter 9, 14, 27, 40, 200–1
li 理 (principles or patterns, dynamically patterned articulation) 7, 62, 104, 140, 143–4
li 禮 (ritual propriety, rites) 36–7, 65, 144, 220–2
li-studies *lixue* (理學) 7, 138, 140–1, 146–7
life 2–3, 9, 19–20, 22, 31, 40, 42, 46, 52–4, 56, 57, 60–1, 64, 71–5, 77, 81–2, 84, 90–1, 98–100, 114, 129–30, 132–3, 137, 139, 142–4, 149, 160–1, 180–4, 186–7, 190, 192, 193–4, 198, 201, 208, 210–2, 215, 217, 221–5, 228, 232, 235, 237–8, 242, 245, 250–1, 255
   as a work of art 75
   good life 5, 69–70, 76, 79–80, 106–7
   inner-life 41–2, 91
   life and death 1, 3, 18, 59–60, 72, 114, 227, 236–8
   life experience, lifeworld 41, 51, 112, 224
   life narrative 41
   lifelines 62
   perspectival life 216
   vibrant life 137, 139
   will to or of life, 100, 192
lithophilia 7, 62, 125, 129, 133, 137, 138, 146, 148
logos 2, 228, 230
   bio-logos 149

Macht (power) 229
   as necessary to create (*machen*) 229
mana, Polynesian idea of 250

materialism 138, 141
  new materialism 7, 137–8, 144, 146–7
  vital materialism 138–9, 141, 143
matter 54, 141
  dull matter 137
  physical, dead, soulless matter 4, 8, 49, 50–2, 137, 139, 140
  vital energy-matter, condensed *qi* 50, 137–41, 145
melody 56–7, 93, 150–1
metaphysics 13, 60, 99–100, 145, 148–9, 168, 173, 202, 228
mimesis 163
mind vii, 8, 15, 17, 33, 39, 40, 41, 47, 50, 52, 54, 56, 77–8, 82, 90, 97, 129, 131, 139, 144, 150, 151–5, 159–60, 163, 165–6, 168, 185, 199
  beginner's mind (shoshin 初心) 131
  emptying the mind (or soul), of projections 58, 185, 189, 195, 222, 225, 227
  social mind 155
ming 命 (fate or circumstance) 199
míng, 明 (illumination, clarity) 217
mood (*Stimmung*) 56, 88–90, 102, 128
mu 無 198
multipolar 44–5
music vii, 3–6, 10, 31, 33–5, 42, 52, 55–7, 59–61, 64–6, 70–1, 75–6, 80, 87–100, 102, 105–6, 118–19, 135, 151, 155–9, 164–6, 168–70, 172–5, 219, 229, 252, 254–5
  centrality of musicality 159
  heart-mind of the music 105
  Heidegger's reticence to music 5, 6, 87–91, 93–8
  music of earth 55, 57, 64
  musical innovation 105
  musical improvisation, improvised music (sonic/aesthetic ecology) 6, 105–6
  yue 樂 220, 222

narratives 14, 21, 40–1, 45, 143, 146, 171, 173, 186, 187, 205
  metanarratives 186, 204
  modern-postmodern narrate 143
  narratives-within-narratives 41

nature 1, 4, 38, 41, 49–51, 55–64, 67, 76, 104, 134, 141, 147–8, 171, 175, 180–7, 189, 194, 196, 200, 202–4, 210, 213, 215, 218, 250
  as origin 185, 187, 204, 214
  homo natura 4, 49, 56, 58–61, 221
  natural, cosmic cycle 38–9, 201
  "self-so-ing" (*ziran* 自然) 40, 55, 59, 219, 221, 233
nihilism 27, 188, 195, 200, 207, 210–1, 213–14, 217, 241
nirvāṇa 23, 190–1, 193, 208
non-attachment 58
nonduality 103–4

ontology, ontological 43, 137, 152–3, 159, 161–2, 166–7, 192, 226, 230
  ontological gestalt shift 43
  pre-ontological 153
  theo-ontological thinking 46
Ouroboros 200–1
overhuman, overman (Übermensch) 18, 22, 24–7, 58, 129, 182–3, 185, 200, 223, 225, 234, 236–7, 239

panpsychism 134, 137, 139
persons, personhood 6, 31–2, 39–41, 43–7, 51, 56–7, 90–1, 95, 101, 106–7, 111, 118, 156–7, 162, 164–5, 167, 190, 198, 218–19, 222, 225, 227, 236
  cultivating persons 8, 31, 32–3, 36–7, 40, 43, 79, 143–4, 227, 236, 243, 252
perspective 2–3, 9, 15–16, 18–22, 25, 31, 40–2, 44, 49, 52–5, 58–62, 67, 77–8, 83–4, 93, 95, 102, 107, 117, 121–2, 132, 134, 152, 154, 159–61, 163, 166, 169, 171, 173, 175, 180–3, 185–8, 190, 194–5, 202, 204, 209, 213–14, 216–24, 226, 229, 235, 240–1, 249
  Apollonian perspective 151
  perspectival perspective 223, 241
  possibility of perspectiveless seeing, experience 20, 77, 182–3, 185, 194
perspectivism 16–17, 20, 103, 180, 182–5, 188–90, 216

pharmakon (cure, poison) 228, 236
poetry 5, 14, 73, 78, 89, 91–4, 98–100, 119, 165
polycentric 44–5, 101, 250
    Zarathustra's polycentric psyche 237
practice vii, 7–8, 50–1, 54–60, 62–4, 67, 69, 72, 74, 76–80, 101–9, 112–13, 118–20, 127–9, 131–3, 138, 142–5, 153–4, 160, 166, 171, 173, 186, 206, 209, 212, 218, 249, 256
    Buddhist practice 107, 127
    caregiving practices 160, 164
    continuous practice (*shugyō*) of mountains 132
    embodied practice, body-mind practices 6, 101, 104–5, 144
    linguistic practice 6, 111
    practice of fengshui (風水) 59, 137, 142
    practice of philosophy vii, 7, 112, 125, 132, 249
    practice of yoga 193
pratītyasamutpāda (mutual conditioning, doctrine of interdependence) 103, 190
predicament resolution 102
primeval forces (*Ursachen*) 238
psychotherapy 53, 63, 66, 172

qi (vital stuff, psychophysical stuff)
    qi-studies (*qixue* 氣學) 7, 138, 141, 144
    qi-realism 8, 138–41, 146, 148
quietism 56, 58

relations 19, 32, 36, 41, 43, 45, 50–2, 55, 158, 229
    pattern of relationships 41, 43
    phase relationships 164, 167–8
    relationships 35, 154–5
    to the world 77
rendao 人道 or 仁道 36
resonating, the (*der Anklang*) 89, 170
responsibility 4, 31–3, 45, 220–2
ressentiment 190, 217
rhythm vii, 6, 8, 31, 33, 55, 93, 149–52, 155–60, 163–70, 174–5, 223, 226, 239, 241
    interpersonal rhythms 163
    music rhythms 3, 33, 57, 93, 96, 119, 150, 170, 175

    polyrhythms 163
    rhythmic entraining 159
    rhythmic pattern 158
    synrhythmic regulation 160
rocks 1, 4, 7, 8, 10, 28, 49, 51, 59, 62, 77, 125, 126, 128–32, 138–9, 201, 218, 250–1
    as kernels of energy 51–2, 130, 132, 135, 138–9, 146, 148
    rock gardens 34, 79, 83, 84, 126–8, 130, 132, 135, 138–9, 142, 146, 148, 198

scenic phantasmata 6, 111
Seiltänzer (tightrope dancer) 183
self 1, 4–5, 15, 18, 23–4, 26, 41–2, 57, 64, 79, 131, 145, 149–150, 170, 173, 195, 208, 221, 223, 224, 226–9, 234, 241–2, 245, 253
    becoming aware of the self 195, 226, 235
    forgetting the self 194, 221–2, 225, 227
    no-self (*anātman*) 190–1, 193, 208, 225
    participatory self 1–3, 10 (fractal), 26, 54, 102 (of intimacy), 129, 197, 217, 219, 222, 223, 243 (fractal), 245 (fractal)
    self and beauty 4, 76, 157, 184
    self and other 65, 162, 219
    self and world 2, 3, 13, 25, 40, 95, 104, 180, 197, 220
    self as multiplicity 20–1, 91, 101, 195, 218, 227, 229, 237
    self-improvement 77
    self-overcoming 80, 131–2, 195, 200, 210–1, 213, 226, 233–4, 243
    self-so-ing (*ziran* 自然) see nature
shengren 聖人 (Daoist sage) 1, 25, 38, 184, 196, 217, 220, 226–7, 232–3, 238, 243, 245
shi 事 (each thing or situation) 103
social coordination 152, 155, 159, 164
Sociopragmatists 153, 166
soul (*psuchē*) vii, 1, 3–4, 6, 9, 22–4, 27–8, 34, 46, 48–9, 53–4, 59, 63, 72, 76, 81–2, 84, 91, 101–2, 104–6, 108–9, 114, 127, 132–5, 149–50, 167–8, 174, 180, 191, 193–4,

199, 109, 212–13, 217–18, 221–6, 229–43, 245–6, 251
anima mundi, world soul 52–4, 59, 63–4, 66–7, 223
sound (*Klang*) 89
    musical sound, rhythm in sound 89, 93, 119, 233
    sound and image 117
    sound in general 3, 27, 55–7, 70, 79, 89, 97–100, 106, 114–15, 120, 137, 166, 200, 220
śūnyatā (emptiness) 103, 129, 131, 191, 195, 199, 209
Sympathy 163–4, 166, 172, 224
    intersubjective sympathy 164
    primal bodily sympathy 166
    rhythmic sympathy 164

taiji quan 51
taṇhā (thirst or craving) 191, 194
tathata (suchness) 193, 209
tea ceremony 77
    Zen-cha-Roku 禅茶録 (*Zen Tea Record*) 74
temporality 120, 150–1
    of the body 152, 164
"this" and "that" 57, 189, 216–17, 219, 221, 223–4
Tian 天 (heaven) 27, 58–9
tiandao 天道 (way of heaven) 36
tiandi 天地 (heaven and earth) 54–5, 181
time, intensity, and form 160
tragedy 20, 70, 80, 84, 99–100, 151, 186–8, 191, 197, 201, 205, 212, 214
    tragedy of the commons, 63
    Zarathustra as a tragedy 197
truth 4, 61, 90–2, 104, 112, 128, 134 (of impermanence), 139, 152, 180, 185, 187–8, 192, 204, 206, 209, 213, 218, 223, 234 (Four Noble Truths), 241, 243, 245, 253
    as happening 91–2
    fixed truth 107
    "woman-truth" in Nietzsche 186

Unendlich, limitless 188
Unendlichkeit infinity, limitlessness 188, 189

vijñapti-mātra 193–4, 209
vitalism 137, 139–41, 143

wander (*you* 遊)
wanwu 萬物 (the ten-thousand things, everything that is happening)
water 1, 3, 9, 15, 73, 76, 83, 126, 127 (as emptiness), 134, 144 (image for clear mirror), 215, 226
    in Dōgen 194–5, 209–10, 212, 213
    in the Daoism 15, 22, 27, 38, 181, 183, 231–3, 242
    in *fengshui* 50–1, 55, 60, 61–3, 67, 137, 141, 144, 147–8
    in Heraclitus 15, 17, 228
    in Nietzsche 58, 60, 197, 225
will to power 9, 15, 18, 20, 22, 24, 27–8, 58, 60–1, 65–6, 80, 84, 101–4, 109, 131, 134, 180–1, 191–4, 196–7, 200, 205, 207–9, 213, 215, 229, 240–2, 245–6
Wonne (delight) 225, 241
    as intoxicating bliss 225
worldview 9, 51, 61
    Cartesian-Newtonian worldview 50
    Chinese worldview 4, 32, 38, 50, 234
wu 無-forms 33, 35, 46

xiaoren 小人 (the small person) 236

yi 義 (appropriateness, morality) 221, 236
yin and yang 17, 23, 60, 65, 92, 189–90, 206–7, 214, 218–19, 227, 230, 238, 242, 246
Yogācāra 193–4, 208–9, 212
you 遊 (wander, rambling) 35, 55, 183
yóu 游 (to dance, float, swim about in water) 183

Zazen 131, 251
Zhi 知 (understanding, realizing) 33, 35, 220, 240

Name Index

Abe, Masao 78, 83, 108, 134
Aeschylus 205–6
Ames, Roger (as referenced by other book authors) 26, 28, 45, 46, 48, 61, 63, 67, 137, 140, 145–7, 202–3, 206, 212–14, 226, 240, 242–5, 253
Aristotle 15–16, 43, 252

Bashō 78
Bataille, Georges 129, 133, 134
Bateson, Mary Catherine 165, 173
Bennett, Jane 137–47
Bergson, Henri 139–40
Buddha 103–4, 107, 130–1, 133, 190–5, 208, 212, 214
Byers, Paul 163, 164, 172

Christ (Jesus), 25, 252

Davis, Bret 104, 108–9, 134, 191–2, 194–5, 207–10, 212, 256
Deleuze, Gilles 200, 212, 241, 245
Derrida, Jacques 204, 210, 211–2
Dewey, John 41, 46
Dōgen 77–9, 82–3, 125, 127, 129, 131–5, 149, 194–5, 199, 202, 209–10, 212–13, 251, 256
Driesch, Hans 139–40, 143
Dukes, Edwin Joshua 142

Emerson, Ralph Waldo 34, 82, 84

Furong Daokai 125, 129, 132

Gadamer, Hans-Georg 87, 168
Gardner, Daniel K. 137, 145–7
Gibson, James 163, 172–3

Hall, David L. 137, 140, 226, 243
Hayman, Ronald 208, 211–2

Hegel, Georg Wilhelm Friedrich 23, 90–1, 95, 138
Heidegger, Martin vii, 5–6, 10, 19, 46, 87–100, 152–3, 166, 233–43, 245, 250–1, 253
Heraclitus 1–2, 8–9, 10, 13–15, 17–18, 27, 61, 186, 201, 204, 211–2, 215, 222, 225, 228, 230, 234–6, 240
Hershock, Peter (as referenced by other book authors) 43–4, 47–8, 108
Hillman, James 4, 53–4, 59, 63, 64–6
Horwicz, Adolf 42
Huiri 130–1, 133
Husserl, Edmund 6, 111, 118, 120, 170

Ii, Naosuke 74

James, William 41, 42, 47, 48
Jōdo Shinshū 131
Jones, David (as referenced by other book authors) 202, 207–8, 212, 241, 243, 245

Kamo no Chōmei 73
Kim, Yung Suk 140, 146, 148
Kubota, Fujitarō 126–7, 133
Kūkai 103, 108

Lampert, Lawrence 196, 204–5, 210, 212
Large, Edward 159, 168, 170, 173–4
Latour, Bruno 137, 142–3
Lee, Christopher 158–9, 169–70, 175
Lingis, Alphonso 221, 229, 240, 242, 245
Livingston, Paisley 7, 112, 114–22
Loehr, Janeen 159, 168, 170, 174
Lohmar, Dieter 6, 111, 118, 121–2
Lusthaus, Dan 209, 212
Lyotard, Jean François 204, 212

Malloch, Stephen 163, 166, 172–4
Mead, George Herbert 41

Mencius, Mengzi 47
Merleau-Ponty, Maurice 63, 152–3, 161, 163, 170–1, 174
Metzinger, Thomas 243, 245
Mishima, Yukio 71–2
Moeller, Hans-Georg 34–5, 46, 48

Nietzsche vii–viii, 2–6, 8–10, 13–20, 22–28, 31–32, 34, 45–46, 48, 56–61, 63–67, 69–84, 87, 95, 98–100, 101–104, 106–109, 117, 119, 128–129, 131–135, 138, 140, 146, 148–151, 167, 174, 180–215, 217–230, 232–233, 236, 239–246, 249, 251–254, 256
Nietzsche and Zhuangzi vii, 2, 3, 9, 13, 15, 17–19, 24, 56, 64, 67, 87, 101, 149, 181–183, 185, 188–189, 200, 203–204, 212–214, 218, 220–222, 226, 229–230, 238–239, 241, 245
Nishitani, Keiji 195, 199–200, 210–1, 213
Nitobe, Inazō 72, 81, 84

Okumura, Shokaku 194, 209, 212

Palmer, Caroline 159, 168, 170, 174
Plato, Platonism 8, 14–15, 17, 23, 25, 53, 76, 81–4, 97–8, 100, 129, 150–3, 167, 182, 186–7, 202, 204–5, 214, 217, 224, 228, 242–3, 246, 252

Rancière, Jacques 137
Reddy, Vasudevi 161–2, 165, 170–1, 173–4
Rojcewicz, Richard 89, 98, 100
Rorty, Richard 46
Rowles, Jimmy 156–7

Sallis, John 205, 214, 243, 245, 253
Sartre, Jean-Paul 75, 80, 84, 117, 161, 170, 174
Schopenhauer, Arthur 99, 103, 151, 190, 224
Sellars, Wilfred 46
Sen no Rikyū 75
Shibayama, Zenkei 211, 214
Snyder, Joel 159, 170, 173

Socrates 3, 14, 18, 117, 151, 180, 186, 205, 228
Sophocles 205
Spinoza, Baruch 140
Stambaugh, Joan 89, 97, 100
Stern, Daniel 160–1, 164–6, 170, 172–3, 175
Su Zhe 蘇轍 38
Suzuki, Shunryū 131, 133, 135

Tang Junyi 45
Theseus 199, 230
Thoreau, Henry David 131, 134–5
Todd, Niel 158–9, 169–70, 175
Trevarthen, Colwyn 160–1, 163–6, 170–5
Tu, Wei-Ming 51, 63

Unno, Mark 131
Unno, Taitetsu 130–1

Vallega-Neu, Daniela 89, 98, 100
Van Gogh 91, 98
Vasubhandu 208
Vattimo, Gianni 207, 214
Vogel, Steven 201, 214

Wang, Robin 206–7, 214, 242, 246
Wenders, Wim 179–80, 201
Wheeler, Michael 89, 97, 100
White Jr., Lynn 202, 214
Whitehead, Alfred North 140
Whiteside, Abby 157–8, 175

Yan Hui 20, 27, 54–6, 221–2, 224–6, 229–30
Young, Julian 45, 88, 93–4, 98
Yunmen Wenyan 132

Zhao, Tingyang 43–5, 47–8
Zhu Xi 138, 140–1, 144, 146–8
Zhuangzi vii–vii, 2–4, 8–9, 13–28, 35, 48, 54–59, 63–65, 67, 85, 101, 104, 149, 179, 181–185, 188–190, 199–200, 203–204, 206, 211–223, 225–227, 229–231, 233–246, 251
Ziporyn, Brook 27, 63–4, 67, 206, 214, 216, 239–40, 246
Ziqi 55

www.ingramcontent.com/pod-product-compliance
Lightning Source LLC
Chambersburg PA
CBHW071814300426
44116CB00009B/1316